Emerging Crop Pest Problems: Redefining Management Strategies

By

P. PARVATHA REDDY

Former Director

Indian Institute of Horticultural Research, Bangalore

SCIENTIFIC PUBLISHERS

Published by:

SCIENTIFIC PUBLISHERS (INDIA)
5A, New Pali Road, P.O. Box 91
Jodhpur 342 001, India
E-mail: info@scientificpub.com
Website: www.scientificpub.com

Print: 2020

ISBN: 978-93-86652-24-9
eISBN: 978-93-88172-08-0

Printed in India

Foreword

Dr. PREM NATH AGRICULTURAL SCIENCE FOUNDATION (PNASF)

To promote agricultural education, research and sustainable development with focus on food and nutrition security

No. 9, 1st Cross, 1st Main, 1st Block, Rajmahal Vilas Extension 2nd Stage,
Bangalore - 560 094, Karnataka State, INDIA
Tel: +91 - 80 - 23415188; Fax: +91 - 80 - 23411555
E-mail: drpremnath@vsnl.net

Dr. PREM NATH

Chairman, PNASF
Chairperson, VEGINET
Former Asst. Director General,
Food & Agriculture Organization of the
United Nations (FAO-UN)

With the development of agricultural technology as well as biotic and abiotic factors, the crop losses and pest problems are changing continuously. Chronic, emerging and invasive pests are considered as one of the major threat to food security. There is a need to explore certain reformative measures to narrow down these losses. The climate change will affect the pest's distribution and status as well as the pest management practices.

Increased movement of people, plants and products in the globalized economy on the one hand, and the concentration and intensification of production systems on the other, have accelerated and enlarged redistribution of plant pests with a clear tendency to expand to all regions of the globe. In addition, climate change is creating new ecological niches for the (re)emergence and spread of pests and diseases. As a result, the impact of crop pests has considerably increased.

The truly scary possibility is that a new or re-emerging crop pest could decimate one of the few crops — rice, wheat, corn — that the global diet is based on. We have already had a few close calls. Wheat rust, which is caused by a fungus, devastated wheat crops in Africa, and is poised to spread to other major wheat-producing countries. It does not help that over the years farmers have narrowed the genetic diversity of commodity crops, which limits our ability to respond if a new pest or disease takes hold. That is why we need to support seed banks, which store a variety of strains within a crop, to ensure that farmers have weapons to respond to a new plant plague. They will need those bullets in a warmer world.

Emerging pests have increased in incidence, geographical or host range; have changed damage potential; have been discovered or newly recognized. Interest in emerging pests has focused on those affecting humans, livestock and wildlife. Emerging crop pests impact negatively on human wellbeing through agricultural and economic loss, and also have consequences for biodiversity conservation.

Thus, emerging pests have themselves become new drivers of global environmental change. Emerging pests can cause extinction of endangered species; alter the ratios of predators, prey, competitors, and recyclers necessary for healthy, well-functioning ecosystems; and alter habitat already threatened by fragmentation and global climate change.

Hence, there is an urgent need to modify crop protection measures with changed climate in order to attain the goal of food security.

In this context, the book written by **Dr. P. Parvatha Reddy** is very timely and comprehensively deals with changing pest scenario with respect to insects, mites, diseases and nematodes affecting field, fruit, vegetable, ornamental, medicinal, plantation, tuber, and forest crops. The management strategies for the emerging pests have been discussed in detail and the practical recommendations have been outlined. The pests which are likely to become serious threats in future due to changes in the ecosystems and habitats are discussed. The possible technical and policy responses, and policy considerations to solve the problems of emerging pest problems have also been outlined.

I compliment Dr. Reddy for his meticulous contribution on a very potential topic of emerging crop pest problems and redefining their management strategies. This book will be of immense value to scientific community in agriculture as a whole and who are involved in crop protection in particular. The material can also be used for teaching post-graduate courses. The book can also serve as a very useful reference to policy makers and practicing farmers.

(Dr. PREM NATH)

Date: May 1, 2017
Place: Bangalore

Preface

Crop pest introduction, establishment, re-emergence and outbreaks have resulted in major food problems either directly through yield reductions of food crops, or indirectly through the reduction of yields of cash crops and loss of consumer confidence, e.g. potato late blight and locusts.

In recent years, with changes in the cropping systems and climate, and introduction of highly input intensive high yielding varieties/hybrids, a shift in pest status has been observed. In light of national concerns for crop biosecurity, new, emerging, and threatening plant pests have received more attention from all the interested parties. Thus, emerging pests have themselves become new drivers of global environmental change. They are responsible for causing extinction of endangered species; altering the ratios of predators, prey, competitors, and recyclers necessary for healthy, well-functioning ecosystems; and altering habitat already threatened by fragmentation and global climate change. New and re-emerging plant pests have raised fear of their potential impact on livelihoods, food security and global markets.

The incidences of several insect pests like mealy bugs, particularly *Phenacoccus solenopsis* on cotton; sugarcane woolly aphid, *Ceratovacuna lanigera* on sugarcane; *Pieris brassicae* on crucifers; and tobacco caterpillar, *Spodoptera litura* on several crops; have shown an increasing trend. There was a decline in the pest status of bollworm (*Helicoverpa armigera)* in cotton, whereas the sap feeders, *viz.* aphids, jassids, mirids and mealy bugs are emerging as serious pests. Recently, the occurrence of resistance in *Bt* cotton to *Helicoverpa zea* in different regions had been reported.

The truly scary possibility is that a new or re-emerging disease pathogen could decimate one of the few crops — rice, wheat, corn — that the global diet is based on. For example, wheat rust fungus, *Puccinia graminis* devastated wheat crops in Africa and is poised to spread to other major wheat-producing countries. Cassava mosaic virus transmitted by white flies, has caused enormous losses in cassava in sub-Saharan Africa, where it is a staple in the diet of millions. As new strains of *Phytophthora infestans* evolve, new outbreaks of the disease emerge, causing devastating epidemics globally; such as the virulent fungicide-resistant strain US-8, which emerged during 1992. In 1972, Karnal bunt fungus, *Tilletia indica* became globally important following its discovery in Mexico and in USA during1996. Recent interest in the disease has surrounded the potential for its use as a biological weapon and exports of wheat from many regions with Karnal bunt have been banned, leading to severe economic loss for affected countries.

There has been an outbreak of serious root-knot nematode problems in the recent years in vegetable crops grown under protected cultivation systems and fruit crops such as pomegranate and guava through dissemination of nematodes with infected planting materials. Rice root-knot nematode has also emerged as a national problem, and the problem is getting accentuated. The detection of potato cyst nematodes in northern India is a serious concern with ramifications on export and quarantine issues.

A focused attention by experts like plant pathologists, entomologists, nematologists, horticulturists, policy makers, pesticide industries, and farmers is the need of the hour for in-depth discussion at one platform in order to develop a broad strategy for tackling these problems.

The present book on "**Emerging Crop Pest Problems: Redefining Management Strategies**" comprehensively deals with the rapid and accurate detection, diagnosis, and development of management recommendations for the emerging crop pests. The book is divided into five sections. The first section deals with an overview of emerging crop pest scenario including drivers of pest emergence, impacts of emerging pests, and management of emerging pests. The emerging insect and mite pests on field, fruit, vegetable, plantation, tuber, and forest crops; and strategies for their management are dealt in section two. The third section deals with emerging bacterial, fungal and viral diseases of field, fruit, vegetable, ornamental, spice, and tuber crops and their management. The emerging nematode scenario on field, fruit, vegetable, ornamental, medicinal, spice, and tuber crops and strategies for their management are dealt in section four. The final section deals with pests likely to become serious threats in future, and potential impact and anticipated effect of climate change on emerging pests. The possible technical and policy responses, policy considerations and the road map ahead are also discussed in this section. The book is extensively illustrated with excellent quality photographs enhancing the quality of publication. The book is written in lucid style, easy to understand language along with adoptable management recommendations involving eco-friendly practices.

This book will be of immense value to scientific community involved in teaching, research and extension activities related to emerging crop pest problems and their management strategies. The material can be used for teaching post-graduate courses. The book can also serve as a very useful reference to policy makers and practicing farmers. Suggestions to improve the contents of the book are most welcome (E-mail: reddy_parvatha@yahoo.com). I am very much thankful to Dr. Prem Nath for kindly writing Foreword to my book.

The publisher, Scientific Publishers, Jodhpur, India, deserves commendation for their professional contribution.

P. Parvatha Reddy

34, UAS Layout, Ist Main, 7th Cross

Sanjay Nagar, Bangalore-560 094

Bangalore

May 5, 2017

Contents

Foreword iii
Preface v

Section I. INTRODUCTION

1. Emerging Crop Pest Problems – An Overview 1

1.1. Introduction 1
1.2. Crop Losses Due to Pests 1
1.3. Emerging Pest Problems 4
1.3.1. What is an Emerging Pest? 4
1.3.2. Insect Pests 5
1.3.3. Disease Pathogens 6
1.3.4. Nematode Pathogens 8
1.4. Invasive Pests 9
1.5. Impact of Climate Change on Emergence of Pests 11
1.5.1. Expansion of Geographic Ranges 12
1.5.2. Increased Overwintering Survival 13
1.5.3. Increase in Number of Generations 13
1.5.4. Physiological and Ecological Impact 13
1.5.5. Loss of Ecological Biodiversity 14
1.5.6. Changes in Insect Feeding 14
1.5.7. Breakdown of Host Plant Resistance 14
1.5.8. Introduction of Transgenic Crops 14
1.5.9. Changes in Tillage Technology 14
1.5.10. Other Effects 14
1.6. Drivers of Pest Emergence 15
1.6.1. Insect Pests 16
1.6.2. Disease Pathogens 16
1.7. Impacts of Emerging Pests 18
1.7.1 Food Security 18
1.7.2. Human Health and Welfare 19
1.7.3. Biodiversity 19
1.8. Management of Emerging Pests 20
1.8.1. Insect Pests 20
1.8.2. Disease Pathogens 22
1.9. Conclusions 22

Section II. INSECT AND MITE PESTS

2. Emerging Insect Pest Problems in Field Crops 25

2.1. Introduction 25

2.2. Cotton, *Gossypium* spp. 25

2.2.1. Mealy Bug, *Phenacoccus solenopsis* (Hemiptera: Pseudococcidae) 27

2.2.2. Tobacco Caterpillar, *Spodoptera litura* (Lepidoptera: Noctuidae) 31

2.2.3. Mirid Bug, *Creontiades biseratense* (Miridae: Hemiptera) 32

2.2.4. Stink Bugs, *Acrosternum hilare, Nezara viridula, Euschistus servus* 34

2.2.5. False Chinch Bug, *Nysius raphanus* (Hemiptera: Lygaeidae) 35

2.2.6. Sugarcane Beetle, *Euetheola rugiceps* 35

2.2.7. Three-Cornered Alfalfa Hopper, *Spissistilus festinus* 35

2.2.8. Tea Mosquito Bug, *Helopeltis bryadi* (Hemiptera: Miridae) 36

2.2.9. Flower Bud Maggot/Gall Midge, *Dasineura gossypii* (Cecidomyiidae: Diptera) 36

2.3. Soybean, *Glycine max* 38

2.3.1. Tobacco Caterpillar, *Spodoptera litura* 38

2.4. Sunflower, *Helianthus annus* 39

2.4.1. Tobacco Caterpillar, *Spodoptera litura* 39

2.5. Sugarcane, *Saccharum officinarum* 39

2.5.1. Woolly Aphid, *Ceratovacuna lanigera* (Aphididae: Hemiptera) 39

2.6. Rice, *Oryza sativa* 41

2.6.1. Brown Plant Hopper, *Nilaparvata lugens* (Hemiptera: Delphacidae) 41

2.6.2. Swarming Caterpillar, *Spodoptera mauritia* (Lepidoptera: Noctuidae) 43

2.7. Sorghum, *Sorghum bicolor* and Pearl Millet, *Pennisetum glaucum* 46

2.7.1. Grey Weevil, *Myllocerus* spp. 46

2.7.2. Leaf Hopper, *Pyrilla perpusilla* 46

2.7.3. Leaf Roller, *Cnaphalocrosis trapezalis* (Lepidoptera: Crambidae) 47

2.8. Pulse Crops 48

2.8.1. Changing Pest Scenario 48

2.8.2. Gram Caterpillar, *Helicoverpa armigera* 48

2.8.3. Spotted Pod Borer, *Maruca vitrata* (Pyralidae: Lepidoptera) 49

2.8.4. Pod Wasp, *Tanaostigmodes cajaninae* 50

2.9. Jute, *Corchorus* spp. 51

2.9.1. Mealy Bug, *Phenacoccus solenopsis* 51

2.10. Subabul, *Leucaena leucocephala* 51

2.10.1. Leucaena psyllid, *Hereopsylla cubana* (Order: Homoptera, Family: Psyllidae) 51

2.11. Conclusions 52

3. Emerging Insect Pest Problems in Fruit Crops 53

3.1. Introduction 53

3.2. Gauva, *Psidium guajava* 54

3.2.1. Spiraling Whitefly, *Aleurodicus dispersus* (Homoptera: Aleyrodidae) 54

3.3. Papaya, *Carica papaya* 55

3.3.1. Mealy Bug, *Paracoccus marginatus* 55

3.4. Sapota, *Manilkara zapota* 58

3.4.1. Seed Borer, *Trymalitis margarias* (Lepidoptera: Tortricidae) 58

3.5. Banana, *Musa* spp. 60

3.5.1. Root Mealy Bug, *Geococcus citrinus* 60

3.6. Mango, *Mangifera indica* 61

3.6.1. Fruitfly, *Bactrocera caryeae* 61

3.6.2. Red Banded Fruit Borer, *Deanolis sublimbalis* (Lepidoptera: Pyralidae) 62

3.6.3. Leaf Weevil, *Rhynchaenus mangiferae* (Coleoptera: Curclionide) 63

3.7. Amla, *Embelica officinalis* 64

3.7.1. Anar butterfly, *Deudorix isocrates* (Lepidoptera: Lycaenidae) 64

3.8. Litchi, *Litchi chinensis* 65

3.8.1. Key Facts 65

3.8.2. Red Weeveil, *Apoderous blandus* (Coleoptera: Curculionidae) 65

3.8.3. Fruit Borer, *Conopomorpha cramerella* (Lepidoptera: Gracillariidae) 66

3.9. Conclusions 67

4. Emerging Insect Pest Problems in Vegetable Crops 68

4.1. Introduction 68

4.2. Recent Changes in the Pest Scenario 69

4.3. Emerging Vegetable Pests 70

4.3.1. Solenopsis Mealy Bug, *Phenacoccus solenopsis* (Pseudococcidae: Homoptera) 70

4.3.2. Serpentine Leaf Miner, *Liriomyza trifolii* (Agromyzidae: Lepidoptera) 71

4.3.3. Phytophagous Mites, *Tetranychus* spp., *Polyphagotarsonemus latus* 72
4.3.4. Fruit Flies, *Bactrocera* spp. 73
4.3.5. Whiteflies, *Bemisia tabaci* 74
4.3.6. Giant African Snail, *Achatina* (*Lissachatina*) *fulica* 75
4.4. Tomato, *Solanum lycopersicum* 76
4.4.1. Silver Leaf Whitefly, *Bemisia argentifolii* 76
4.4.2. Pinworm, *Tuta absoluta* (Lepidoptera: Gelechiidae) 77
4.4.3. Western Flower Thrips, *Frankliniella occidentalis* (Thripidae: Thysanoptera) 79
4.4.4. Serpentine Leaf Miner, *Liriomyza trifolii* (Agromyzidae: Lepidoptera) 80
4.5. Brinjal/Egg Plant 81
4.5.1. Mealy Bug, *Phenacoccus solenopsis* (Pseudococcidae: Hemiptera) 81
4.5.2. Gall Midge, *Asphondylia capparis* 82
4.6. Chilli, Capsicum annuum 82
4.6.1. Gall Midge, *Asphondylia capparis* 82
4.7. Potato, *Solanum tuberosum* 83
4.7.1. Changing Pest Complex 83
4.7.2. Aphid, *Myzus persicae* 83
4.8. Cauliflower, *Brassica oleracea* var. *botrytis* 84
4.8.1. Diamondback Moth, *Plutella xylostella* 85
4.9. Cowpea, *Vigna unguiculata* 86
4.9.1. Spotted Pod Borer, *Maruca vitrata* (Pyralidae: Lepidoptera) 86
4.9.2. Hadda beetles, *Henosepilachna vigitioctopunctata* and *Epilachna dodecastigma* (Curculionidae: Coleoptera) 87
4.10. Field Bean, *Lab lab purpureus* var. *lignosus* 88
4.11. Conclusions 89

5. Emerging Insect Pest Problems in Plantation and Tuber Crops 90
5.1. Introduction 90
5.2. Plantation Crops 90
5.2.1. Coffee, *Coffea canephora* / *Coffea arabica* 90
5.2.2. Coconut, *Cocus nucifera* 93
5.2.3. Oil Palm, *Elaeis guineensis* 97
5.2.4. Cashew, *Anacardium occidentale* 97
5.3. Tuber Crops 99
5.3.1. Taro, *Colocasia esculenta* 99
5.4. Conclusions 100

6. Emerging Insect Pest Problems of Forest Crops **101**
6.1. Introduction 101
6.2. Coral Tree, *Erythrina spp.* 101
6.2.1. Gall Wasp, *Quadrastichus erythrinae* (Hymenoptera: Eulophidae) 101
6.3. Eucalyptus, *Eucalyptus teriticornis* 103
6.3.1. Gall wasp, *Leptocybe invasa* (Hymenoptera: Eulophidae) 103
6.4. Oak, *Quercus* spp. 105
6.4.1. Processionary Moth, *Thaumetopoea processionea* 105
6.5. Pine, *Pinus* spp. 105
6.5.1. Processionary Moth, *Thaumetopoea pityocampa* 105
6.5.2. Mountain Pine Beetle, *Dendroctonus ponderosae* 105
6.5.3. Southern Pine Beetle, *Dendroctonus frontalis* 105
6.6. Conclusions 106

Section III. DISEASE PATHOGENS

7. Emerging Disease Problems in Field Crops **108**
7.1. Introduction 108
7.2. Groundnut, *Arachis hypogea* 109
7.2.1. Peanut Bud Necrosis Virus 109
7.2.2. Peanut Yellow Spot Tospovirus (Tospovirus: Bunyaviridae) 112
7.3. Wheat, *Triticum spp.* 113
7.3.1. Karnal Bunt, *Tilletia indica* 113
7.3.2. Blast, *Magnaporthe oryzae* 116
7.4. Conclusions 120

8. Emerging Disease Problems in Fruit Crops **121**
8.1. Introduction 121
8.2. Pomegranate, *Punica granatum* 121
8.2.1. Bacterial Blight, *Xanthomonas axanopodis* pv. *punicae* 121
8.2.2. Disease Complex, *Fusarium* spp., *Rhizoctonia bataticola* 124
8.3. Guava, *Psidium guajava* 125
8.3.1. Fusarium Wilt, *Fusarium oxysporum* f. sp. *psidii* 125
8.4. Papaya, *Carica papaya* 130
8.4.1. Papaya Ringspot Virus (PRSV) 130
8.5. Conclusions 133

9. Emerging Disease Problems in Vegetable Crops **134**
9.1. Introduction 134
9.2. Potato, *Solanum tuberosum* 134
9.2.1. Late Blight, *Phytophthora infestans* 134

9.3. Tomato, *Solanum lycopersicum* 138
- 9.3.1. Tomato Yellow Leaf Curl Virus (TYLCV) 138
- 9.3.2. Tomato Spotted Wilt Virus (TSWV) 142

9.4. Onion, *Allim cepa* 145
- 9.4.1. Iris Yellow Spot Virus (Iris Yellow Spot Disease) 145

9.5. Watermelon, *Citrullus lanatus* var. *lanatus* 149
- 9.5.1. Watermelon Bud Necrosis Virus (WBNV) 149

9.6. Conclusions 151

10. Emerging Disease Problems in Ornamental, Spice and Tuber Crops 152

10.1. Introduction 152

10.2. Ornamental Crops 152
- 10.2.1. Ornamental Plants 152
- 10.2.2. Rose, *Rosa* spp. 154

10.3. Spice Crops 157
- 10.3.1. Ginger, *Zingiber officinale* 157

10.4. Tuber Crops 159
- 10.4.1. Cassava, *Manihot esculenta* 159
- 10.4.2. Taro, *Colocasia esculenta* 162
- 10.4.3. Elephant Foot Yam, *Amorphophallus poeniifolius* 168

10.5. Conclusions 172

Section IV. NEMATODE PATHOGENS

11. Emerging Nematode Problems in Field Crops 174

11.1. Introduction 174

11.2. Rice, *Oryza sativa* 174
- 11.2.1. Root-Knot Nematode, *Meloidogyne graminicola* 174
- 11.2.2. White Tip Nematode, *Aphelenchoides besseyi* 183

11.3. Wheat, *Triticum* spp. and Barley, *Hordeum vulgare* 186
- 11.3.1. Cereal Cyst Nematode, *Heterodera avenae* (Molya Disease) 186

11.4. Pigeon Pea, *Cajanus cajan* 188
- 11.4.1. Cyst Nematode, *Heterodera cajani* 188

11.5. Groundnut, *Arachis hypogaea* 191
- 11.5.1. Root-knot Nematodes, *Meloidogyne arenaria*, *M. javanica* 191

11.6. Conclusions 195

12. Emerging Nematode Problems in Fruit Crops 196

12.1. Introduction 196

12.2. Guava, *Psidium guajava* 196
- 12.2.1. Root-Knot Nematode, *Meloidogyne enterolobii* (=*M. mayaguensis*) 196

12.2.2. Guava Decline Disease Complex, *Meloidogyne enterolobii, Fusarium solani* 203
12.3. Pomegranate, *Punica granatum* 205
12.3.1. Root-Knot Nematode, *Meloidogyne incognita* 205
12.3.2. Wilt Disease Complex (Fungi, Nematodes and Insect pests) 208
12.4. Mulberry, *Morus* spp. 212
12.4.1. Root-knot Nematodes, *Meloidogyne incognita* 212
12.4.2. Root-knot Nematode, *Meloidogyne incognita* and Root Rot, *Macrophamina phaseolina* Disease Complex 214
12.5. Citrus, *Citrus* spp. 215
12.5.1. Root-knot Nematode, *Meloidogyne indica* 215
12.6. Conclusions and Recommendations 216

13. Emerging Nematode Problems in Vegetable Crops 218

13.1. Introduction 218
13.2. Potato, *Solanum tuberosum* 218
13.2.1. Cyst Nematodes, *Globodera rostochiensis, G. pallida* 218
13.3. Nematode Problems on Polyhouse Vegetables 229
13.3.1. Main Causes 230
13.3.2. Management 230
13.3.3. Cropping Sequences 231
13.3.4. Recommendations 231
13.4. Tomato, *Solanum lycopersicum* 232
13.4.1. Root-knot Nematodes, *Meloidogyne incognita, M. javanica* 232
13.4.2. Root-knot Nematode, *Meloidogyne incognita* and Wilt, *Fusarium oxysporum* f. sp. *lycopersici/F. solani* Disease Complex 233
13.5. Bell Pepper, *Capsicum annuum* 234
13.5.1. Root-knot Nematode, *Meloidogyne incognita* 234
13.5.2. Root-knot Nematode, *Meloidogyne incognita* and Bacterial Wilt, *Ralstonia solanacearum* Disease Complex 236
13.6. Cucumber, *Cucumis sativus* 238
13.6.1. Root-knot Nematodes, *Meloidogyne incognita* 238
13.7. Conclusions 239

14. Emerging Nematode Problems in Ornamental and Medicinal Crops 240

14.1. Introduction 240
14.2. Ornamental Crops 240
14.2.1. Tuberose, *Polianthes tuberosa* 240
14.3. Polyhouse Flower Crops 243
14.3.1. Preparation of Beds 244

14.3.2. Enrichment of Organic Amendments (FYM/ Neem/ Pongamia/ Mahua Cake/ Vermicompost) 244
14.3.3. Application of Enriched Organic Amendments 244
14.3.4. Recommendations 245
14.4. Carnation, *Dianthus caryophyllus* 245
14.4.1. Root-Knot Nematode, *Meloidogyne incognita* 245
14.4.2. Root-knot Nematode, *Meloidogyne incognita* and Wilt, *Fusarium oxysporum* f. sp. *dianthi* Disease Complex 247
14.5. Gerbera, *Gerbera* spp. 248
14.5.1. Root-knot Nematode, *Meloidogyne incognita* 248
14.5.2. Root-knot Nematode, *Meloidogyne incognita* and Foot Rot, *Phytophthora parasitica* Disease Complex 249
14.6. Medicinal Crops 249
14.6.1. Coleus, *Coleus forskohlii* 249
14.6.2. Aswagandha, *Withania somnifera* 255
14.7. Conclusions 259

Section V. FUTURE THRUSTS

15. Pests likely to become Serious Threats in Future 261
15.1. Introduction 261
15.2. Potential Impact of Climate Change on Emerging Pests in Future 262
15.2.1. Increased CO_2 Concentration 262
15.2.2. Rising Temperature 262
15.2.3. Precipitation 263
15.3. Anticipated effect of Climate Change on Pests 263
15.3.1. Insect Pests 263
15.3.2. Disease Pathogens 269
15.3.3. Weeds 271
15.3.4. Nematodes 272
15.4. Conclusions 272

16. The Way Forward 274
16.1. Introduction 274
16.2. Possible Technical and Policy Responses 275
16.2.1. Data for Projecting Risk 275
16.2.2. Early Warning and Prevention Strategies 276
16.2.3. Eradication, Containment, and Impact Reduction 276
16.2.4. Information Options 276
16.2.5. Food Trade Industry 277
16.2.6. Government Constraints 277
16.2.7. Government Priorities 277
16.2.8. Ecosystem Processes 278
16.2.9. Global Frameworks 278

16.3. Policy Considerations 278
16.3.1. Disaster Prevention 279
16.3.2. Capacity Development 279
16.3.3. Protecting Livelihoods 279
16.4. Roadmap Ahead 280
16.5. Conclusions 280

References **282**
Acronyms **329**
Subject index **330**

Chapter 1. PLATES

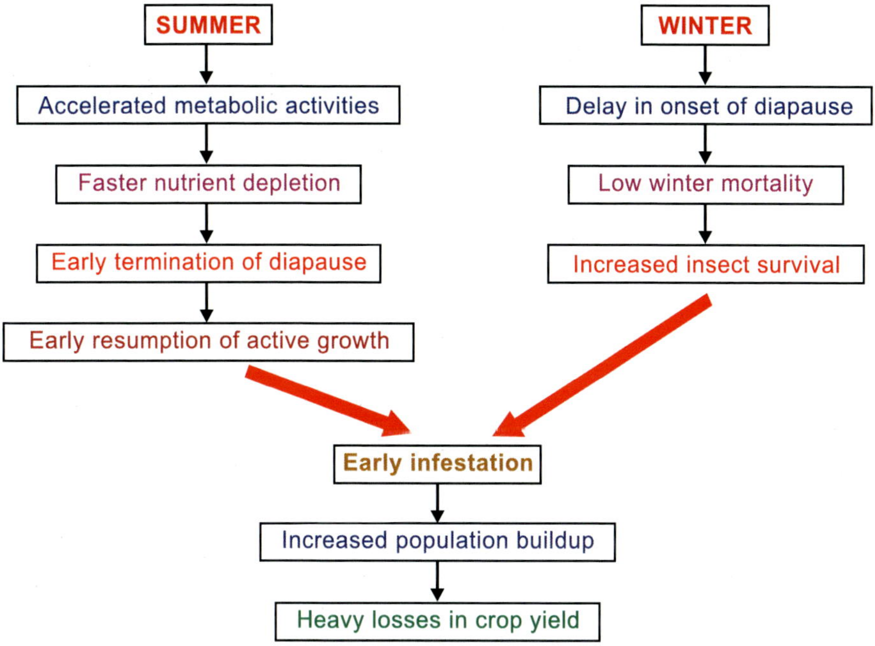

Plate 1.1. Impact of increasing temperature on insect population build up

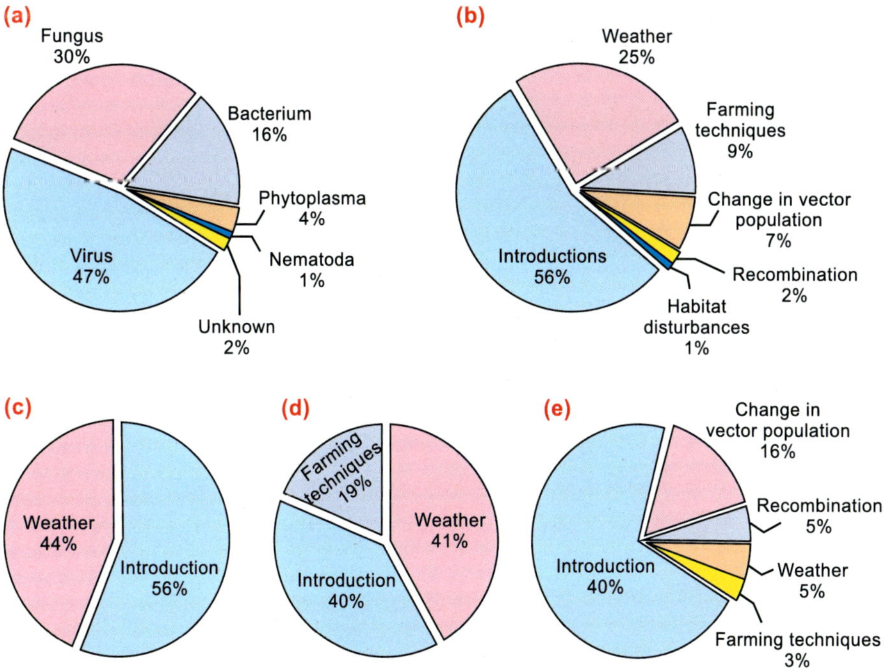

Plate 1.2. Drivers of emerging plant diseases

Chapter 2. PLATES

Plate 2.1. Infestation of mealy bug, *Phenacoccus solenopsis* on cotton

Plate 2.2. Predators of cotton mealy bug. Left - *Aenasius bambawalei*, Middle - *Cryptolaemus* sp. and Right - *Zygogrmma bicolorata*

Plate 2.3. Larva and adult of *Spodoptera litura* on cotton leaves

Plate 2.4. The nymph of mirid bug, *Creontiades biseratense* on cotton boll

Plate 2.5. Cotton boll damaged by stinkbugs

Plate 2.6. False chinch bug, *Nysius raphanus*

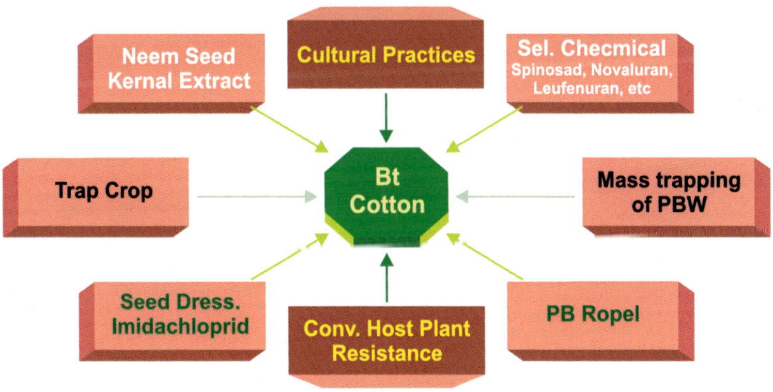

Plate 2.7. Sustainable IPM based on transgenic *Bt* Cotton

Plate 2.8. *Spodoptera litura* on soybean

Plate 2.9. Tobacco caterpillar and adult, *Spodoptera litura* on sunflower leaves

Plate 2.10. Wooly aphid, *Ceratovacuna lanigera* infestation on sugarcane leaf

Plate 2.11. Hopper burn symptoms caused by the rice brown plant hopper.

Plate 2.12. The brown plant hopper has two forms, short winged 'brachypterous' (A) and long-winged 'macropterous' (B) which migrates.

Plate 2.13. Rice swarming caterpillar, Spodoptera mauritia.
Left – Damage to rice crop. *Middle* –Caterpillar. *Right* – Adult

Plate 2.14. Grey weevil, *Myllocerus* spp. on sorghum and pearl millet

Plate 2.15. Leaf hopper, *Pyrilla purpusilla* on sorghum and pearl millet

Plate 2.16. Sorghum leaf roller, *Cnaphalocrosis trapezalis*. *Left*– larva, *Right*– adult

Plate 2.17. Larva and adult of spotted pod borer, *Maruca vitrata* on pigeon pea

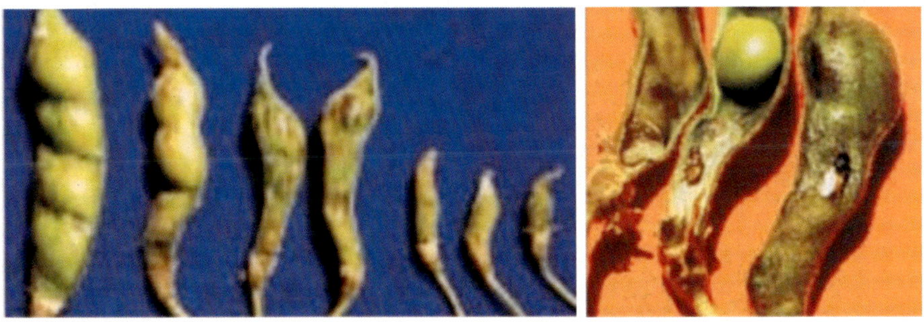

Plate 2.18. *Tanaostigmodes cajaninae* damage on pigeon pea pods

Plate 2.19. Jute field, severely infested with mealy bug. a) basal part, b) terminal part, and c) infested plant in field

Plate 2.20. Leucaena psyllid, *Hereopsylla cubana* infestation on Subabul

Chapter 3. PLATES

Plate 3.1. Spiraling whitefly, *Aleurodicus dispersus* on guava leaves

Plate 3.2. Papaya mealy bug, *Paracoccus marginatus* infestation on leaves and fruits

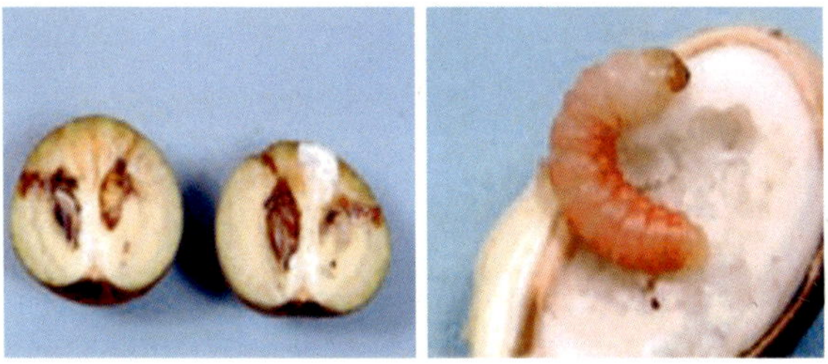

Plate 3.3. Sapota seed borer, *Trymalitis margarias*

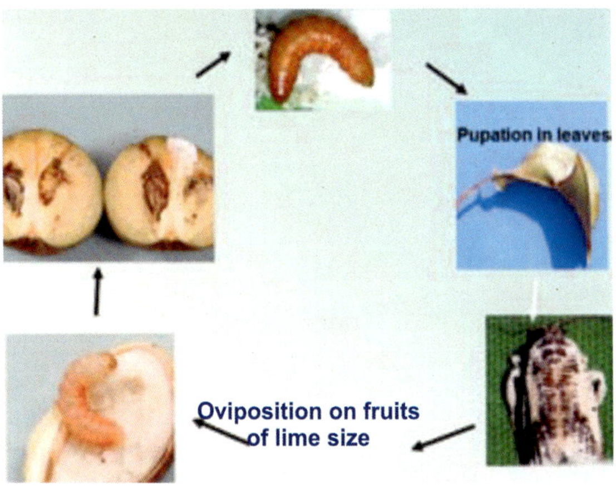

Plate 3.4. Biology of sapota seed borer, *Trymalitis margarias*

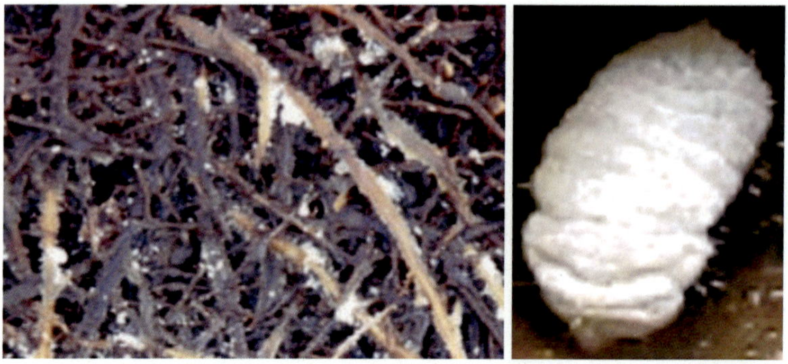

Plate 3.5. Banana root mealy bug, *Geococcus citrinus*

Plate 3.6. Fruit borer, *Deonolis albizonalis* damage on mango fruits

Plate 3.7. Mango leaf weevil, *Rhynchaenus mangiferae*

Plate 3.8. Amla fruits damaged by pomegranate butterfly

Plate 3.9. Red weevil, *Apoderus blandus* on litchi

Plate 3.10. Fruit borer, *Conopomorpha cramerella* on litchi

Chapter 4. PLATES

Plate 4.1. Serpentine leaf miner, *Liriomyza trifolii* on tomato, cowpea, cucurbits and gerbera

Plate 4.2. Giant African Snail, *Achatina fulica*

Plate 4.3. Silver leaf whitefly, *Bemisia argentifolii*

Plate 4.4. Pin worm damage on leaf and fruit of tomato

Plate 4.5. Western flower thrips feeding damage to tomato leaf. Dark specks are frass (excrement).

Plate 4.6. Silver reflective mulch used to repel thrips from tomato.

Plate 4.7. Minute pirate bug, a predator of thrips

Plate 4.8. *Left* – Mining of tomato leaf by serpentine leaf miner, *Liriomyza trifolii*. *Right*– Oviposition by adult

Plate 4.9. Mealy bug, *Phenacoccus solenopsis* infestation on brinjal

Plate 4.10. *Top* – Symptoms of gall midge damage on brinjal flower and fruits.
Bottom – Adult gall midge

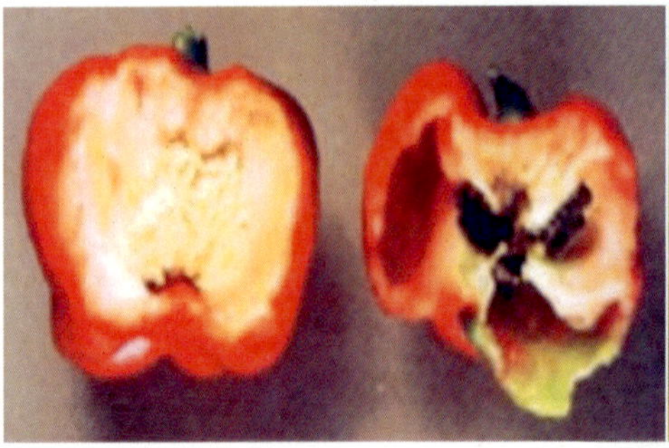

Plate 4.11. Gall midge, *Asphondylia capparis* damage on bell pepper and chilli

Plate 4.12. Potato leaf infested with aphids

Plate 4.13. Diamondback moth, *Plutella xylostella* larva and adult

Plate 4.14. IPM in cauliflower using Indian mustard as a trap crop

Plate 4.15. Spotted pod borer damage on cowpea flowers and pod

Plate 4.16. Hadda beetle, *Henosepilachna vigitioctopunctata*

Plate 4.17. Field bean plume moth, *Sphenarches caffer* larva and adult

Chapter 5. PLATES

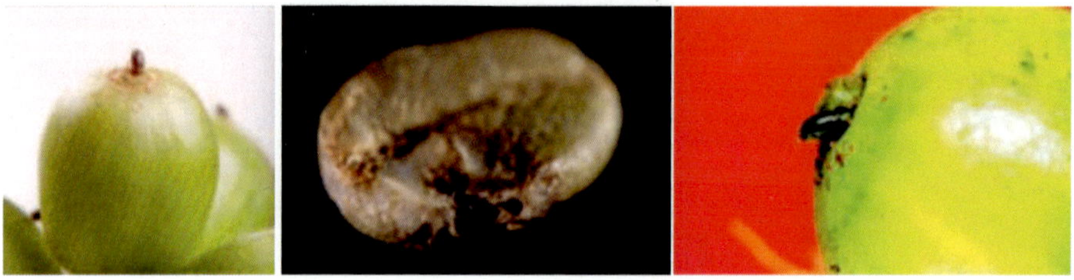

Plate 5.1. Coffee berry borer, *Hypothenemus hampei* damage on coffee beans and adult beetle

Plate 5.2. Mite, *Aceria guerreronis* damage to coconut

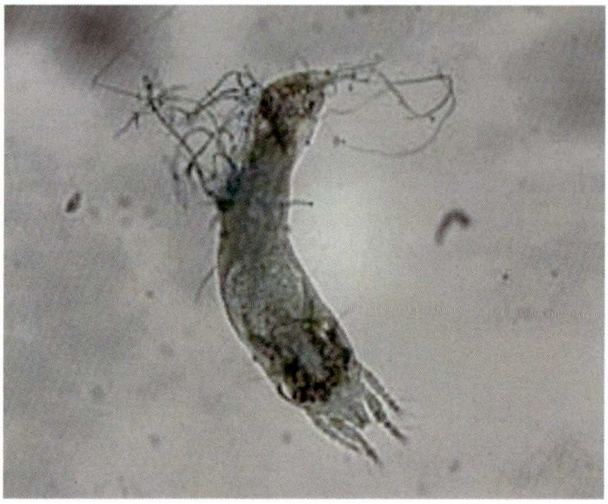

Plate 5.3. *Hirsutella thompsonii* infecting coconut mite

Plate 5.4. Leaf beetle, *Brontispa longissima* damage on coconut plant and leaf

Plate 5.5. Life cycle of the coconut leaf beetle

Plate 5.6. *Left*– Infection of the beetle by *Metarrhizium anisopliae*. *Right*– Coconut leaf beetle parasitoid, *Asecodes hispinarum*

Plate 5.7. *Left*– Rhinoceros beetle damage on base of immature palm, *Right*– Rhinoceros beetle adult

Plate 5.8. *Left*– Pheromone trap for rhinoceros beetle. *Right*– Virus infected larva.

Plate 5.9. Typical damage (necrosis on the shoot) of tea mosquito bug, *Helopeltis antonii* on a young shoot of cashew (*Courtesy:* A. M. Varela, ICIPE).

Plate 5.10. Tea mosquito bug, *Helopeltis antonii* on cashew leaf and nut

Plate 5.11. Weaver ants foraging on cashew apples and nuts (*Courtesy:* A. M. Varela, ICIPE).

Plate 5.12. Corm borer, *Aplosonyx chalybaeus* on colocasia

Plate 5.13. Earwigs feeding on the eggs of corm borer

Chapter 6. PLATES

Plate 6.1. Deformed leaves of erythrina, Galls formed due to wasp infestation and adult gall wasp

Plate 6.2. Tall or windbreak Erythrina 12 weeks after injection with (*Left*) and without (*Right*) imidacloprid

Plate 6.3. Symptoms of eucalyptus gall wasp, *Leptocybe invasa* on leaves and stem

Plate 6.4. Galls on stem, petiole and midrib of Eucalyptus

Plate 6.5. *Aprostocetus gala* female and male

Plate 6.6. Parasitoids of *Leptocybe invasa*

Chapter 7. PLATES

Plate 7.1. Symptoms of peanut bud necrosis virus disease.
Left– Chlorotic mottle or specks on leaves, *Right*– Necrosis of the terminal bud.

Plate 7.2. Peanut bud necrosis virus. *Left*– Healthy and infected plants, *Right*– Healthy and infected pods and kernels.

Plate 7.3. Transmition of PBNV by *Thrips palmi*

Plate 7.4. Symptoms of peanut yellow spot *Tospovirus*

Plate 7.5. Karnal bunt infected wheat seeds showing varying severities of disease.

Plate 7.6. Symptoms of blast disease in wheat. **a-** A completely bleached wheat spike with traces of gray from blast sporulation at the neck (*arrow*) of the spike. **b-** Complete bleaching of a wheat spike above the point (*arrow*) of infection. **c-** Two completely bleached spikes with traces of gray (*upper arrow*) and a lesion (*lower arrow*) from blast sporulation at the base. **d-** Typical eye-shaped lesion (*arrow*) and dark gray spots on a severely diseased wheat leaf. **e-** Mild blast disease-affected slightly shrivelled wheat seeds. **f-** Severe blast-affected shrivelled and pale wheat seeds. **g-** A severely infected rachis with dark gray blast sporulation at the neck (*arrow*) and severely damaged spikelets. **h-** Micrograph of two conidia isolated from the infected spike of wheat (Scale bars in e and f = 1 cm and in h = 10 μ)

Chapter 8. PLATES

Plate 8.1. Bacterial blight disease on pomegranate leaves and fruits

Plate 8.2. Symptoms of pomegranate wilt

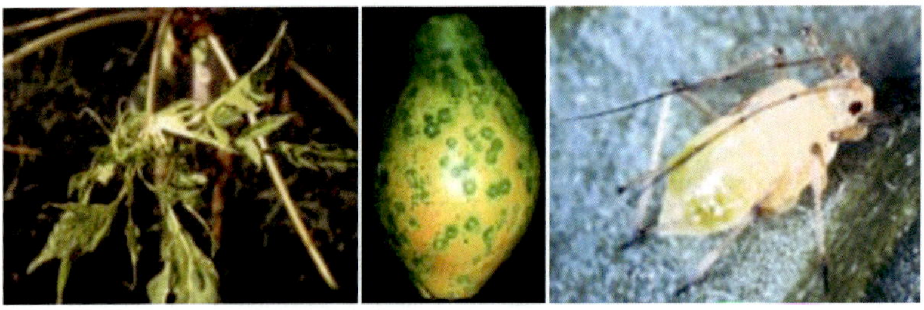

Plate 8.3. PRSV-P symptoms on papaya leaves and fruit, and aphid vector

Plate 8.4. PRSV-W symptoms on squash leaf and fruits.

Plate 8.5. GM papaya (Rainbow) resistant to PRSV (*Right*) and susceptible papaya (*Left*)

Chapter 9. PLATES

Plate 9.1. Potato field devastated by *Phytophthora infestans.*

Plate 9.2. Symptoms of late blight on potato leaves and tuber.

Plate 9.3. *Left*– Symptoms of Tomato yellow leaf curl virus (leaf deformation, yellowing and stunting). *Right*– whitefly vector

Plate 9.4. *Left*– Symptoms of TSWV on tomato plant. *Middle*– necrotic streaks on petiole; *Right*– chlorotic spots on fruit

Plate 9.5. Western flower thrips (WFT) *Frankliniella occidentalis* is a vector of TSWV in tomatoes

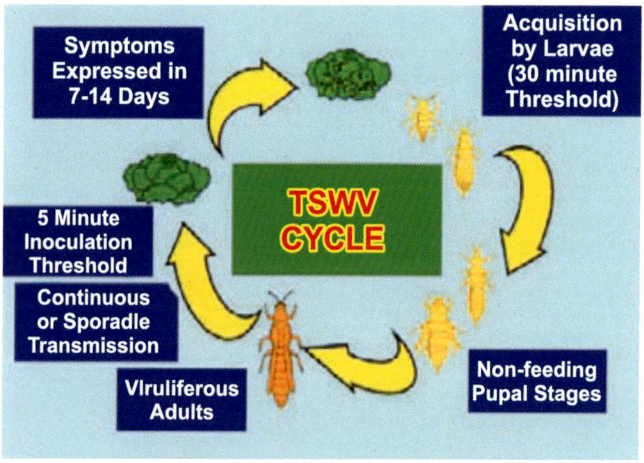

Plate 9.6. The transmission cycle for Tospovirus by thrips (*Photo:* Ron Mau, University of Hawaii).

Plate 9.7. Effect of rouging on the incidence of TSWV (*Left*– Rouging – Removal of symptomatic seedlings before transplanting and within 45 days after planting. *Right*– No rouging - Transplanting all seedlings (no removal of symptomatic seedlings)

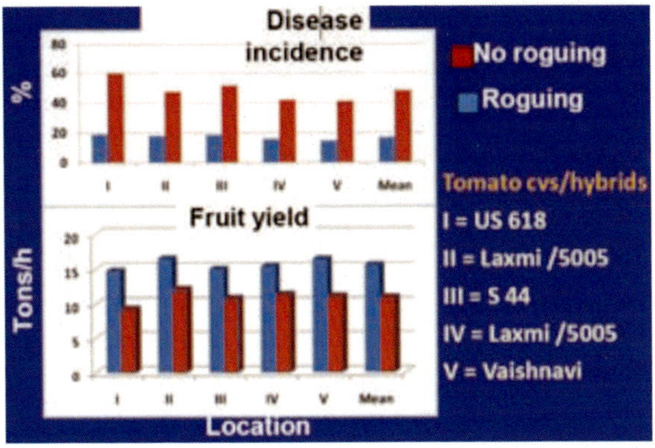

Plate 9.8. Effect of rouging on the incidence of TSWV and fruit yield

Plate 9.9. Tomato in net house to avoid thrips entrance

Plate 9.10. Sticky traps for thrips monitoring

Plate 9.11. Predatory mites (*Left– Transieus montdorensis* and Right– *Amblyseius cucumeris*)

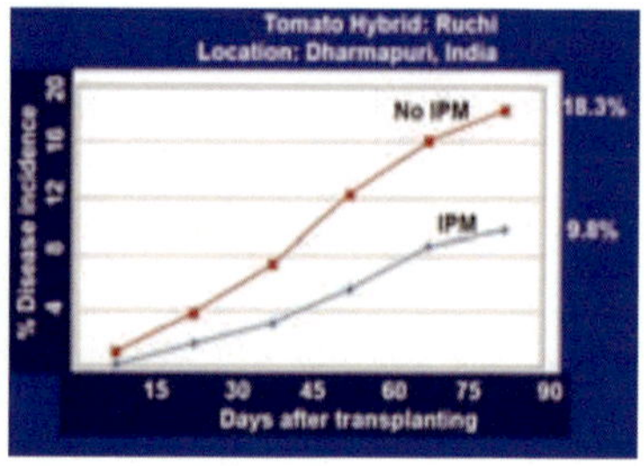

Plate 9.12. Effect of IPM on the incidence of TSWV

Plate 9.13. Iris yellow spot virus symptoms on onion (diamond shaped necrotic lesions on the onion stipes, and collapse and loss of the seed head)

Plate 9.14. Bud necrosis virus disease symptoms on watermelon.

Plate 9.15. Thrips vector *Thrips palmi*

Chapter 10. PLATES

Plate 10.1. INSV symptoms (ringspots) on garden Impatiens (*Left*) and Gloxinia (*Right*)

Plate 10.2. INSV symptoms Begonia (veinal necrosis on leaves and circular pigmentation patterns on petals) (*Right*)

Plate 10.3. *Left*– Typical rose leaf with downy mildew. *Right*– Rose buds infected with downy mildew.

Plate 10.4. Symptoms of bacterial wilt on mature field of ginger (*Left*) and on young immature ginger plant (*Right*)

Plate 10.5. Severely rotten ginger rhizomes infected by bacterial wilt (*Left*) and oozing of the white, milky bacterial stream from the discoloured, necrotic centre (*Right*)

Plate 10.6. Symptoms of cassava mosaic disease (CMD) infection on leaves (mosaic and leaf distortion)

Plate 10.7. Reduction in tuber size caused by CMD (*Left*– Healthy, *Right*– Infected with CMD)

Plate 10.8. Production of CMD-free plants through meristem culture (*Left*) and hardening of plants (*Right*)

Plate 10.9. Symptoms of leaf blight on taro foliage

Plate 10.10. Collar rot symptoms on elephant foot yam plant (*Left*) and sclerotial bodies at the base of stem (*Right*)

Chapter 11. PLATES

Plate 11.1. Field symptoms of rice root-knot nematode – patches of stunted and chlorotic rice plants.

Plate 11.2. *Left*– Rice roots showing terminal and horse-shoe galls caused by *Meloidogyne graminicola. Middle*– Characteristic hooked, root tip galls on rice. *Right*– Female nematodes and eggs inside rice root gall.

Plate 11.3. Wheat crop as a green bridge supporting the *M. graminicola* population. *Left*– Field symptoms on wheat. *Right*– Galling on wheat roots due to *M. graminicola*

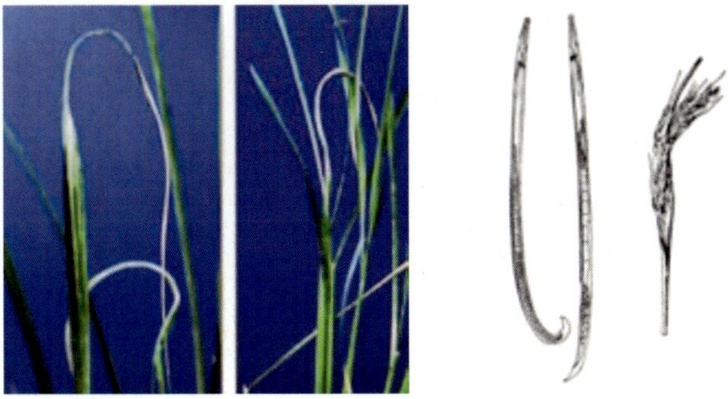

Plate 11.4. *Left*– Symptoms of white tip of rice.
Right– Male and female white tip nematodes

Plate 11.5. Distinct patches of yellowed and stunted plants caused by the cereal cyst nematode.

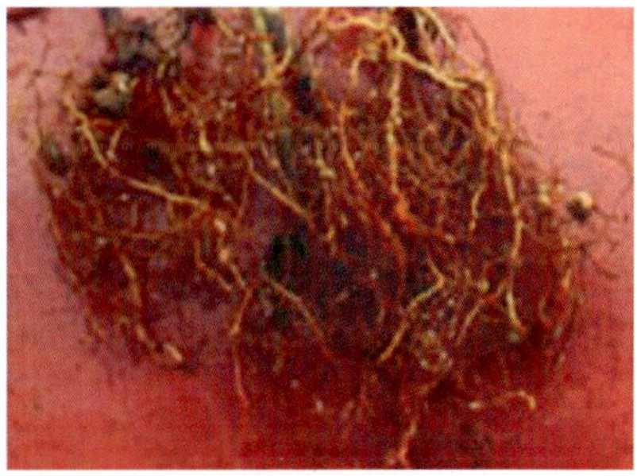

Plate 11.6. Symptoms of pearly root caused by *Heterodera cajani* in pigeon pea

Plate 11.7. Peanut fields exhibiting symptoms of extensive damage due to a severe infestation of the root-knot nematode *Meloidogyne arenaria.*

Plate 11.8. Galling of peanut pods and pegs due to infection by *Meloidogyne arenaria.*

Plate 11.9. Comparison of growth of the root-knot nematode resistant peanut cultivar NemaTAM (Left) with the growth of the susceptible cultivar Florunner (Right) in a field infested with *Meloidogyne arenaria.*

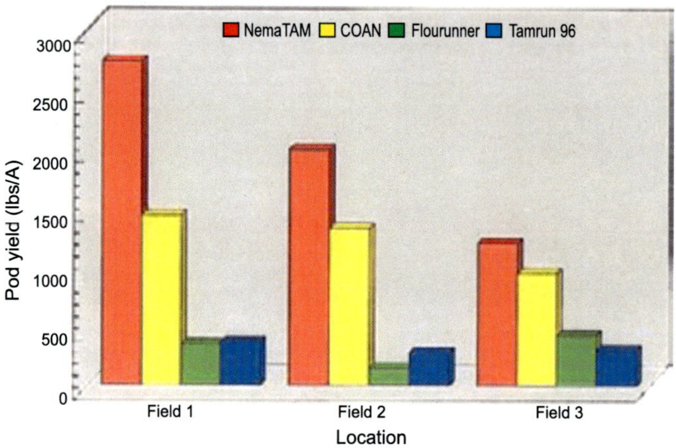

Plate 11.10. Comparison of pod yields of the root-knot nematode resistant cultivars COAN and NemaTAM with the yields of the susceptible cultivars Florunner and Tamrun 96 in three fields infested with *Meloidogyne arenaria.*

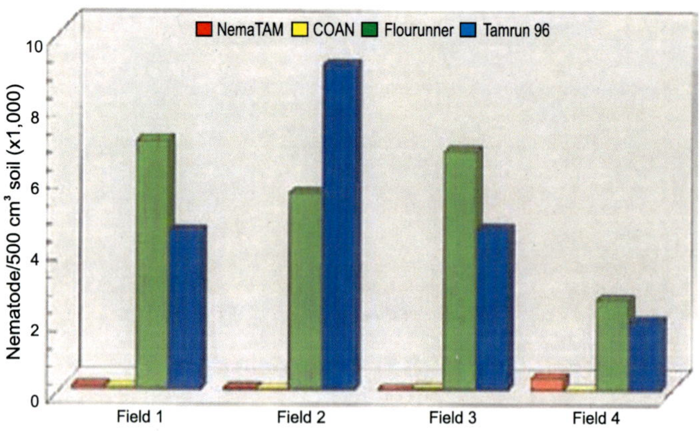

Plate 11.11. Comparison of population densities of *Meloidogyne arenaria* at harvest of the resistant cultivars COAN and NemaTAM with the population densities at harvest of the susceptible cultivars Florunner and Tamrun 96.

Chapter 12. PLATES

Fig. 12.1. Symptoms of root-knot nematode on guava

Plate 12.2. Effect of inoculation of *Meloidogyne enterolobii* and *Fusarium solani* alone or in combination on the root system of guava. From left to right: Uninoculated control; inoculated with Fusarium solani alone; inoculated with *F. solani* alone combined with physical injury of the roots with a knife; inoculated with *M. enterolobii* only; and inoculated with *M. enterolobii*, followed 21days later with *F. solani*.

Plate 12.3. Effect of *Meloidogyne enterolobii* and *Fusarium solani* on development of the root and aerial part of guava plants with root system divided into two plastic container (I and II), in greenhouse A) both halves of the root system non-inoculated; B) half inoculated with the fungus / and half non-inoculated; C) half inoculated with nematode / and half non-inoculated; D) half inoculated with nematode / and half inoculated with the fungus; E) half inoculated with nematode and 21 days later inoculated with the fungus / and half non-inoculated; F) detail of half of the root system inoculated only with *F. solani*; G) detail of half of the root system inoculated only with *M. enterolobii*; H) detail of half the root system inoculated with *M. enterolobii* and 21 days later with *F. solani*.

Plate 12.4. Symptoms of root-knot nematode on pomegranate. *Left*– yellowing and stunting of pomegranate plant. *Right*– Wilted pomegranate plant.

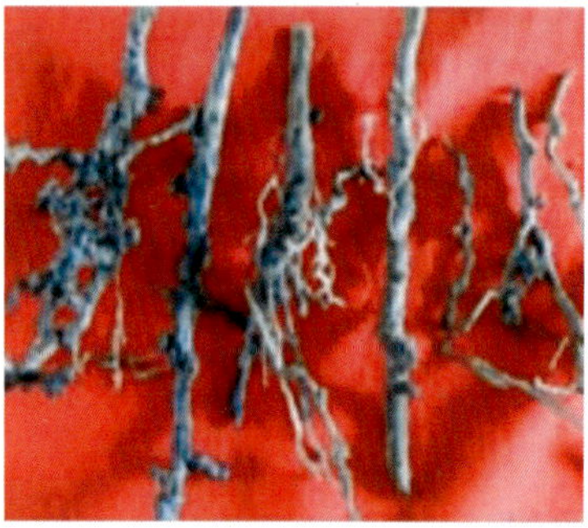

Plate 12.5. Severe root galling on pomegranate

Plate 12.6. Wilted plants due to *Ceratocystis fimbriata*

Plate 12.7. Brown dicolouration of vascular tissue and wood due to *Ceratocystis fimbriata*

Plate 12.8. Shot hole borer with larva and pupa

Plate 12.9. Severe root infestation of pomegranate with root-knot nematodes

Plate 12.10. Mulberry roots infected with root-knot nematode

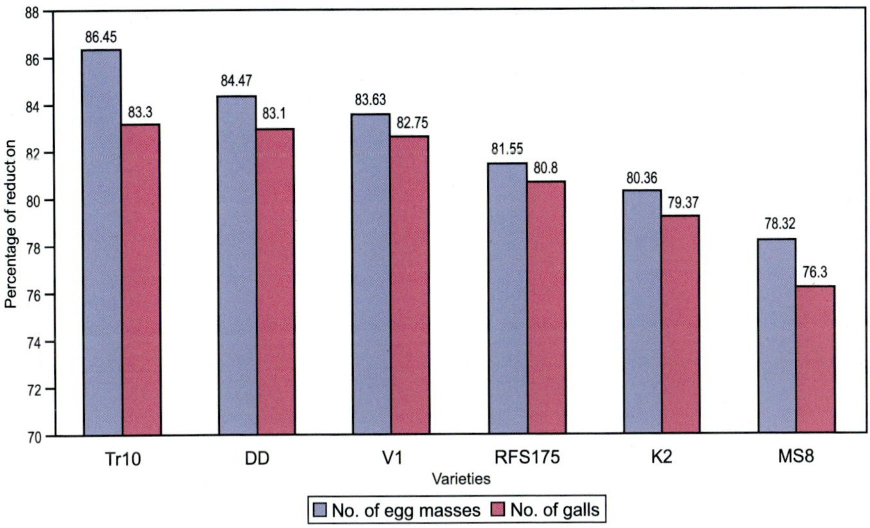

Plate 12.11. Integrated management of root knot nematode *Meloidogyne incognita* in different mulberry varieties

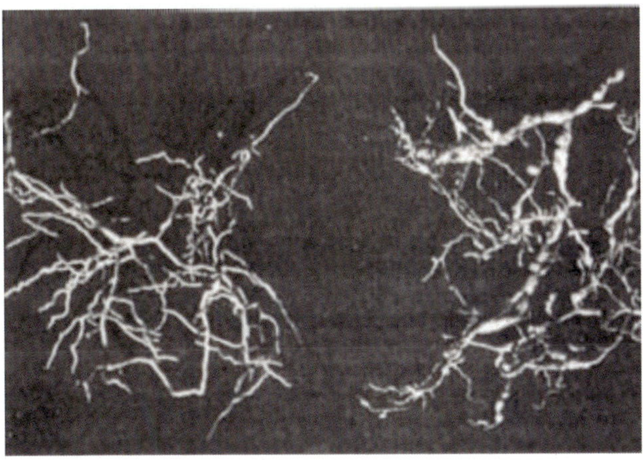

Plate 12.12. Acid lime roots infected with *Meloidogyne indica*. *Left*– Healthy roots; *Right*– Infected roots.

Chapter 13. PLATES

Plate 13.1. *Left*– A patch of potato field infected with PCN. *Right*– Cyst nematode on potato. *Left*– Infected, *Right*– Healthy.

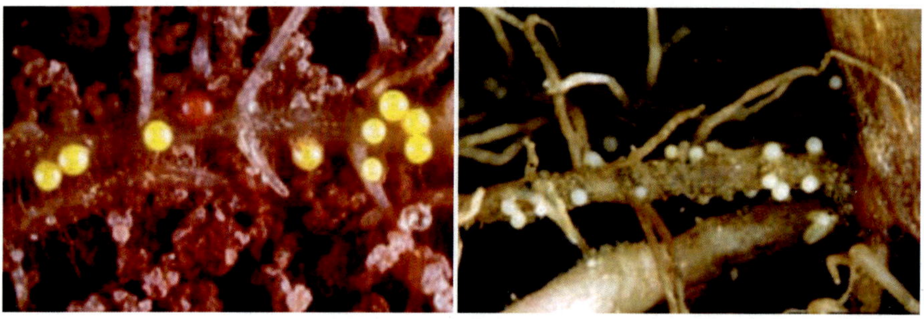

Plate 13.2. Potato roots infected with *Globodera rostochiensis* (Left) and *G. pallida* (right)

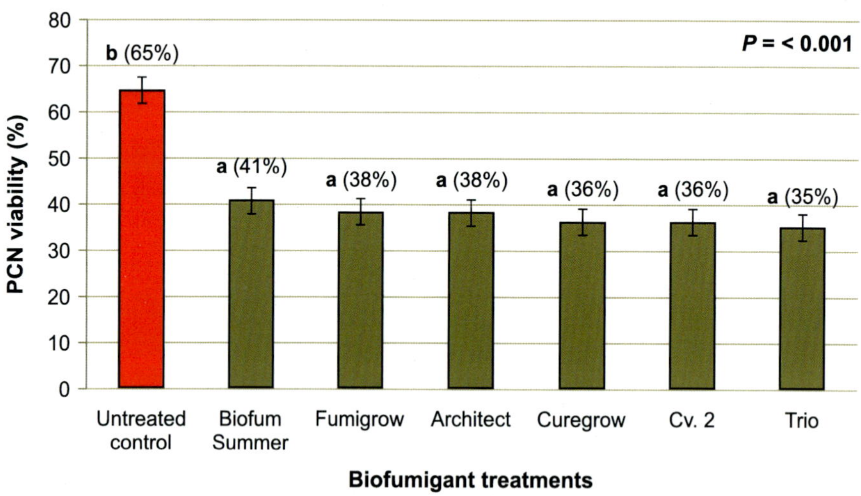

Plate 13.3. Biofumigation effects on PCN viability (**Biofum Summer** - *Rhaphanus sativus*, *Brassica carinata* and *Sinapis alba* blend; **Fumigrow** - *B. juncea, S. alba* & *Eruca sativa* blend; **Architect** – *S. alba*; **Curegrow** – *S. alba* & *B. juncea* blend; **Cv. 2** – *B. juncea*; **Trio** – *E. sativa*).

Plate 13.4. *Left*– Golden potato cyst nematode (*Globodera rostochiensis*) damage to potato crop. *Right*– Resistant variety

Plate 13.5. Tomato under protected cultivation

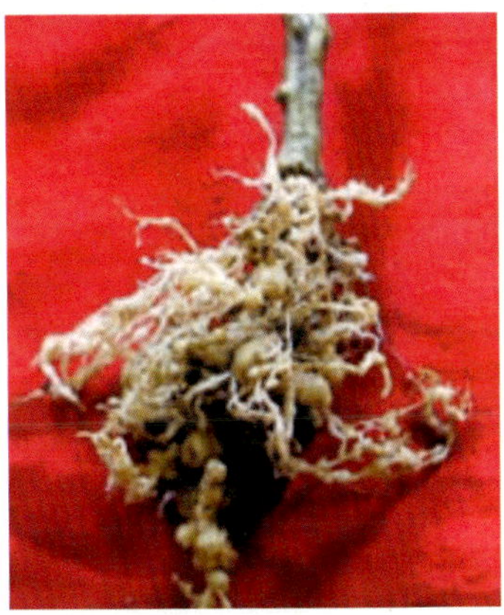

Plate 13.6. Heavy galling of tomato roots infected with *Meloidogyne incognita*

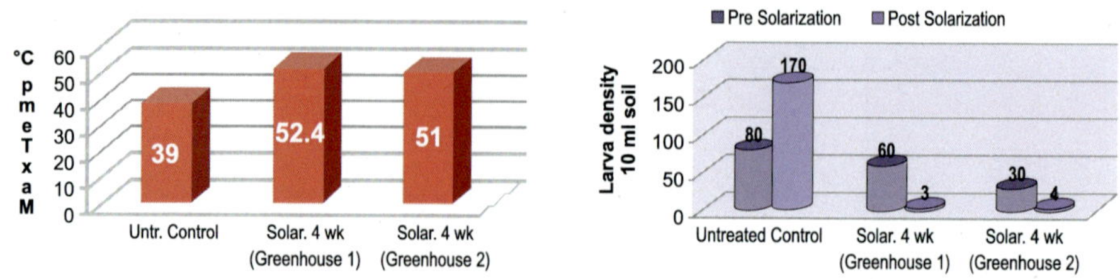

Plate 13.7. Effect of soil solarization on temperature increase and reduction in nematode population. *Left*– Maximal soil temperatures during soil solarization. *Right*– Effect of soil solarization on the root-knot nematode population densities in soil.

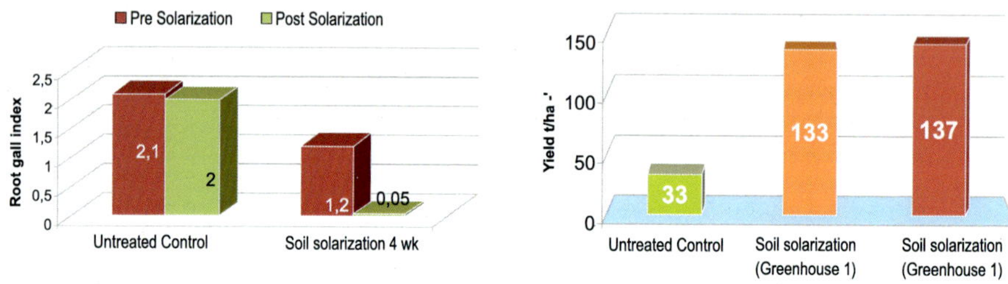

Plate 13.8. Effect of soil solarization on root gall index and fruit yield. *Left*– Effect of soil solarization on root galling due to root-knot nematode. *Right*– Effect of soil solarization on tomato fruit yield

Plate 13.9. The simultaneous occurrence of both root-knot nematode (*Meloidogyne* spp.) and Fusarium wilt disease (*Fusarium oxysporum*) causing enhanced disease development and tomato yield loss

Plate 13.10. Coloured capsicum under greenhouse

Plate 13.11. Root-knot nematode infestation on bell pepper

Plate 13.12. Cucumber under protected cultivation

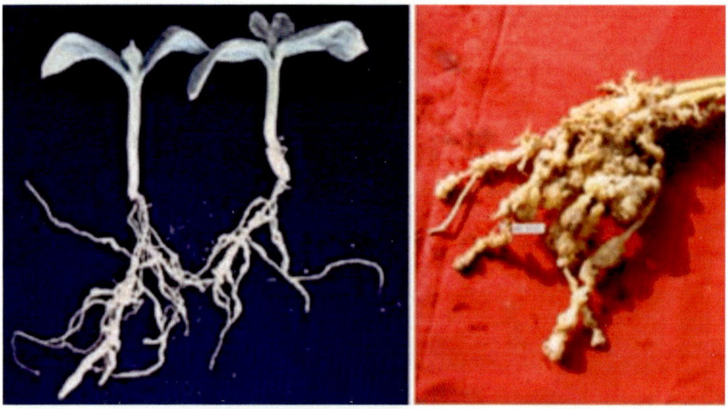

Plate 13.13. Root-knot nematodes on cucumber

Plate 13.14. Effect of different treatments on root growth of cucumber plants infested by *M. incognita*

Plate 13.15. Comparison of root galling due to *M. incognita* in *Paecilomyces lilacinus* treated plant with control

Chapter 14. PLATES

Plate 14.1. Symptoms of *Aphelenchoides besseyi* damage on tuberose flowers.

Plate 14.2. Enrichment of organic amendments with biopesticides

Plate 14.3. Carnation roots infected by root-knot nematodes

Plate 14.4. Management of root-knot nematodes on carnation using organic amendments enriched with biopesticides (*Left*– untreated, *Right*– treated) (Shylaja, 2005)

Plate 14.5. Root-knot nematode and Fusarium wilt disease complex in carnation

Plate 14.6. Symptoms of root-knot nematode infection on gerbera

Plate 14.7. Root-knot nematode infested plant and roots of coleus

Plate 14.8. Symptoms of root-knot and wilt disease complex on coleus

Plate 14.9. Efficacy of talc formulation of PGPR strains against root-knot nematode on coleus under field conditions (*Left*– RB 50. *Right*– Control).

Plate 14.10. Root-knot and collar rot disease complex on coleus. *Left*– Infected, *Right*– Healthy

Plate 14.12. Root knot infested plants and roots of ashwagandha

Plate 14.11. Sequential inoculation of *M. incognita* 7 days prior to inoculation fungal pathogens. 1 – Control, 2 – Nematode → *Fusarium*, 3 – Nematode → *Rhizoctonia*, 4 –Nematode → *Sclerotium*, 5 –Nematode → *Fusarium* + *Rhizoctonia* + *Sclerotium*.

Plate 14.13. Symptoms of root-knot and root rot complex on ashwagandha. *Left* to *Right*– Rotting, decaying and disintegration of roots

Plate 14.14. Efficacy of talc formulation of PGPR strains in ashwagandha under field conditions (*Left* - RB 50. *Right* – Control)

Section I
INTRODUCTION

1

Emerging Crop Pest Problems – An Overview

1.1. INTRODUCTION

Agriculture faces two major challenges i.e. production of sufficient food to feed the growing population, and prevention of environmental degradation. In order to meet the projected demands of a growing population expected to exceed 9 billion by 2050, farmers in the developing world must double food production, a challenge complicated by the effects of climate change and growing competition for land, water and energy. Pests (insect and mite pests, disease pathogens, nematodes and weeds) are one of the important limiting factors in crop production and productivity. Crop protection always aims to avoid or prevent crop losses or to reduce them to economically acceptable level but losses due to pests had been enormously increased. Food plants of the world are damaged by more than 10,000 species of insects, 30,000 species of weeds, 100,000 disease pathogens (caused by fungi, viruses, bacteria and other microorganisms) and 1000 species of nematodes (Hall, 1995; Dhaliwal *et al.*, 2007). However, less than 10 per cent of the total identified pest species are generally considered as major pests.

Plant pests reduce the availability of food and affect quality. Although quantification of losses and potential losses due to pests is limited, historically, plant pest introduction, establishment, re-emergence and outbreaks have resulted in major food problems either directly through yield reductions of food crops, or indirectly through the reduction of yields of cash crops and loss of consumer confidence, e.g. potato late blight and locusts.

Damage by crop pests is one of the major constraints to increase food production. Therefore, reduction in losses caused by pests is the obvious strategy for increased food supply.

1.2 CROP LOSSES DUE TO PESTS

Global crop losses due to pests are estimated at more than 50 per cent of potential crop output. Insect pests are estimated to have caused the destruction of 15 per cent of crops, and 13 per cent each by pathogens and weeds, and post-harvest pest infestations another 10 per cent. Losses of 100 per cent occur in specific circumstances, creating enormous variability in productivity and risks to the livelihoods of farm families.

In spite of a variety of control measures applied against pests, crop losses have consistently shown an increasing trend (Dhaliwal and Koul, 2010). Despite the annual investment of US $35,000 for the application of there million metric tons of pesticides, plus the use of various biological and other non-chemical controls worldwide, global crop losses remain a matter of concern (Pimentel, 2007; 2009). Presently, 35-42% of growing and stored crops are lost to pests, pathogens, and weeds annually, amounting to losses of US$ 244 billion worldwide annually (Pimental, 1997).

Although crop protection aims to avoid or prevent crop losses or to reduce them to economically acceptable levels, the availability of quantitative data on the effect of different categories of pests is very limited.

Since crop production technology and especially crop protection methods are changing, Oerke (2006) updated the crop loss data for major food and cash crops for the period 2001-03. The actual losses due to various pests have been estimated as 26-29 per cent for soybean, wheat and cotton, and 31, 37 and 40 per cent for maize, rice and potatoes, respectively (Table 1.1).

Table 1.1. Current global losses (%) due to various categories of pests in major crops

Crop	Animal pests	Weeds	Pathogens	Viruses	Total
Cotton	12.3	8.6	7.2	0.7	28.8
Maize	9.6	10.5	5.8	5.2	31.3
Potato	10.9	8.3	14.5	6.6	40.3
Rice	15.1	10.2	10.8	1.4	37.5
Soybean	8.8	7.5	8.9	1.2	26.4
Wheat	7.9	7.7	10.2	2.4	28.2
Average	**10.8**	**8.8**	**10.0**	**2.5**	**32.1**

Source: Oerke (2006)

Losses due to insect pests in Indian agriculture have been estimated from time to time (Pradhan, 1964; Dhaliwal *et al.*, 2004). In general, the losses in post-green revolution era (Dhaliwal *et al.*, 2004) have shown an increasing trend than in the pre-green revolution era (Pradhan, 1964). Overall, the losses increased from 7.2 per cent in early 1960s to 23.3 per cent in early 2000s (Table 1.2). The maximum increase in loss occurred in cotton (18.0 to 50.0 %), followed by other crops like sorghum and millets (3.5 to 30.0 %), maize (5.0 to 25.0 %) and oilseeds (other than groundnut) (5.0 to 25.0 %).

Table 1.2. Crop losses (%) due to insect pests during pre- and post-green revolution in India

Crop	Pre-green revolution (early 1960's) (1)	Post-green revolution (early 2000's) (2)	Changes in loss (2-1)
Cotton	18.0	50.0	+32.0
Groundnut	5.0	15.0	+10.0
Other oilseeds	5.0	25.0	+20.0
Pulses	5.0	15.0	+10.0
Rice	10.0	25.0	+15.0
Maize	5.0	25.0	+20.0
Sorghum and millets	3.5	30.0	+26.5
Wheat	3.0	5.0	+2.0

Crop	Pre-green revolution (early 1960's) (1)	Post-green revolution (early 2000's) (2)	Changes in loss (2-1)
sugarcane	10.0	20.0	+10.0
Average	**7.2**	**23.3**	**+16.1**

Source: Dhaliwal *et al.* (2004)

Plant-parasitic nematodes constitute a major constraint to agriculture. Estimates of their crop-losses are important for establishing research, extension, and budget priority. Regulatory policy action, pesticide impact assessment, resource allocation, and programme prioritisation are usually contingent upon crop loss data.

In the tropical and sub-tropical climates, crop production losses attributable to nematodes were estimated at 14.6% compared with 8.8% in developed countries. Perhaps more importantly, only ~0.2% of the crop value lost to nematodes is used to fund nematological research to address these losses (Sasser and Freckman, 1987). It is estimated that the potato cyst nematodes (*Globodera rostochiensis* and *G. pallida*) alone account for an estimated ~$70 million per annum in UK or 9% of UK production (DEFRA, 2010). In the USA a survey of 35 States on various crops indicated nematode-derived losses of up to 25% (Koenning *et al.*, 1999). More recently Handoo (1998) estimated global crop losses due to nematode attack in the region of $80 billion.

Estimated overall average annual yield loss of the world's major horticultural crops due to damage by plant parasitic nematodes is estimated to be 13.54% (Table 1.3). For the 7 horticultural crops (left-hand column) that stand between man and starvation (life sustaining horticultural crops), an estimated annual yield loss of 12.77% is reported (Sasser and Freckman, 1987).

Table 1.3. Estimated annual yield losses due to damage by plant parasitic nematodes – world basis

Life sustaining horticultural crops	Loss (%)	Economically important horticultural crops	Loss (%)
Banana	19.7	Cocoa	10.5
Cassava	8.4	Citrus	14.2
Coconut	17.1	Coffee	15.0
Field bean	10.9	Cowpea	15.1
Potato	12.2	Eggplant	16.9
Sugar beet	10.9	Grapes	12.5
Sweet potato	10.2	Guava	10.8
		Melons	13.8
		Okra	20.4
		Ornamentals	11.1
		Papaya	15.1
		Pepper	12.2
		Pineapple	14.9
		Tea	8.2
		Tomato	20.6
		Yam	17.7
Average	**12.77**	**Average**	**14.31**
	Overall average – 13.54%		

Source: (Sasser and Freckman, 1987)

Recent questionnaire results of the study for world-wide estimates of crop losses due to plant-parasitic nematodes are presented herein. For the 20 life-sustaining crops, an average worldwide crop loss due to nematodes is estimated to be 12.6% which equalled $215.77 billion annual yield loss. The other crops that represent a miscellaneous group important for food or export value have an average annual yield loss of 14.45% which equalled $142.47 billion. Losses for the total 40 crops average is 13.5%. So, worldwide crop losses due to nematodes on 37 crops are estimated at $358.24 billion annually based on the 2010-2013 production figures and prices. These figures are staggering, and the real figure, when all crops throughout the world are considered, probably exceeds such estimations. Our current estimates far exceed previous ones probably because of challenging issues such as reduced number of effective nematicides available and limitation in their use due to environmental issues, increased adoption of intensive agriculture, climate change, occurrence of resistance-breaking pathotypes on economically important crops, and potential introduction of quarantine-nematodes. Basic and applied nematological research should be more oriented to provide better management of plant-parasitic nematodes in an economically and environmentally beneficial manner (Abd-Elgawad, 2014).

1.3 EMERGING PEST PROBLEMS

Crop production provides about 84 per cent of global food, feed and fibre needs and virtually all other human endeavours depend on food security. Green revolution led to increased food supplies but caused several ecological, environmental and socio-economic problems. This technology relied on use of dwarf, semi-dwarf, high yielding varieties, and increased use of agrochemicals and irrigation. These practices favoured build-up of pests and increased their intensity. Several minor pests became major and new pest problems appeared. Misuse of pesticides led to the problem of pesticide resistance, resurgence and contamination of environment. Biotechnology has added new dimension in pest management but its potential and effect on humans are debatable. In addition, the climate change has profound effect on the intensity of pest problems (Ramamurthy *et al.,* 2009). Hence, there is considerable shift in pest problems and crop losses. Many pests having minor or secondary status are becoming major problems. The severity of pest problems has been changing with the developments in agricultural technology and modifications of farming practices. Crop pests, including diseases, insects, nematodes, and weeds should be overcome to meet future food needs.

1.3.1. What is an Emerging Pest?

So what are the characteristics of emerging plant pests? They have increased in incidence, geography or host range, they have changed in pathogenesis, or they are newly evolved and they may have been discovered or newly recognized.

- The pest reported from an area from a particular crop whose population has been increasing over a period of time considerably causing or likely to cause economic damage above economic injury level (EIL).
- An already known pest whose incidence or geographical distribution has been increasing notably, but it can also be a newly described indigenous or invasive species.

- The pest, whose status has been changing from minor to major or secondary to primary.

In recent years, with changes in the cropping systems and climate, and introduction of highly input intensive high yielding varieties/hybrids, a shift in pest status has been observed.

In light of national concerns for crop biosecurity, new, emerging, and threatening plant pests have received more attention from all the interested parties. Thus, emerging pests have themselves become new drivers of global environmental change. They are responsible for:

- Causing extinction of endangered species.
- Altering the ratios of predators, prey, competitors, and recyclers necessary for healthy, well-functioning ecosystems.
- Altering habitat already threatened by fragmentation and global climate change.

1.3.2. Insect Pests

The change of insect pest scenario in Indian agriculture was more influenced by the introduction of new agricultural production technology based on high yielding varieties and increased use of inputs like irrigation water, fertilizers and pesticides during the last four decades. Many ecological attributes, which are favourable for plant growth, also favour multiplication of pests. Basic principles of herbivore were highlighted in several pestilence episodes. Most of these intense pestilences in various crops could be mitigated through regulated plant nutrition system. The quality of plant nutrition is the determinant of the intensity and build-up of agricultural pests, which are capable of adapting to large changes in the environment. Consequently, there is a change in the insect pest scenario in almost all the crops due to changes in agro-ecosystem.

The *desi* tall varieties of rice were found to suffer from only five insect pests, viz. yellow stem borer, hispa, rice bug, rice grasshopper, and surface grasshopper. Following the introduction of high yielding dwarf varieties and many river command-area projects, and also the associated chemical intensive technology, there has been a considerable increase in area under rice. Extensive sole cropping of rice is very common in many river and tank command areas, which favoured the development and multiplication of multitude of pests. Similarly, leaf folder has emerged as a serious pest in Kerala, Tamil Nadu, Andhra Pradesh, Karnataka, Gujarat, Punjab and Haryana. Gall midge has been observed to occur during *rabi* in Godavari delta of Andhra Pradesh and coastal Maharashtra. Gundhi bug, which was earlier confined to north and north-eastern states of India, has been frequently causing serious damage to rice in parts of Tamil Nadu and Andhra Pradesh. There is a need to cultivate mosaics of varieties of crop plants in monocropping systems in large geographic tracts to prevent generation and invasions of various pests as a tool to avoid enormous biotic stresses.

The incidences of several insect pests like mealy bugs, particularly *Phenacoccus solenopsis* on cotton; sugarcane woolly aphid, *Ceratovacuna lanigera* on sugarcane; *Pieris brassicae* on crucifers; and tobacco caterpillar, *Spodoptera litura* on several crops; have shown an increasing trend. There was a decline in the pest status of bollworm (*Helicoverpa armigera)* in cotton, whereas the sap feeders, *viz.* aphids, jassids, mirids and mealy bugs are emerging as serious pests. Recently, the occurrence of resistance in *Bt* cotton to *Helicoverpa zea* in different regions had been reported.

New and re-emerging plant pests have raised fear of their potential impact on livelihoods, food security and global markets (Table 1.4).

Table 1.4. Newly emerging major pests during last decade in Northern Plain Zone in India

Crop	Pest/s
Cotton	Mealy bug, whitefly, tobacco caterpillar
Wheat, barley, oat	Aphid species
Rice	Brown plant hopper (BPH), white backed plant hopper (WBPH), leaf folder
Maize, sorghum	Shoot fly, Pyrilla
Vegetable crops	Cole crops: Cabbage caterpillar, tobacco caterpillar
	Tomato, cucurbits: Leaf miner
	Okra, Cole crops, chilli: *Helicoverpa armigera*
	Brinjal, okra: Spider mites
Oil seeds	Tobacco caterpillar, cabbage caterpillar
Fruit crops	Fruit moth (citrus), mealy bugs, fruit flies

1.3.3. Disease Pathogens

Plant disease epiphytotics have been a major cause to change agricultural patterns and even food habits in many parts of the world. For example, cereal rye was replaced by potato due to ergot (*Claviceps purpurea*) disease, and potato was replaced by wheat due to late blight (*Phytophthora infestans*) in Europe. Some other similar devastating diseases which had far reaching impact on agriculture were: wheat rust in Mexico; coffee rust in Sri Lanka and Brazil; southern corn blight in the USA; cassava mosaic and maize rust in Africa; Panama disease (*Fusarium oxysporum* f sp *cubense)* of banana in South America and Liberia; bunchy top of banana in Australia; swollen shoot of cocoa in Ghana; bacterial rot (*Pseudomonas solanacearum*) of potato in Kenya and many others. In India, the brown spot of rice (*Cochliobolus miyabeanus*) caused the great Bengal famine in 1943; red rot of sugarcane caused severe ephiphytotics in Uttar Pradesh and Bihar during 1938-42; wheat rust in Uttar Pradesh and Madhya Pradesh in 1946-47; *Helminthosporium* blight of wheat and barely in Uttar Pradesh, Bihar during 1979-81; leaf curl of cotton in Punjab and Rajasthan during 1994-95; tungro disease of rice in Punjab during 1998-99; and necrosis disease of sunflower in Karnataka are just a few examples of the serious problems caused by plant diseases. Ironically most of these diseases have caused destruction in crops grown under rainfed agriculture, indicating vulnerability of such cropping systems.

In wheat, rusts were the most serious problems until the mid-seventies, but currently with the wide use of rust-resistant varieties (although there is overwhelming evidence of the yellow rust resistant wheat variety, PB 343 appear to be susceptible to this dreadful disease in the 'green revolution area' of the country), a minor disease of the past, namely, Karnal bunt has assumed serious-proportions. Karnal bunt in some wheat growing states, breakdown of resistance in PB 343 variety for yellow rust race, 78 S 84 have emerged as a serious disease problems. The rice tungro virus and the bacterial leaf blight of rice are the most devastating diseases in the new varieties. Sheath blight has become serious on rice in the non-conventional areas. Maize and millets are now devastated by downy mildews. Blight of cotton became a major problem when the indigenous diploid cottons were replaced by exotic cotton tetraploid cottons. The new exotic cotton varieties are also highly susceptible to a new disease known as 'parawilt of cotton' of unknown etiology. During the past few years the viruses of cotton (particularly whitefly transmitted leaf curl) have become dangerous and required special efforts to check the spread to different areas. Such examples of changing disease scenario are available for pulses, oilseeds, vegetables, fruits, etc. i.e. the crops in which productivity has increased tremendously. Certain diseases of complex or unknown etiology, e.g., parawilt of cotton, coconut root wilt, dieback in citrus, mango malformation, crown rot of oil palms, brown-bast disease of rubber, etc need special efforts to develop management practices to minimize the losses.

Plant diseases can have a huge impact on global food security and national food security. One of the characteristics of pathogens and drivers of emerging plant diseases, a large percentage of introductions are due to viruses because they are not easily detected in the plant material. The largest way that pathogens move into plant material is through the introduction of plant species that are infected with the pathogen. Plant introductions are a major way in which bacteria, viruses and fungi are introduced, but weather has a larger impact on the fungal and the bacterial plant pathogens. So pathogens move in trade and they move in infected plant material, and globalized trade has been a major driver for the increased frequency of plant pathogen introductions.

Viruses, fungi and bacteria are the major pathogens causing emerging plant diseases. Viruses cause just under half (47%) of the reported emerging plant diseases, fungi represent a moderate proportion (30%), and bacteria cause a lower proportion (16%).

The truly scary possibility is that a new or re-emerging plant disease could decimate one of the few crops — rice, wheat, corn — that the global diet is based on. For example, wheat rust, which is caused by a fungus, devastated wheat crops in Africa, and is poised to spread to other major wheat-producing countries.

Cassava mosaic virus (family of geminiviruses) carried by white flies (Anderson and Morales 1993), has caused enormous losses in cassava (manioc, yucca, or tapioca) in sub-Saharan Africa, where it is a staple in the diet of millions.

As new strains of *Phytophthora infestans* evolve, new outbreaks of the disease emerge, causing devastating epidemics globally; such as the virulent fungicide-resistant strain US-8, which emerged during 1992 (Edwards, 1996).

In 1972, Karnal bunt, a fungal disease of wheat, durum, rye and triticale (a hybrid of wheat and rye) caused by *Tilletia (Neovossia) indica*, became of global importance following its discovery in Mexico and, in 1996, it was reported for the first time in the USA (Bonde *et al.,* 1997). Recent interest in the disease has surrounded the potential for its use as a biological weapon and exports of wheat from many regions with Karnal bunt have been banned, leading to severe economic loss for affected countries (Gewin, 2003).

1.3.4. Nematode Pathogens

Plant parasitic nematodes, the unseen enemies of the farmer, have been recognized as serious perpetual problems in agricultural production all over the world. Favourable weather and almost continuous availability of host crops in the tropical and subtropical regions including India favour their build up. Being soil-borne and with wide host range, nematodes are one of the toughest pests to control. International estimates have put crop losses due to nematodes at over $100 billion (Sasser and Freckman, 1987). The root-knot nematodes cause intense pestilence problem in vegetables, fruit crops, rice, pulses, fibre and oilseed crops.

Extensive nematode quarantine and control efforts in developed countries have paid rich dividends. India suffers heavy quantitative and qualitative losses in various food, fibre and commercial crops due to nematodes. A number of new nematode problems have emerged in the intensive cropping systems.

Rice revolution in the country has brought about many second-generation nematode problems such as a new root-knot nematode species *Meloidogyne triticoryzae*, which damage to rice and wheat in the north-western plains with rice-wheat cropping system. *Meloidogyne graminicola* infestation was intense in rice crop of Mandya district in Karnataka. *Molya* disease in wheat caused by the cereal cyst nematode, *Heterodera avenae* in Haryana and Bihar has emerged as a serious disease problem.

Numerous challenges in the discipline of Nematology have opened up new opportunities to prove nematodes as important limiting factors in agricultural production: root-knot nematode problem on rice, pomegranate, and guava; cyst nematodes problem on potato, and many others; protected horticulture; impact of climate change; threat to introduction of new nematode pests under globalization of agricultural produce; lack of management strategy in standing crops etc. are some of the challenges in near future (Table 1.5).

Table 1.5. Emerging nematode problems in India

State	Nematode problem/s
Gujarat	*Aphelenchoides besseyi* on rice. *Meloidogyne indica* on acid lime. *Meloidogyne javanica* on pomegranate.
Haryana	*Meloidogyne graminicola* on rice. Causes 10-80% yield loss in Sirsa, Bhiwani and Rewari districts.
Himachal Pradesh	*Meloidogyne incognita* on polyhouse crops - bell pepper and carnations.
Karnataka	*Meloidogyne incognita* on pomegranate and mulberry. *Meloidogyne enterolobii* on guava. *Meloidogyne graminicola* on rice.
Kerala	*Meloidogyne graminicola* on rice.
Maharashtra	*Meloidogyne incognita* on pomegranate in Solapur, Sangli, Nasik and Ahmednagar districts. *Meloidogyne incognita* on polyhouse crops in Pune, Satara, Sangli and Kolhapur districts.
Orissa	*Aphelenchoides besseyi* on tuberose.
Tamil Nadu	*Meloidogyne graminicola* on rice. *Meloidogyne enterolobii* on guava.

State	Nematode problem/s
West Bengal	*Meloidogyne graminicola* on rice. *Aphelenchoides besseyi* on tuberose.

The detection of potato cyst nematode in northern India is a serious concern with ramification on export and quarantine issues.

There has been an outbreak of serious nematode problems in the recent years in crops grown under protected cultivation and fruit crops such as pomegranate and guava through dissemination of nematodes with infected planting materials.

Rice root-knot nematode, *Meloidogyne graminicola* has also emerged as a national problem and is getting accentuated. It can be maintained below damaging thresholds through good water management. However, with reduced availability of water following climatic changes and/or competition for urban use, reduced quality of water management, or the introduction of water saving mechanisms such as direct wet seeding is favouring the development of high populations of *M. graminicola*, drastically raising its economic significance as a damaging pest (de Waele and Elsen, 2007).

1.4 INVASIVE PESTS

Invasive or exotic or trans-boundary pests are organisms that are introduced into an area beyond their natural range and become pests in the new environment. They are also referred to as alien, non-native, or introduced pests, which threatens ecosystems, habitat or plant species. Invasive plant pests have been affecting crop production from time immemorial.

Alien species means, with respect to a particular ecosystem, any species, including its seeds, eggs, spores, or other biological material capable of propagating that species, that is not native to that ecosystem. Ecosystem means the complex of a community of organisms and its environment.

The characteristics of invasive pest species include:

- Most introductions have been unintentional and accidental. Having evolved in a different ecosystem, these non-native species may have few natural enemies in their new locations, which can often lead to population increases that can overwhelm native species by out-competing them for resources (e.g., food, water, light, space).
- An invasive species is a species that does not occur naturally in a specific area and whose introduction does or is likely to cause economic (including agricultural) or environmental harm or harm to human health.
- Common traits of invasive pests and pathogens include rapid reproduction, fast growth, wide dispersal, altering of growth or form to suit a particular habitat, tolerating a wide range of environmental conditions and the ability to feed on a variety of different foods.
- All species have vectors that can move them to new areas to colonize, but for invasive species it is usually human activity of some kind (such as foreign trade and travel) that has transported them unknowingly. Often the best single predictor of invasive ability is whether a species is already known to be an invasive pest somewhere else.

- Occur outside their natural adopted ranges and dispersal protection.
- Some establish, invade and out-compete native species.
- Potential to upset natural biodiversity.

The movement of plant pests and disease pathogens across physical and political boundaries threatens food security and creates a global public concern across all countries and all regions (Table 1.6). Countries allocate large resources to limit the spread and control of trans-boundary pests and diseases. They also adapt plant health services and activities and cooperate regionally and globally for prevention, early warning and control.

Table 1.6. Recent invasions of insect pests in India

Common Name	Scientific Name	Family Order	Introduced From	Host Plant
Serpentine leaf miner	*Liriomyza trifolii*	Diptera Agromyzidae	USA (Florida)	Castor, greens, cucurbits, tomato and ornamental plants
Coffee berry borer	*Hypothenemus hampei*	Coleoptera Scolytidae	Northeast Africa	Coffee
Spiraling whitefly	*Aleurodicus dispersus*	Homoptera Aleyrodidae	Caribbean region and Central America	Guava and papaya
Silver leaf whitefly	*Bemisa argentifolii*	Homoptera Aleyrodidae	—	Tomato
Coconut mite	*Aceria guerreronis*	Acari Eriophyidae	Mexico	Coconut
Coconut leaf beetle	*Brontispa longissima*	Coleoptera	—	Coconut
Eucalyptus gall insect	*Leptocybe invasa*	Eulophidae	Australia	Eucalyptus
Papaya mealy bug	*Paracoccus marginatus*	Pseudococcida e	Sri Lanka	Papaya, cassava and mulberry
Cotton mealy bug	*Phenococcus solenopsis*	Hemiptera Pseudococcida e	—	Cotton
Tomato leaf miner	*Tuta absoluta*	Lepidoptera Gelechiidae	—	Tomato
Western flower thrips	*Frankliniella occidentalis*	Thysanoptera Thripidae	—	Fruits and vegetables
Golden apple snail	*Panacea canaliculata*	Gastropoda Ampullariidae	Brazil	Apple
Rice water weevil	*Lissorhoptrus oryzophilus*	Coleoptera Curculionidae	Eastern USA and Cuba	Rice
Mango fruit borer	*Deanolis albizonalis*	Lepidoptera Pyralidae	—	Mango

Common Name	Scientific Name	Family Order	Introduced From	Host Plant
Thai-sac brood disease on honey bees	—	—	—	Honey bees
Gall midge of chili and bell pepper	*Asphondylia capparis*	Ceceidomyidae	—	Brinjal, Bell pepper
Erythrina gall wasp	*Quadrastichus erythrinae*	Hymenoptera Eulophidae	Florida	Erythrina
Sapota seed borer	*Trymalitis margarias*	Lepidoptera Tortricidae	Sri Lanka	Sapodilla
Leucaena psyllid	*Hereopsylla cubana*	Homoptera Psyllidae	---	Leucaena
Litter beetle	*Luprops tristis*	Tenebrionidae	—	Rubber

Now-a-days, the movement of plant pests and diseases across boundaries has given rise to global threats to food security and agricultural development and trade and often creates global concern especially when such diseases and pests affect commercial crops. Increased movement of people, plants and products in the globalized economy on the one hand, and the concentration and intensification of production systems on the other, have accelerated and enlarged redistribution of plant pests with a clear tendency to expand to all regions of the globe. In addition, climate change is creating new ecological niches for the (re)emergence and spread of pests and diseases. As a result, the impact of plant pests has considerably increased. There are indications that the poorest countries with vulnerable sanitary regulation and infrastructure carry a larger burden than others. The list of imported exotic diseases and pests has grown while no substantial progress has been made in the control of local entrenched pests. Poor countries and producers may perceive different risks and incentives associated with plant pests, and it is essential to recognize these differences in the design and implementation of pest prevention and control measures. It is in the interest of the international community to avoid creating country or regional reservoirs for plant pests that will maintain the threat at global level.

Quarantine pests and diseases can be introduced to a country mainly by trade and movement of people. Fruit flies, for example, aphids, cassava mosaic virus, banana wilt, are increasingly important as trans-boundary pests. Economic losses due to fruit flies alone are estimated at more than US$ 1 billion per year.

The larger grain borer originally arrived from Central America in the 1980s, first in the United Republic of Tanzania and then Kenya. During the last 20 years, this storage pest has spread over many countries in West, Central and East Africa. The beetle is causing devastating post-harvest losses of up to 90 per cent on stored maize and on dried cassava, both primary staple foods in most African countries.

1.5 IMPACT OF CLIMATE CHANGE ON EMERGENCE OF PESTS

Climate change has added new dimensions and challenges in the agricultural system. Elevated temperatures (Plate 1.1), droughts, rising sea levels, floods, changes in wind patterns and abnormal weather have been causing changes in the ecosystem balance, particularly in pests and diseases behaviours and occurrence. Changes in climate may trigger changes in geographical distribution, increased overwintering, changes in population growth rates, increases in the number of generations, extension of the development season, changes in crop-pest synchrony, changes in interspecific interactions, pest biotypes, activity and abundance of natural enemies, species extinction, increased risk of invasion by migrant pests and efficacy of crop protection technologies.

1.5.1. Expansion of Geographic Ranges

The insect-pests are expected to extend their geographic range from tropics and subtropics to temperate regions, which lead to increased abundance of tropical insect species and sudden outbreaks of insect-pests (Kannan and James, 2009).

There is clear evidence that climate change is altering the distribution, incidence and intensity of plant pests and diseases whose distribution is shifting north in the United Kingdom and other parts of Europe. For example, migrant moths of the Old World bollworm *(Helicoverpa armigera)* had a phenomenal increase in the United Kingdom from 1969-2004 and there have been outbreaks at the northern edge of its range in Europe. Similarly, the cottony cushion scale *(Icerya purchasi)* populations appear to be spreading northwards perhaps as a consequence of global warming. Likewise, the cottony camellia scale *(Pulvinaria – Chloropulvinaria – floccifera)* has become much more common in the United Kingdom, extending its range northwards in England and increasing its host range in the last decade or so, which is almost certainly in response to climate change. In Sweden, this species was previously only known as a greenhouse species, but is now established as an outdoor species. *Nezara viridula*, a tropical and subtropical crop pest, is gradually moving northward in southwestern Japan, possibly due to global warming, replacing the more temperate species *Nezara antennata.*

Tea mosquito bug, *Helopeltis antonii* is a serious constraint in cashew (west coast-Kerala, Karnataka, and east coast-Tamil Nadu). Cashew tracts of Andhra Pradesh, West Bengal and Orissa are free from this pest. The hot spot areas of the bug across the cashew tract of the whole country have been demarcated, taking into consideration the optimum temperature during flushing and following stages. The pest may spread to new areas under current scenario of climate change and states like Andhra Pradesh, West Bengal and Orissa may come under this pest attack in the changed situation (Gupta *et al.* 2009).

New England hemlock trees are under assault from the woolly adelgid, an aphid-like insect that has migrated northward with warmer winters.

The burrowing nematode, *Radopholus similis* occurs only below ~1,400 m altitude in the East African Highlands where it is a principal pest of banana and plantain, a regional key starch staple for over 20 million people. A small raise in temperature due to climate change would result in *R. similis*, which is cold-sensitive, infecting millions more bananas grown at higher altitudes.

If crop pests continue to march polewards as the earth warms, the combined effects of a growing world population and the increased loss of crops to pests will pose a serious threat to global food security.

1.5.2. Increased Overwintering Survival

The temperature in India is expected to increase by 1-5^{0}C within next 100 years (Arora and Dhawan, 2013). Insects undergoing a winter diapause are likely to experience the most significant changes in their thermal environment (Bale and Hayward, 2010). Increase in temperature in the range of 1-5^{0}C would increase insect survival (Sharma *et al.*, 2010).

1.5.3. Increase in Number of Generations

With every 2° C rise in temperature, multivoltine insects may have 1-5 additional generations (Yamamura and Kiritani, 1998). Rice stem borer, *Chilo suppressalis* would extend its distribution range in Japan and produce two generations a year after 2^{O} C warming (Morimoto *et al.*, 1998).

In Alaska, spruce bark beetles have denuded 4 million acres of conifers on the Kenai Peninsula, as warming allows the beetles time for an extra generation each year (Kerlin, 2002). Dead stands are then vulnerable to fire.

1.5.4. Physiological and Ecological Impact

Some of the species that are able to adapt to the warmer climates may become major pests (Dillon *et al.*, 2010). Differences in the pattern of response to temperature changes would disrupt synchronization in phenology between insects and host plants or natural enemies (Kiritani, 2006).

Most threatening are outbreaks of migratory pests because of their sudden emergence, often without or with limited warning, triggered by changing ecological conditions or agricultural practices which favour an explosive increase in the pest population and its rapid spreading into neighbouring areas. The unexpected appearances of such massive pest outbreaks hit the countries which in most cases are ill prepared, with no or only poor capacities to effectively cope with rapidly evolving and fast moving threats.

Various species of mealy bugs have started appearing in serious proportions on field crops, vegetables, fruits and ornamentals (Tanwar *et al.*, 2007). In fact, mealy bugs have become indicator insects for the current ecosystem alterations due to slow changes in climate during the period from 2002 to 2005. Among these, *Phenacoccus solenopsis* on cotton and *Paracoccus marginatus* on papaya have become quite serious. The papaya mealy bug, *P. marginatus,* has become quite alarming in Tamil Nadu, challenging the pesticides or other IPM measures. *Maconellicoccus hirsutus* causes extensive damage to ornamentals, though its host range extends to 76 families and over 200 genera, including beans, chrysanthemum, citrus, coconut, coffee, cotton, croton, cucumber, grape, groundnut, guava, *Hibiscus,* maize, mulberry, pumpkin and rose (Tanwar *et al.*, 2007; Rajendran, 2009).

Some of the insect pests of vegetable crops have become major and are gradually attaining the major pest status in different regions of the country due to changes in the ecosystem and habitats. Fruit borer, *Helicoverpa armigera* in tomato; whitefly, *Bemisia tabaci* and serpentine leaf miner *Liriomyza trifolii* on tomato and cucurbitaceous crops; fruit fly on fruits and vegetables; mealy bugs on several horticultural crops; gall midge on brinjal; okra stem fly; bitter gourd leafhopper red spider mite on okra, brinjal, cowpea, Indian bean; and nematodes on several vegetable crops are some of the examples. In recent times, there were outbreaks of gall midge (*Asphondylia capparis*), which was known to be a minor pest and gradually becoming a regular problem in chilli, bell pepper and brinjal in the states of Andhra Pradesh and Karnataka, and on brinjal in Chhattisgarh. Various species of mealy bugs in cotton, vegetables and papaya have intensified their severity of occurrence in different parts of the country and have become indicator insects for the current ecosystems due to slow changes in climate in the last one decade. Several national and international polyphagous pests like termite, white grub, hairy caterpillar, gram pod borer and *Spodoptera litura* are also becoming severe and expanding their host-horizon. This envisages that there is a need to relook into the cropping systems for reorientation and vegetables should be given importance with adequate insect pest management measures. In this context, integrated pest management (IPM) system was developed out of the need for sustainable crop protection strategy against the backdrop of increasing pesticide use and deleterious effect of residues in the environment.

1.5.5. Loss of Ecological Biodiversity

In the present day scenario, many butterfly species are under a real threat due to depletion of the natural vegetation for various anthropogenic developmental activities (Sidhu and Mehta, 2008).

1.5.6. Changes in Insect Feeding

Insects have been shown to feed more when plants are grown in elevated levels of carbon dioxide (CO_2) (Hamilton *et al.*, 2005). Sap-feeders were the only functional group to show positive responses to elevated CO_2 with increased food consumption and higher build up at higher CO_2 levels (Whittaker, 1999).

1.5.7. Breakdown of Host Plant Resistance

Severe yield loss in sorghum may occur in India due to breakdown of resistance against midge, *Stenodiplosis* sp. and spotted stem borer, *Chilo partellus* (Sharma *et al.*, 2006).

1.5.8. Introduction of Transgenic Crops

Introduction of transgenic crops decline the status of major pest but the population of minor pest increases. The pest scenario in cotton ecosystem is changing and sap feeder like aphids, jassids, mirids and mealy bugs are emerging as serious pests (Vennila, 2008). *Spodoptera litura* and other minor pests like thrips and weevils are also becoming serious pests of cotton (Sarode *et al.*, 2009).

1.5.9. Changes in Tillage Technology

Zero tillage, rotary tillage, and bed planting contributed to new pest problems in wheat. Direct seeded rice may also result in change in the pest complex of rice.

1.5.10. Other Effects

There are indications of shift of insect pests of plantation crops to new crops and new areas. The rhinoceros beetle, *Oryctes rhinoceros* is an established pest of coconut palms in India. Quite recently, it has become a problem for oil palm in southern states of India. The incidence of the pest in oil palm plantations closer to natural forest areas has been noticed at many places.

Recent locust outbreaks in Central Asia, Southeast Asia, and Central and Southern Africa surfaced raising concerns with regard to locust species other than the desert locust, putting the livelihoods of more than 15 million people at risk in Central and Southern Africa alone.

Oaks in New Orleans, Louisiana, are bristling with termites (CNN, 1997), as killing frosts became less frequent in the 1990s. The rice blast fungus, present in scores of countries, has migrated to wheat, and has become a scourge of farmers in Brazil.

Habitat is also being subjected to the dual threats of climate change and emerging infectious diseases. In California, several species of trees are infected with *Phytophthora*, a fungus related to the one responsible for the Irish potato famine (Davidson *et al.*, 2003). Extreme weather weakens the hosts and emboldens the pest attack.

1.6. DRIVERS OF PEST EMERGENCE

Major drivers for the spread of plant pests and diseases are international trade and traffic (except for migratory pests). Plant pests and diseases are not evenly distributed over the globe, often because they are limited by physical barriers such as mountains, seas and deserts. The increase in movement of people, animals, plants, goods and conveyances has accelerated the redistribution of plant pests and diseases and climate change will create new ecological niches allowing for the establishment and spread of pests and diseases into new geographical areas and from one region to another. This expansion will continue to result in huge financial losses and require large eradication programmes and control measures. Among the major occurrences are the spread of coffee leaf rust throughout the world, soybean rust into the Americas, and citrus tristeza virus in South and Central America and now in the Mediterranean.

The major factors that have contributed towards the emergence of pest problems include:

- Favourable prevailing weather conditions.
- Extensive cultivation of high yielding varieties.
- Growing of varieties lacking resistance to major pests.
- Abundant food supply due to mono-cropping practices.
- Imbalanced use of fertilizers, particularly application of high levels of nitrogen.
- Non-judicious use of insecticides resulting in pest resistance to pesticides.
- Absence of natural enemies.
- Expansion of irrigation facilities.
- Modifications of cultural practices.
- Climate change.
- Change of genotypes.
- Intensified crop cultivation throughout the year providing constant niches for pest multiplication,
- Resurgence of pests and outbreaks of minor pests (emergence of sucking pests in *Bt* cotton).

In addition, unforeseen emergence of "new" diseases and pests has been relatively common. New vectors, selection and recombination of disease genotypes may occur when plant species and varieties mix or when insect pests and vectors are introduced without their natural enemies. Change in climate resulting in changes in species composition and interactions will augment the emergence of unexpected events, including the emergence of new diseases and pests.

1.6.1. Insect Pests

The range of the oak processionary moth (*Thaumetopoea processionea)* has extended northward from central and southern Europe into Belgium, Netherlands and Denmark due to improved synchrony of egg hatch and reduction of late frosts. The massive population build up of mountain pine beetle (*Dendroctonus ponderosae)* and its northward progression in the North American Pacific Northwest has most likely been due to a combination of warmer winter temperatures, reduced episodes of under bark mortality and increased drought which weakened the trees. Kirtani (2007) reported on the polar extension of several plant pests in Japan over the period 1965 to 2000. Similarly, about 40 of the 250 butterfly species in Japan have exhibited northward range extensions in recent years.

The expansion of maize production driven by climate change will make more areas vulnerable to entry, establishment and spread of the corn root worm (*Diabrotica*). The range of tephritid fruit flies will alter considerably with climate change with corresponding changes in phytosanitary regulations and international trade opportunities. Mountain pine beetle (*Dendroctonus ponderosae*), a pest of North American forests, is expected to decrease in generation time and winter mortality, which will increase the risk of range extension into vulnerable ecosystems. Conversely, some pests will be less damaging because climate suitability will decrease and through interactions with natural enemies and plant defences.

Migratory plant pests, in particular locusts, are totally dependent on rain, temperature and vegetation and their habitats change rapidly. The desert locust (*Schistocerca gregaria*), like other locusts, can change its behaviour and physiology from solitary grasshoppers to gregarious stages that form swarms. Solitary desert locusts occur at low density in the recession area, which covers North Africa, the Sahel, the Red Sea countries and parts of India, Pakistan, Iran and Afghanistan. The outbreak area reaches from Mauritania to India and from southern Europe to Cameroon and Tanzania. Outbreaks originate in the recession areas when there are several cycles of good breeding conditions. Although the effects of climate change on this system are difficult to judge, climate scenarios with more winter rain in the Sahel may provide better breeding conditions.

1.6.2. Disease Pathogens

Pathogen introduction is the most important driver of emerging plant diseases. Weather conditions are also important, which might be related to the sensitivity of plants to humidity and moisture levels and the responses of plant pathogens to weather events. Recombination process was identified as the cause of emergence for 2% of the plant diseases (Plate 1.2).

1.6.2.1. Pathogen Introduction:

It is defined as the anthropogenic movement of pathogens (parasites) outside their natural geographical or host-species range. It can occur through anthropogenic change resulting in a parasite crossing an evolutionary boundary, such as geographical or ecological separation (Cunningham *et al.*, 2003). The introduction of alien pathogens or hosts that leads to disease emergence is the most commonly cited driver of emerging infectious diseases. The increasing volume of globalized trade has driven increased frequency of introduction events. This is combined with a heightened impact that introduced pathogens often have on native susceptible host populations (Cunningham *et al.*, 2003). The result is a series of high-profile introduced emerging diseases. Given the predicted continued rise of global air travel and trade volume, the number of introduced emerging diseases is likely to also increase (Daszak and Cunningham, 2003).

1.6.2.2. Climate Change as a Driver of Emerging Infectious Diseases of Plants:

Climate change can lead to disease emergence through gradual changes in climate (e.g. through altering the distribution of invertebrate vectors or increasing water or temperature stresses on plants) and a greater frequency of unusual weather events (e.g. dry weather tends to favour insect vectors and viruses, whereas wet weather favours fungal and bacterial pathogens) (Agrios, 1997). Thus, climate change can lead to the emergence of pre-existing pathogens as major disease agents or can provide the climatic conditions required for introduced pathogens to emerge.

Harvell *et al.* (2002) suggested that milder winters, higher nocturnal temperatures and higher overall temperatures will enable increased winter survival of plant pathogens, accelerated vector and pathogen life cycles, and increased sporulation and infectiousness of foliar fungi. Because climate change will enable plants and pathogens to survive outside their historic ranges, Harvell *et al.* (2002) predicted an increase in the number of invasive pathogens. The ranges of several important crop insects, weeds and plant diseases have already expanded northward (Rosenzweig *et al.*, 2000). Range expansion of the grey leaf blight of corn, caused by the fungus *Cercospora zeae-maydis*, was first noticed during the 1970s, and, in the past two decades, has become the major cause of corn yield loss in the USA (Rosenzweig *et al.*, 2000).

1.6.2.3. Factors Cited as the Cause of Disease Emergence by Pathogen Group:

Analysis of the factors cited as the cause of disease emergence for the three most significant taxonomic groups of pathogens [bacteria (Plate 1.2c), fungi (Plate 1.2d) and viruses (Plate 1.2e)] shows that, although introduction is the most, or second most, important driver for each pathogen group, the percentage of emerging diseases driven by introduction declines proportionately with size of pathogen, being lowest for fungi and highest for viruses. Weather conditions are major drivers of bacterial and fungal plant emerging diseases, but are relatively unimportant for plant emerging diseases that are caused by viruses, where changes in vector populations are the most important driver after pathogen introduction. Interestingly, although agricultural changes were identified as important drivers of plant emerging diseases caused by fungi and viruses, they were not mentioned as drivers of bacterial diseases.

1.6.2.4. Agricultural Changes and Disease Emergence:

Agricultural changes such as diversification introduce new hosts, and globalization new pathogens, promoting the spread of alien diseases and vectors. For example, during the early 1970s, expansion of soybean (*Glycine max*) cultivation in Brazil and a seven-fold increase of soybean acreage in Argentina led to the emergence of two major disease threats to bean (*Phaseolus vulgaris*) production, with yield losses of up to 85%. In southern Brazil, Bean golden mosaic virus (BGMV), a minor crop pathogen before intensification of soybean cultivation, emerged to become the limiting factor in dry season bean production (Costa, 1975). In Argentina, Bean dwarf mosaic virus (BDMV) threatened the collapse of bean production (Morales and Anderson, 2001). Both viruses are transmitted by the whitefly *Bemisia tabaci*, which reproduces on soybean plants. Thus, intensive cultivation of the host plant of this vector led to disease emergence in other species.

Large irrigation projects enabled rice to be extensively cultivated, and native grass species to be widely disseminated beyond their natural ranges. This resulted in build up of Rice yellow mottle virus and its beetle vectors, leading to disease emergence throughout East Africa.

1.7. IMPACTS OF EMERGING PESTS

1.7.1 Food Security

Food security is defined as "When all people, at all times, have physical, social and economic access to sufficient, safe and nutritious food which meets their dietary needs and food preferences for an active and healthy life" (World Food Summit, 1996). Components of food security are food availability, food access, food utilization and food stability.

1.7.1.1. Availability of Food:

Plant pests and diseases reduce the availability of quantities of food of appropriate quality, whether supplied through domestic production or imports. Overall quantifications of losses and potential losses due to plant pests and diseases are limited. Entry and establishment, emergence and outbreaks of plant pests and diseases have historically resulted in major food problems either directly through yield reductions of food crops, or indirectly through yield reduction of cash crops (e.g. potato blight, locusts). Climate change will result in a higher volatility and, therefore, is likely to cause additional crises in local agricultural production, in particular for small farmers and those involved in subsistence agriculture with different consequences for socio-economic groups and genders.

1.7.1.2. Access to Food:

Plant pest and disease regulations are designed to facilitate trade while reducing the risk of international movement of restricted organisms whose introduction could require expensive eradication or control operations. Plant pests and diseases reduce food access through reduction of income from reduction in yields of food and cash crops, reduction in forest productivity, as well as increased costs of control. Indirect effects are reduced access to international markets due to the occurrence of quarantine for plant pests.

1.7.1.3. Utilization of Food:

Food utilization in relation to plant diseases and pests, chiefly concerns food safety. Climate change may lead to redistribution of plant pests and changes in pest incidence and intensity resulting in additional and inappropriate pesticide use. Consequently, there may be higher and even unacceptable levels of pesticide in food. Mycotoxins in food are a growing problem. Changes in rainfall, temperature and relative humidity may favour the growth of fungi that produce mycotoxins and thus may make food such as groundnuts, wheat, maize, rice and coffee unsuitable for human and animal consumption.

1.7.1.4. Stability of Food:

To be food secure, populations, households or individuals must have access to adequate food at all times. They should not risk losing access to food as a consequence of sudden shocks such as economic or climatic crises or cyclical events such as seasonal food insecurity. The concept of stability refers to both the availability and access dimensions of food security. Introduction or emergence of new plant pests and diseases, and outbreaks of migratory pests may have substantial effects on the stability of food supply through direct losses as well as through the reduction of income and will also influence the stability of the production system.

1.7.2. Human Health and Welfare

Although crop production is affected by established as well as emerging diseases, epidemics caused by emerging plant diseases infecting new hosts with little resistance or tolerance to the emerging pathogen can be particularly devastating. The latest available data suggest that pre-harvest pest and disease damage in the eight most important food and cash crops accounts for about 42% of the attainable crop production, or a production value of more than US$ 300 billion (Pimentel *et al.*, 1993; Oerke *et al.*, 1994). In Uganda, cassava production plummeted from a high of 3.5 million tons during 1989 to 2.25 million tons during 1996 as a result of the emergence of cassava mosaic virus (EACMV-Ug), with losses estimated at US$ 60 million yr^{-1} between 1992 and 1997 (USAID, 2000). In the north and east of Uganda, the epidemic led to famine. Food insecurity resulting from this emerging disease still threatens regions of East, Central and West Africa (Otim Nape *et al.*, 1997).

1.7.3. Biodiversity

Emerging diseases have not coevolved with the host or ecosystem in which they emerge and, as such, are more likely than are endemic diseases to pose a threat to biodiversity through biomass loss, changes in host species complements and via the extinction of host species.

The consequences of the Dutch elm disease pandemics are profound, with the second pandemic (which began during the 1940s) resulting in the loss of hundreds of millions of elms and causing closed canopy forests to become open habitat (Brasier, 2001), leading to changes in bird community composition (Canterbury and Blockstein, 1997). The deaths of mature elms resulted in increased mortality rates of sympatric trees, such as paper birch *Betula papyrifera*, as for example, they became more exposed to storm damage. The American chestnut was an important component of the southern Appalachians (Vandermast and Van Lear, 2002) before its removal by chestnut blight. This disease-induced effective extinction has had serious consequences for the ecology of the region (Real, 1996). Aggregations of dead wood from affected trees have environmental repercussions even decades after the initial epidemic, such as influencing the structure and function of Appalachian streams (Wallace *et al.*, 2001).

The negative impact of emerging plant diseases on biodiversity might reduce the potential for the discovery of new pharmaceuticals, new crops or disease-resistant strains of existing crop species as rare strains of crop plants or wild plants are lost. The increased use of pesticides to combat emerging diseases can have toxic effects on humans and ecosystems. Before the 1990s, metalaxyl was used as a therapeutic fungicide to manage potato late blight. Following metalaxyl-resistant potato blight emergence in the USA and Europe, fungicides are now used as a prophylactic and primary management tool, with a predicted 15–25% increase in fungicide use (Fry and Goodwin, 1997).

1.8. MANAGEMENT OF EMERGING PESTS

1.8.1. Insect Pests

1.8.1.1. Strategies to Manage Emerging Pests

- Study the biology and ecology of known insect pests and their natural enemies under changing climatic conditions.
- Study the ecology and genetic make-up of the invasive alien species (IAS).
- Careful tracking of geographical distribution of pest.
- Developing cultivars resistant to insect pests.
- Judicious use of insecticides to prevent resistance and resurgence development.
- To identify, conserve and augment natural enemies of emerging insect pests.
- Modifying crop management practices.
- Developing suitable integrated pest management strategies.
- Phyto-sanitary regulations to prevent or limit the introduction of risky insect pests.
- Better weather forecasting.

1.8.1.2. Sucking pests:

These pests inflict direct as well as indirect losses by acting as vectors of plant viruses.

- Being vectors, it is difficult to establish ETL for these pests.
- Thick coating of waxy secretion, dust deposits over webs makes chemical pesticides ineffective; hence wax dissolving organic solvent with insecticides is required.
- Hidden infestation makes it difficult to understand the start of infestation as in case of woolly apple aphid, mealy bugs, mites etc.
- It is difficult to apply pesticides in case of dense plant canopy crop like sugarcane that requires application of pesticides at proper site and time.

- Control measures needs to be taken up by all cultivators on campaign basis, especially in case of pests spread by wind.
- Proper management of nitrogenous fertilizers and water to avoid vulnerability of crop and favourable climate to the pest.
- Use of optimum plant density, inter or mixed cropping to facilitate better aeration, pesticide application and movement of natural enemies.
- Conservation and augmentation of generalist predators is necessary. The specialist predators and parasitoids of sucking pests may be identified, protected, collected, reared and released on routine basis.
- Farmers' training in respect of identification of pests and natural enemies to protect them and to avoid unnecessary application of pesticides.
- Due to wide host range of sucking pests, it is very difficult to control as all host plants cannot be sprayed.
- Presence of preferred host throughout the year.
- Systemic insecticides lead to residue problem while synthetic pyrethroids lead to outbreaks of different pests. Therefore frequent applications of these pesticides are not advisable.

1.8.1.3. Internal Feeders/Soil Pests:

These pests require careful monitoring for accessing pest status and for implementation of management strategy.

- Being internal feeders, it is difficult to understand the start of infestation. By the time of appearance of external symptoms, damage is already done; hence there is need to develop some effective system of monitoring the externally visible stages. Control measures needs to be applied on community basis.
- Clean cultivation, collection and destruction of infested fruits and plant parts throughout the year on regular basis. e.g. fruit flies, mango fruit fly.
- Information on realistic or potential yield loss and package of nematode management is lacking. There is a need to develop eco-friendly approaches. Emphasis should be given on development of integrated nematode management programmes.
- Nematode may transmit plant viruses or may interact with other pathogens to cause disease complexes that are worse than disease caused by pathogens alone.
- Nematode infestation may increase host plant susceptibility to other environmental stresses such as higher temperature or moisture. Nematode infestation may nullify host plant resistance to pathogens.
- Some nematode diseases are vectored by insect pests. e.g. Pine wilt nematode - long horn beetle, red ring of coconut - palm weevils

1.8.1.4. Polyphagous, Sporadic and Cosmopolitan Pests:

As these pests have wide host range, common approach should be made.

- Repeated application of an insecticide should be avoided as they have developed resistance against large number of insecticides. Insecticides of different chemical group may be used to avoid resistance development.

- It is difficult to apply pesticides on trees due to tallness and thick canopy e.g., mango fruit fly.
- These pests have large number of natural enemies. Care should be taken to protect those. Botanical or microbial pesticide may be used to avoid the bad impact of chemical pesticides.
- Augmentation of indigenous species of natural enemies by undertaking mixed cropping with maize, cowpea and sorghum in cotton; green gram, black gram, and sorghum in pigeon pea; mustard in reducing the *Helicoverpa* menace. While sparse planting of castor in groundnut results in reduced infestation of *Spodoptera.*

The introduction of gene technology has added a new dimension to pest management. There are varied claims and counter-claims about the potential of gene technology in shaping the crop protection scenario in the twenty-first century (Dhaliwal, 2008; James, 2009).

1.8.2. Disease Pathogens

The past few decades have seen the development of an increasingly secure global system of plant health inspection and quarantine (phytosanitary system) that balances agriculture protection against international trade (McGee, 1997). However, our understanding of the risk of pathogen introductions is still poor and there is a need to improve surveillance strategies. Currently, the United Nations Food and Agricultural Organization (FAO) mandates countries to report plant pests and diseases to the International Plant Protection Convention (FAO, 1995), but economic consequences often hinder accurate reporting. There is a lack of standardization in testing protocols for seed-borne diseases and significant variation in seed protection treatments among countries. Strengthening the global phytosanitary system (i.e. international agreements on the inspection and quarantining of plants to prevent the spread of pathogens and diseases) requires improved information, standardized seed testing, improved access to seed treatment protocols and revised pathology research priorities (McGee, 1997), but funding for adequate diagnostic facilities and programmes is often lacking. Emerging infectious diseases (EIDs) research must change from passive cataloguing to analysis and interpretation of the factors contributing to disease emergence. Spatial and temporal analyses of anthropogenic drivers of EIDs, coupled with an understanding of pathogen biology, evolutionary and environmental factors are required to predict and prevent emergence (Anderson, 1994).

1.9. CONCLUSIONS

The main findings on emerging crop pest problems and their management strategies include:

- Insect pests on an average are estimated to cause 15-20 per cent yield losses in principal food and cash crops.
- Climate change and other agronomic factors have affected diversity and abundance of insect pests and extent of losses.
- Chronic, emerging and invasive pests are considered as one of the major threat to food security.

- Sucking insect pests and defoliators have emerged as major pests in recent times like mirid bug, mealy bug, whitefly, tobacco caterpillar on cotton; BPH, WBPH, leaf folder on rice; maize shoot fly on maize; cabbage butterfly on crucifers; tobacco caterpillar on Cole crops; and aphids on wheat crop.
- Need to modify current management strategies and devise appropriate IPM measures to manage emerging pests and mitigate the otherwise incalculable losses.

With the development of agricultural technology as well as biotic and abiotic factors, the crop losses and pest problems are changing continuously. There is a need to explore certain reformative measures to narrow down these losses. The climate change will affect the pest's distribution and status as well as the pest management practices. There is an urgent need to modify crop protection measures with changed climate in order to attain the goal of food security.

This story of Earth's ills is not a cheery one, but systems are resilient, and unstable systems can be re-stabilized. Unearthing root causes raises the urgency of conserving natural systems, and deciphering the pattern of consequences can guide us toward local and global solutions. Hence, there is an urgent need for rapid and accurate detection, diagnosis, and development of management recommendations for these new pests and diseases. A focussed attention by experts like plant pathologists, entomologists, nematologists, horticulturists, policy makers, pesticide industries, and farmers is the need of the hour for in-depth discussion at one platform in order to develop a broad strategy for tackling these problems.

Section II
INSECT AND MITE PESTS

2

Emerging Insect Pest Problems in Field Crops

2.1. INTRODUCTION

Various species of mealy bugs have started appearing in serious proportions on field crops (Tanwar *et al.*, 2007). Among these, *Phenacoccus solenopsis* on cotton has become quite serious. *Maconellicoccus hirsutus* causes extensive damage to cotton, groundnut, and maize. Cottony cushion scale (*Icerya purchasi*) populations on sugarcane appear to be spreading northwards perhaps as a consequence of global warming. Likewise, cottony camellia scale (*Pulvinaria – Chloropulvinaria – floccifera*) has become much more common in the United Kingdom, extending its range northwards in England and increasing its host range in the last decade or so, which is almost certainly in response to climate change. The expansion of maize production driven by climate change will make more areas vulnerable to entry, establishment and spread of the corn root worm (*Diabrotica*). There is clear evidence that climate change is altering the distribution, incidence and intensity of plant pests that are moving north into more temperate zones of Europe.

2.2. COTTON, *Gossypium* spp.

The pest scenario in cotton ecosystem is changing and sap feeder like aphids, jassids, mirids and mealy bugs are emerging as serious pests while decline in the pest status of bollworms was observed (Vennila, 2008). Also, some of the minor pests like thrips, *Thrips tabaci*; shoot weevil, *Alcidodes affaber* and stem weevil, *Pempherulus affinis* are becoming serious on *Bt* cotton (Sarode *et al.*, 2009). American and spotted bollworms attained secondary pest status in the cotton ecosystems while tobacco caterpillar, pink bollworm, mirids and mealy bugs are emerging as major pests. The mealy bug, *P. solenopsis* was observed for the first time on cotton in USA (Fuchs *et al.*, 1991). Recently, mirid bugs, *Ragmus* spp. and *Creontiades biseratense* appeared in epidemic form in Dharwad (Karnataka) and Coimbatore (Tamil Nadu.). There are also indications of development of resistance to *H. armigera* in parts of China (Wang *et al.*, 2009) and occasional reports of survival of *H. armigera* on *Bt* cotton in Australia (Downes *et al.*, 2007) and India (Kranthi *et al.*, 2009) (Table 2.1).

Table 2.1. Emerging insect pest species on *Bt* cotton in India

Emerging insect pests	Remarks
Cotton leaf hopper, *Amrasca biguttula biguttula*	Major pest of cotton with increased incidence across India.
Cotton aphid, *Aphis gossypii*	Emerging pest of *Bt* cotton in South-Central India.
Whitefly, *Bemisia gossypii*	Major pest across cotton growing regions of India; with outbreaks in 2010 in North and South-Central India.
Red cotton bug, *Disdercus* spp.	Major pest of cotton with increased incidence.
Green stink bug, *Nezara viridula*	Emerging pest of *Bt* cotton in South-Central India.
Dusk cottony bug, *Oxycaraenus leatus*	Major pest of cotton with increased incidence.
Mealy bugs, *Phenacoccus solenopsis* and *Maconellicoccus hirsutus*	Emerged as major pest of cotton including *Bt* cotton since 2006.
Mirid bugs, *Creontiodes biseratense*	Recorded as serious pest of *Bt* cotton in Karnataka.
Cotton stem weevil, *Alcidodes affaber*	Serious pest of cotton in Southern India since 2003.
Cotton stem weevil, *Phemperulus affinis*	Major pest of cotton in Southern India since 2006.
Grey weevil, *Myllocerus* spp.	Major foliage pest of cotton since 2002.
Blister beetle, *Zonabris pustulata*	Emerging pest of *Bt* cotton.
Tobacco caterpillar, *Spodoprera litura**	Emerged as foliage pest of cotton since 2007.
Pink bollworm, *Pectinophora gossypiella**	Resistance to Cry 1Ac *Bt* cotton reported in 2009.
Serpentine leaf miner, *Liriomyza trifolii*	Major seedling pest of *Bt* cotton in South-Central India since 2006.
Thrips, *Thrips* spp.	Emerging as major foliage pest of cotton since 2005.

* Effectively controlled by Bollgard II. *Source:* Dhillon *et al.* 2011; Gujar *et al.* 2010.

Some experts are attributing the increase in non-traditional cotton pests to the rapid adoption of *Bt* cotton varieties, as well as the expansion in both conservation tillage acreage and areas included in the boll weevil eradication programme. Although some of these "unusual" pests may have existed in cotton fields in the past, they are becoming more prominent now that *Bt* varieties are controlling tobacco budworm and bollworm infestations and the eradication programme is steadily wiping out the boll weevil.

A number of non-caterpillar pests are also emerging as problem pests, specifically in *Bt* cotton. Among the insects threatening to reduce Delta lint yields are stink bugs, false chinch bugs, sugarcane beetles, and three-cornered alfalfa hoppers.

2.2.1. Mealy Bug, *Phenacoccus solenopsis* (Hemiptera: Pseudococcidae)

2.2.1.1. Key Facts

- First observed in *Gossypium hirsutum* in central Gujarat in 2000.
- Polyphagous pest and capable of causing complete crop loss.
- More than 100 acres of *Bt* cotton in Bhatinda, Punjab, was uprooted because of severe mealy bug infestation.
- Polyphagous pest reported to invade 154 plant species belonging to 53 families (Arif *et al.*, 2009). Besides cotton, this pest attacks sunflower, a number of medicinal crops, brinjal, okra, tomato etc.
- Recent research indicates that this mealy bug is of a single clone and indiscriminate use of insecticides if avoided will drastically reduce the mealy bug infestation.
- In Haryana, the pest was first time reported in 2006 in some villages of Sirsa district, an area adjoining Punjab and spread to more areas during 2007.
- Dhawan *et al.* (2007) reported 30-40% losses due to mealy bug infestation in cotton.
- In Punjab the losses caused by mealy bug were estimated to be Rs. 159 crores to cotton growers during Kharif season 2007.
- Nagrare *et al.* (2009) also reported that mealy bug reduced cotton yields in several parts of Gujarat up to 40-50%.
- Reasons for spread include conducive weather, introduction of *Bt* cotton, and indiscriminate use of insecticides.

Survey across 47 locations of the country between months of late 2007 and early 2008 established the predominance of cotton mealy bug, *P. solenopsis.* Widespread infestation of the pest causing economic damage to cotton growing states of the country had occurred in Punjab, Haryana, Rajasthan, Gujarat, Madhya Pradesh, Andhra Pradesh, Karnataka, Tamil Nadu and Maharashtra during 2008-09 crop season. This pest prefers cotton among all other host plants.

There was a sudden invasion of mealy bugs on *Bt* cotton in Gujarat in 2004, from which it spread to almost the entire country and even neighbouring Pakistan in the next two years, causing substantial economic damage in 2007 (Deshpande, 2009).

In the Malwa belt of rural Punjab, miles after miles of *Bt* cotton fields were under attack by the mealy bug pest during 2006. Punjab's four major cotton-growing districts, Bhatinda, Muktsar, Faridkot and Ferozepur, have been badly affected. The so-called "magic bullet" *Bt* cotton has turned into a bitter pill for farmers who were promised profits but who are now faced with huge losses. During 2006, *P. solenopsis* appeared for the first time on cotton crop in Punjab and caused severe losses in some pockets of Ferozepur, Muktsar and Bhatinda districts (Dhawan and Saini, 2009). Since then this pest has spread to several states like Haryana, Rajasthan, Maharashtra and Gujarat and southern states (Table 2.2) (Atwal and Dhaliwal, 2009; Nagrare *et al.*, 2009).

Table 2.2. Spread of cotton mealy bug to different states

Year of 1st notice	State
2004-05	Punjab
2005-06	Andhra Pradesh, Gujarat
2006-07	Tamil Nadu, Maharashtra
2007-08	Madhya Pradesh, Karnataka, Haryana, Rajasthan

2.2.1.2. Economic Importance:

During 2006, the mealy bug caused economic damage, reducing yields by up to 40-50 per cent in infested fields in several parts of Gujarat (Nagrare *et al.*, 2009). At around the same time, mealy bug infestations were found in all the nine cotton-growing states: Punjab, Haryana, Rajasthan, Gujarat, Madhya Pradesh, Maharashtra, Tamil Nadu, Andhra Pradesh and Karnataka. Severe economic damage was reported in 2007 in four districts of Punjab, two districts of Haryana, and low to moderate damage in parts of Maharashtra, Tamil Nadu and Andhra Pradesh. Nearly 2000 acres of cotton crop were destroyed by the mealy bug by mid-July, and over 100 acres of bug-infested *Bt* cotton was uprooted in Arike-Kalan village in Bhatinda. By the end of the Kharif (autumn) season (June-October), the total damage in 2007 was estimated to range from US$ 400,000 to 500,000 in north India alone.

A report published by the Centre for Agro-Informatics Research in Pakistan in 2006 (Zaka *et al.*, 2006) also stated that the exotic mealy bug *P. solenopsis* had destroyed 0.2 million bales and 50,000 acres (out of the 8 million acres) of cotton area across Pakistan, especially in Punjab and Sindh provinces. It warned that the pest was still increasing, and could result in an epidemic in the cotton-growing areas if unchecked.

The mealy bug has become a major pest in almost all cotton growing states of India and Pakistan. Apart from yield losses, the cost of insecticide application has increased by US$ 250-375 per acre in both India and Pakistan (Nagrare *et al.*, 2009).

2.2.1.3. Damage:

Plants infested with mealy bugs (Plate 2.1) during vegetative phase exhibit symptoms of distorted and bushy shoots, crinkled and/or twisted and bunchy leaves, and stunted plants that dry completely in severe cases. Late season infestations during reproductive crop stage result in reduced plant vigour and early crop senescence.

2.2.1.4. Biology

The developmental period from immature crawler to adult stage was greater for males (18.7 ± 0.9 days) compared to females (13.2 ± 1.8 days), probably due to the additional moult to the pupal stage in males. Survival of second instars was lower (45.5%) than first and third instars (71.4%). Females showed dynamic patterns of fecundity with the number of crawlers produced per female ranging between 128 and 812, with a mean of 344 ± 82. The reproductive period lasted 30.2 ± 8.2 days. Parthenogenesis with ovoviviparity (96.5%) was dominant over the oviparous (3.5%) mode of reproduction. Adult females lived 42.4 ± 5.7 days. Males accounted for less than 5% of the population, and lived 1.5 ± 0.1 days (Vennila *et al.*, 2010).

2.2.1.5. Host Range:

Besides cotton, *P. solenopsis* has been recorded on several economic crops like okra, tomato, brinjal, chilli, grapes, fig, date palm, apple, avocado, banana, citrus, sunflower, and a number of medicinal crops (Mohindru *et al.*, 2009).

2.2.1.6. Management

i) **Cultural Methods:** It is important to remember that mealy bug crawlers spread through human interventions such as spraying, irrigations, frequent movement through the infested area, etc. Therefore disturbing mealy bug affected plants should be avoided. It is important to remember that young cotton plants can overcome mealy bugs and it is better not to resort to chemical sprays on young plants that have slight infestation of the mealy bugs in early vegetative stages of the crop. It has been observed that the mealy bugs were unable to establish colonies on the cotton crop during early vegetative and peak vegetative stages. It is only in rare cases, which is generally possible on a few susceptible genotypes, that mealy bugs colonize plants during vegetative stage.

   The scientists from Central Institute for Cotton Research (CICR), Nagpur, recommended removing the preferred weed *Parthenium hysterophorus* on which the mealy bug multiplies and moves onto crop plants during the season. Planting border rows and strips (every 5-6 rows of cotton) of pigeon pea *Cajanus cajan*, which does not support the growth and multiplication of *P. solenopsis*, would help to control the pest. Timely detection and removal of infested plants along borders of crop will prevent further spread of the pest (Nagrare *et al.*, 2009).

ii) **Biological Methods:** All over the country, several parasitoids (predominantly *Aenasius* spp.) and coccinellid beetle predators are now found to keep mealy bug populations under control, thereby preventing spread and damage. Recently, a nymphal parasitoid, *Aenasius bambawalei* of *P. solenopsis* has been recorded (Hayat, 2009), which caused more than 80 per cent parasitization on cotton (Ram *et al.*, 2009) and 30 per cent on tomato (Mohindru *et al.*, 2009) (Plate 2.2).

   The scientists from CICR, Nagpur, stressed that further studies are essential to detect predators and parasitoids that may be occurring naturally in India to strengthen eco-friendly sustainable mealy bug management. Otherwise it would be necessary to import their natural enemies from the US. Their findings have been confirmed by Hayat (2009) who even isolated a parasitoid, *Aenasius* from the exotic mealy bug pest, *P. solenopsis*.

iii) **Chemical Methods:** Insecticides such as profenophos, chlorpyriphos, monocrotophos etc. which are being commonly used for mealy bug control, destroy the parasitoids and predators and can result in mealy bug outbreaks. Therefore, insecticide applications should be avoided until peak boll formation stage, so as to allow further establishment of the parasitoid and predator complex in the ecosystem. Eco-friendly insecticides such as neem oil based botanicals and buprofezin can be used if necessary in the initial stages so as to keep mealy bugs under check while causing minimum disturbance to the ecosystem.

However during peak boll formation stage, mealy bugs can establish colonies but are initially restricted to a few plants along the border rows, adjacent to the source of infestation and thus can be effectively managed through early detection and initiation of interventions to control early stages of infestation. If timely scouting and appropriate control measures are not initiated, cotton crop is likely to be severely damaged with mealy bugs.

Insecticides should not be applied all over the field to manage mealy bugs. Such a practice disrupts the ecosystem and does not allow naturally occurring parasitoids and predators to establish natural control. Therefore, the following practices are advised:

- Locate infested plants with more than one twig infested completely with mealy bug colonies.
- Do not allow physical contact with the infested plant. Do not disturb the plant vigorously. If possible, the affected twig can be gently detached from the plant, collected in plastic bags and taken far away from fields to be destroyed by burning.
- If at least 10% infested plants exceed grade-2 (at least one stem completely colonized with mealy bugs) in more than 40 plants randomly sampled plants per acre, chemical control measures may be initiated.
- Insecticide application should start first on the neighbouring plants and then as spot application near the root zone, base of the plant and other infested parts.

iv) **Integrated Management:** CICR has evolved an integrated management package for controlling the mealy bug, which consists mainly of symptomatic control without spraying pesticides at early crop stage, continuous monitoring of infestation, destruction of cotton stalks after picking, and use of botanical and biological control formulations like neem seed kernel extract and biopesticides at different stages (Table 2.3).

Table 2.3. Package of practices for managing mealy bugs on cotton

Cultural practices	When and how to do
Use acid delinted seeds for sowing	Fuzzy seeds may harbour infective stages of mealy bug, especially crawlers.
Early crop termination	Removal of cotton crop from the fields immediately after the last picking and maintenance of host-free period.
Destruction of cotton stalks	The dry cotton stalks should be pulled out of the fields or shredded and burnt off *in situ* before ploughing the field.
Clean cultivation: Destroy alternate weed host growing on field bunds, water channels and wastelands in the area	During the crop season and off-season, remove weeds especially congress grass and *Xanthium* which are the most suitable hosts for mealy bugs.
Grow non-hosts like pigeon pea, bajra or maize as border crop wherever possible	Growing 2 rows of densely planted pigeon pea or maize or bajra around the cotton fields and also if possible as intercrop of 1- 2 rows after 5-6 rows of cotton.

Cultural practices	When and how to do
Do not allow free movement of labour/ animals in infested fields	Mealy bugs spread through water, air, human, animal, farm implements, etc.
Botanical and biological control	**When and how to do**
Neem seed kernel extract (NSKE 5%) 50 ml/L + Neem oil 5ml/L + detergent powder 1g/L can be sprayed as spot application on infested stalks.	Spot application restricts the spread of mealy bugs. These formulations are less harmful to natural enemies and thus help in conserving ecosystem.
Fish oil rosin liquid 10 ml mixed with neem oil10 ml/L or Karanj oil 10 ml/L may be sprayed.	These formulations are less harmful to natural enemies and thus help in conserving ecosystem.
Use of *Cryptolaemus montrouzieri* adults /grubs at 10 per infested plants wherever available.	*C. montrouzieri*, is a naturalized predator of mealy bugs and feeds voraciously on *P. solenopsis*. Release the adult beetle during morning or evening hours to avoid direct exposure to hot sunlight.
Spray biopesticides viz., *Verticillium lecanii* (2 x 10^8 cfu/g) 10 g/L and *Beauveria bassiana* (10^8 spores/ml) 10 ml/L.	These formulations disrupt growth and multiplication of mealy bugs by causing disease without harming to the other natural enemies and the environment.
Chemical control	**When and how to do**
Spray less hazardous insecticides, such as Acephate 75 SP 1 g/L, Malathion 50 EC 2 ml/L, Buprofezin 25 SC 1 ml/L	When 10% infested plants observed of grade-2 (at least one stem completely colonized with mealy bugs) in more than 40 plants per acre. Cause less harm to the environment.
As the last option, spray moderately hazardous insecticides: Quinolphos 25 EC 5.0 ml/L, Chlorpyriphos 20EC 3.0ml/L, Profenophos 50EC 5.0 ml/L, Thiodicarb 75WP 5.0 gm/L	When > 10% infested plants observed to exceed grade 2 infestation in more than 40 plants per acre. Cause comparatively less harm to the environment.

2.2.2. Tobacco Caterpillar, *Spodoptera litura* (Lepidoptera: Noctuidae)

2.2.2.1. Key Facts

- The intensity of *S. litura* is likely to further increase under the potential climate change, as it has been found to consume more than 39% cotton leaves at elevated CO_2 levels in cotton crop (Kranthi *et al.*, 2009).
- As moderate toxicity of Cry 1Ac and Cry 2Ab to *S. litura*, it is likely that the pest may be able to feed on Bollgard II (Kranthi, 2012).
- Incidence of *S. litura* has increased in Punjab on cotton crop in recent years probably due to rise in average minimum temperature (0.7°C/year).

Though *Bt* cotton offers inherent toxicity against lepidopteran pests, but many workers have reported that *Bt* cotton with Cry1Ac is not effective against *Spodoptera litura* and has greater potential to survive in the presence of Cry1Ac. It started appearing regularly on cotton crop in Punjab in late 90's and has caused economic damage. As *Bt* cotton (BG-1) (Cry1Ac) does not provide protection against the pest, it inflicts heavy losses in several cotton growing regions of India. The intensity of *S. litura* is likely to further increase under the potential climate change, as it has been found to consume more than 30 per cent cotton leaves at elevated CO_2 levels (Kranthi *et al.*, 2009).

2.2.2.2. Damage:

The larvae feed gregariously on the under surface of the leaves and skeletonize them leaving only the midrib and veins in severe cases. They also attack flowers, buds and squares causing considerable loss (Plate 2.3).

2.2.2.3. Life Cycle:

Females lay eggs in masses that are cream to golden brown in colour. Egg masses are usually covered with body hair, scales and are laid on the underside of the leaf. Egg, larval and pupal periods lasts for 3-4, 3-20 and 8-10 days, respectively. Life cycle is completed in 50-60 days.

2.2.2.4. Management

i) **Biological Methods:** The caterpillars are attacked by numerous predators and parasitoids (e.g. predatory bugs, tachnid flies, braconid wasps and ichneumonid wasps). The use of *Bt* for control will help preserve beneficial insects and also reduce the risk of subsequent whitefly and mite attack.

ii) **Host Resistance:** Among the *Bt* cotton cultivars, highest mortality of 1st and 3rd instar larvae of *S. litura* was observed in BG II cultivars (Cry1Ac + Cry2Ab) when fed on leaves followed by squares and bolls. Highest mortality of 5th instar larvae was recorded on leaves of BG II cultivars.

2.2.3. Mirid Bug, *Creontiades biseratense* (Miridae: Hemiptera)

2.2.3.1. Key Facts

- *Creontiades biseratense* is infesting cotton since 2005 in Central and South India.
- It is emerging as number one pest in Karnataka, Tamil Nadu, Mahrashtra and Andhra Pradesh. In general a high population of mirids was recorded in Haveri (43.85 bugs/25 squares) followed by Belgaum (18.15 bugs) and Gulburga (14.95 bugs) districts of Karnataka in 2011. The estimated yield loss was 449 kg/ha.
- The incidence of *C. biseratense* in Haveri district (10 and 13 bugs/5 squares) followed by Dharwad (6 and 9 bugs/5 squares) during 2009 and 2010, respectively, which were recognized as hot spots.
- Peak incidence of the pest was coinciding with luxuriant reproductive growth of the crop.

Three species of mirid bugs are found infesting cotton in India. None of these have been historically referred as key pests of cotton. It is only after enhanced cultivation of *Bt* cottons, mirid bugs have assumed key status. Three different species of mirid bugs (Miridae: Hemiptera) viz., *Creontiades biseratense*, *Compylomma livida* and *Hyalopeplus linefer* are infesting cotton since 2005. Unlike widespread distribution of mealy bug *Phenacoccous solenopsis* (Nagrare *et al.*, 2009), the mirid bugs have been restricted to Central and South India. The dominant species is *Creontiades biseratense*. It is most potential, widespread and also becoming a production constraint every year. In Karnataka *Creontiades biseratense* mirid bugs are number one pests at present followed by Tamil Nadu. Other states having menace of this pest are Maharastra and Andhra Pradesh. *Compylomma livida* is dominant species in Maharastra. However, it is noticed in Karnataka also.

Recently the cotton mirid bug, *Creontiades biseratense* has emerged as an important insect pest on *Bt* cotton and its incidence in India was recorded during 2006 in Karnataka, Tamil Nadu, Andhra Pradesh and Maharashtra. The study undertaken during 2008-09 on the population level of the bug covering 7 districts as well as other reports showed that *C. biseratense* could be considered as an emerging threat to cotton cultivation in Karnataka. Mirid bug incidence on cotton squares and bolls causes shedding and leads to significant reduction in yield.

2.2.3.2. Economic Importance:

Two application of Acephate 75 SP at 1400 g ai /ha at peak incidence offered maximum protection against mirid bugs. The estimated loss was 449 kg /ha or 18% over zero protection. This was on an average true with both BG-I or BG-II cultivars under moderate insect pressure. There could also be more if incidence is more or avoidable loss may increase with extended spray schedule. However, two spray models could be ideal considering the pest and damage potential nature. The minimum avoidable yield loss with two rounds of protection worked out to be 20.6%.

2.2.3.3. Damage:

Both adults and nymphs damage developing flower buds/ squares and tender bolls. One to two days old bolls with dried intact petals provide a good habitat to the insects for feeding and sheltering (Plate 2.4). The characteristic symptoms of feeding on the flower bud shows oozing out of yellow fluid from the buds and staining of this yellow fluid on the inner surface of the bracts. Infested tender bolls have number of black patches on all sides of the outer surface of boll rind. It was observed shedding of most of the damaged squares and tender bolls. The feeding results in small, dark, sunken lesions on the surface of the boll and in severe cases deformed bolls are formed due to lack of fertilization of some ovules. This symptom is often referred to as "parrot beaking" of bolls. Almost all cultivated *Bt* cotton genotypes are found susceptible to mirid bugs.

Usually the adults are swift fliers and brown in colour. Nymphs are small with yellowish green abdomen and fast moving when disturbed. Feeding on matured bolls leads to parrot beaking and improper opening. It was estimated that damage leading to shedding of tender bolls for two days period revealed a loss of seed cotton yield.

2.2.3.4. Population Dynamics and Severity:

Population dynamics of mirid bugs has been assessed during 2007 to 2009 in Karnataka. Mirid populations ranged from 18 bugs/ 25 squares to 66 bugs/ 25 squares, and reached their peak at all locations in September and continued until December. High population of mirid bugs were recorded in Haveri (44 bugs/25 squares) and Belgaum district (43.8 bugs/ 25 squares). However, in Gulbarga the population was found to be low (18 bugs/ 25 squares). The severity of mirid bug incidence has been also noticed in Raichur and Bellary districts, however in Bidar, Bijapur, Gadag, Mandya and Uttar Kannada its incidence was low to moderate. But in all places irrespective of cultivar it was regularly noticed.

2.2.3.5. Life Cycle:

The female mirid lays egg singly, preferentially on the leaf petiole. Eggs are cylindrical, slightly recurved and laterally compressed; shining white in colour, turning to yellow as it matures. Eggs hatch within 4 to 5 days. There are five nymphal instars, each of about 2 to 3 days duration. The wing pads start to develop at the third instar. Adults are elongated, about 7-9 mm long, with long legs and antennae. Development from egg to adult takes about 15 to 18 days. The adults can live for 3 to 5 weeks and a female can lay up to 80 eggs in the life time.

The females of mirid bug, *C biseratense* preferred to lay eggs in petiole where in 855 eggs were traced. There are five instars in the life history extending total life cycle up to 39-41 days. The details are available in Udikeri *et al.* (2010).

2.2.3.6. Management:

Among the various treatments, Acephate 75 SP at 1.0 g/l was superior in reducing the mirid bug infestation (9.53 mirid/ 25 squares). Next in the order of superiority were acetamiprid (13.60), and imidacloprid (13.60 mirids/ 25 squares). In Australia recent trials have shown that reduced rates of indoxocarb or fipronil combined with salt have given effective control of mirids with reduced negative effects on beneficials (Khan, 2003). Application of Acephate 75 SP at 700 g ai/ha is a good management option presently.

2.2.4. Stink Bugs, *Acrosternum hilare, Nezara viridula, Euschistus servus*

A recent increase in stink bug pest status has been attributed to the reduced use of broad-spectrum insecticides, primarily due to the adoption of *Bt* cotton varieties (for caterpillar control). Three different stink bug species, including the green stink bug *(Acrosternum hilare)*, the brown stink bug *(Euschistus servus)*, and the Southern green stink bug *(Nezara viridula)*, are being reported in Delta cotton fields.

Although all three of these species are shield-shaped and possess the same identifying odor, the green species and the Southern species are bright green. A narrow, orange-yellow line also borders the body of the green stink bug. The brown stink bug is slightly smaller and brownish in colour.

2.2.4.1. Damage:

Stink bugs primarily feed on developing cotton bolls. In cotton, stink bugs prefer to feed on medium-sized bolls (approximately the diameter of a quarter), but are capable of feeding on bolls of any size (Plate 2.5). Although stink bugs may feed on bolls 25 days of age and older, bolls of this maturity are relatively safe from yield loss. Direct yield losses occur due to shedding of young bolls (fewer than 10 days of age) and damage to seeds; indirectly, the transmission of pathogens that cause boll rot affects the yield and fibre quality. For example, excessive stink bug feeding on cotton results in stained lint, poor colour grades, and reductions in physical fibre quality.

2.2.4.2. Management:

When cotton was more aggressively protected during the first, second, and third weeks of bloom, the economic returns were greater than those provided by the static 20 per cent threshold. While the brown stink bug is a little more tolerant to pyrethroids, none of three species is difficult to control with insecticides. Stink bugs overwinter as adults and are highly mobile as they feed on the bolls and seeds of the cotton plant. Stink bug infestations should be treated when the populations of adults or nymphs greater than one-quarter inch in size, exceed one bug per 6 row feet, or five bugs per 100 plants.

Pyrethroids do well against southern green and green stink bugs but are only fair against brown stink bugs. Organophosphates provide good control of all three species. Average sprays for stink bugs are about two with a range of zero to four applications, depending on infestations.

Treatment options suggested include the insecticides like Orthene, Bidrin, Methyl Parathion, Baythroid, Karate, and Scout X-tra.

2.2.5. False Chinch Bug, *Nysius raphanus* (Hemiptera: Lygaeidae)

About one-tenth of an inch long, false chinch bugs are grey or brown in colour with long, narrow bodies. According to the University of California's pest management guidelines, false chinch bugs often hide under plants or dirt clods during the heat of the day. False chinch bugs migrate to cotton when cruciferous weed hosts dry up or are destroyed by cultivation.

2.2.5.1. Damage:

In normal years, false chinch bugs (Plate 2.6) are considered a pest only of seedlings. Individual bugs do little damage, but large migrations can severely injure or kill young plants in a few hours. Damage is usually confined to border rows.

2.2.5.2. Management:

Early-season control of cruciferous weed hosts well before planting will eliminate the probability of this pest occurring in cotton. False chinch bugs are fairly difficult to control. A treatment of Bidrin at a rate of 0.4 to 0.5 pound per acre seems to work most consistently, but it is currently under review by the EPA.

2.2.6. Sugarcane Beetle, *Euetheola rugiceps*

Primarily an infrequent corn pest, the sugarcane beetle, *Euetheola rugiceps* is a black, hard-shelled beetle that is dome-shaped with faint ribbed lines on its back. Sugarcane beetle populations were seen in cotton for the first time in 2000 in a clean conventional tillage field. It looks like a firecracker blew up the root system of the cotton plant. High populations of sugarcane beetles may seem particularly bad, but the good news is that it doesn't seem to do a large amount of damage in most cases.

2.2.6.1. Management:

For control of the sugarcane beetle, application of a pyrethroid, banded, at the full labelled rate is recommended.

2.2.7. Three-Cornered Alfalfa Hopper, *Spissistilus festinus*

Sometimes called the beaver of the crop world, the three-cornered alfalfa hopper, *Spissistilus festinus* is another odd early-season cotton pest. The nymph resembles a beaver, often shows up more in the edges of a field. Adult insects are green, wedge-shaped, and range in length from 6.0 to 6.5 mm.

2.2.7.1. Damage:

Three-cornered alfalfa hoppers pull up plant stems and cause large, knot-like swelling areas on the stems. They girdle stems by their feeding and egg-laying activities. Nymphs and adults weaken the lower stem by piercing it with their needle-like mouths to extract plant juices. As a result, lodging and breaking usually occur weeks after attack.

This insect can mow over a crop if it is not controlled. But, to catch populations in time to treat, you will have to scout heavily and early. Often, when damage by this pest is noticeable it is too late to make control applications.

2.2.7.2. Management:

Pesticides that have been shown to offer control of the three-cornered alfalfa hopper include Orthene at a rate of 0.75 to 1.00 pound per acre and pyrethroids at a mid-range labelled rate.

2.2.8. Tea Mosquito Bug, *Helopeltis bryadi* (Hemiptera: Miridae)

2.2.8.1. Key Facts

- It is a common pest of guava and tea.
- However, it appears rarely on cotton crop.
- First time reported on DCH-32 cotton crop from Davangere district and during 2002 in severe form on the same genotype in Uttara Kannada and Mysore districts, Karnataka.
- Affected genotypes were interspecific hybrids.

Helopeltis bryadi a hemipteran pest has been found damaging the *Bt* cottons severely. It is a common pest on guava / cashew/ tea etc. and are called kajji bug as it causes black lessons on the almost all leaves and young shoots, squares and bolls. However, it appears rarely on cotton crop, first time reported on DCH-32 cotton crop during 1996 in Honnali of Davangere district. Later in 2002, on the same genotype it was noticed in H.D. Kote Taluk of Mysore district. The most severe damage of this pest insect was noticed on *Bt* cotton in Hosalli village of Uttar Kannada district. Affected genotypes are interspecific biotech hybrids.

2.2.8.2. Damage:

Both nymph and adults suck the sap from leaves and young shoots, squares and bolls. As a result of it leaves get rolled at the edge with brown central black lesions particularly near the main veins. Cankers like lesions develop on the lower green bolls. Linear scars with white papery layer appear in tender shoots. Retarded growth leads to gradual dying of shoots. Matured bolls bear blackish brown circular pits towards tip of the bolls. Rotting of boll takes place due to entry of rainwater and such infested dried up bolls/ squares remain intact in plant.

2.2.9. Flower Bud Maggot/Gall Midge, *Dasineura gossypii* (Cecidomyiidae: Diptera)

2.2.9.1. Key Facts

- Appeared as potential pest for the first time in the history of cotton entomology in India during 2009 in Haveri district, Karnataka.

- 90% fruiting body damage was recorded in largely cultivated *Bt* cotton cultivars viz., Kanaka and Neeraj.
- In recent survey, cotton hybrids have shown 50 to 92% of square damage.
- The yield loss attributed was more than 60% during 2009.

During 2009 severe incidence of gall midge, *Dasineura gossypii* was seen in a farmer's field at Hesarur (Savanur Taluk, Haveri District). The crop was sown in first week of June and by August incidence took severe proportions. Later in Kakol, Konanatambagi etc villages (Haveri District) also it has been noticed in severe form. In the research farm (ARS, Dharwad) it has been noticed in negligible proportions earlier, but, since a couple of years its incidence is increasing. In Raichur, Bellary, Belgaum, Haveri, Gulburga and Dharwad districts also the pest has been recorded during survey but, in low proportions.

2.2.9.2. Damage:

The maggots are damaging stages. They feed on anthers and stamen column leading to degradation/ decaying. There will be 3-15 maggots in a flower bud. The infested flower buds fail to grow properly and they will not open as the petals as well as tissue inside dries. Thus comprehensive symptoms give a picture of flower drying through organ degradation and death. The pupation take place inside dried flowers itself. In general, squares formed will not turn into a boll, due to death at flowering stage. Tissue drying and/or death unevenly lead to twisted or contorted stamen column/ anthers. Compared to normal boll, persistent stamina column is seen in affected ones but it needs confirmation. In some cases, where fertilization is not affected, the boll formation is affected. The size remains smaller. Tissue degradation is prominent on boll rind also. The bolls will not reach normal size and no proper filling with fibre.

2.2.9.3. Management:

As the pest remains inside the flower and multiplies faster, it was difficult to manage this pest by the farmers even with 3-4 sprays. The insecticides used viz., Imidacloprid (Confidor 200 SL), Oxydemeton methyl (Metasystox 25 EC), Monocrotophos (Monocil 40 SL), Acephate (Starthene 75 SP), Neem oil etc. could not render satisfactory results. When these insecticides are used along with DDVP (Nuvan 78 EC) as tank mix (at 0.25 ml/l), the spread of incidence was checked to certain extent. In all the locations where the pest has caused wide spread damage, the cultivar was MRC-7351 BG-II (Kanaka) only. In recent survey, other hybrids have also shown varied degree of square damage ranging from 50 to 92%. The yield loss attributed is more than 60%.

Integrated Cotton Pest Management: It is important to develop pest management using a broad spectrum of practices. These include crop rotation, tillage, higher biological diversity in the farming system and promotion of the pests' natural enemies. Taking into account the spatio-temporal dimension of natural population regulatory factors has led to changes in agricultural practices and production systems (Deguine *et al.*, 2008). In cotton, production systems maintaining permanent ground cover are having increasing success. Intercropping and trap cropping have been favourable for the maintenance of beneficial arthropod complexes and unfavourable to the growth of pest populations (Plate 2.7).

2.3. SOYBEAN, *Glycine max*

There was an outbreak of tobacco caterpillar (*Spodoptera litura*) on soybean crop causing disease epidemic in Vidarbha region of Maharastra in August 2008. After the introduction of *Bt* cotton in 2002 and its subsequent rapid adoption, its infestation in these crops has significantly declined in cotton. *H. armigera* has become a menace in the pulse growing regions. As *H. armigera* is not able to survive on *Bt* cotton, its cycle gets disrupted and there is no significant movement of the pest from cotton-to-pigeonpea-to-chickpea (Bambawale *et al.*, 2009).

2.3.1. Tobacco Caterpillar, *Spodoptera litura*

There was an outbreak of *S. litura* on soybean in Kota region of Rajasthan and a loss of Rs 300 crores was estimated. The pest also struck in epidemic form on soybean in Vidarbha region of Maharashtra in August 2008 and caused severe losses in yields to the tune of Rs. 1392 crores.

An investigation was carried out to study the effect of various weather parameters on the occurrence of larva and adult population of *Spodoptera litura* and factors for outbreak of this pest in soybean in Banswara, Rajasthan. The moth populations of *S. litura* were active from August to middle of October and decreased sharply in late October. The peak appearance was observed during September-October months and corresponded with peak activity of egg masses and larval population in soybean contributed to the outbreak of this pest during the reproductive stage of the crop. Overall correlation studies confirmed that the fluctuations in the male moth catches of *S. litura* per pheromone trap is mainly due to weather factors like, maximum temperature, rainfall, sunshine and wind speed. Moth population in pheromone traps will be used as a warning situation indicating the occurrence of larvae in the host for potential attack and to initiate timely management of *S. litura* in soybean (Ramesh Babu *et al.*, 2015).

2.3.1.1. Damage:

The larvae attack the crop normally in the month of August and September during late vegetative and early reproductive stages of the crop. The newly hatched larva, scrap the chlorophyll content and feed voraciously. This gives the appearance of yellowish white web on the leaves. As the caterpillars mature, they completely defoliate the leaves leaving only midribs and stalks. Larval feeding is vigorous as it advances in age. The soft pods are chewed by the larvae and the thick pods are bored and feed on grain. The habit of larvae is to hide under the plants, cracks and crevices of soil debris during the day time and feeding during night hours. The incidence could be noticed by the faecal pellets left on the leaves (Plate 2.8).

2.3.1.2. Management:

The following practices have been recommended for the management of *Spodoptera litura* on soybean:

- Use optimum seed rate of 70-100 kg/ ha.
- Collect and destroy infested plant parts, egg masses and early stage larvae found in clusters.
- Install one light trap (200 W mercury vapour lamp)/ ha in the field to catch the *S. litura* adults.

- Installation of pheromone traps for monitoring incidence of *S. litura.*
- Use of castor as trap crop for tobacco caterpillar.
- Release *Telenomus remus* at 50000/ha against *S. litura.*
- Spray Spodoptera NPV (Nuclear Polyhedrosis Virus) at 250 LE/ha.
- Spray NSKE at 5% to manage early stage larvae.
- Spray *Bacillus thuringiensis* var. *kurstaki*, Serotype H-39, 3b, Strain Z-52 at 0.75 to 1.0 kg/ha.
- Spray methomyl at 2 L/ ha, Ethofenprox 10 EC at 1L/ ha, Triazophos 40 EC at 625 mL/ ha, Quinalphos 25 EC 1.5 L/ ha, Chlorantraniliprole 18.5% SC at 150 mL/ ha, Indoxacarb 15.8% EC at 333 mL/ ha and Lamda-cyhalothrin 50 EC at 300 mL/ ha for controlling tabacco caterpillars.
- Use resistant/tolerant varieties – JS 81-21, PS 564 and PK 472.

2.4. SUNFLOWER, *Helianthus annus*

2.4.1. Tobacco Caterpillar, *Spodoptera litura*

Outbreaks of *S. litura* were noticed in major sunflower growing areas of Central and Southern India. During 2005, the outbreak of *S. litura* led to more than 90 per cent defoliation of sunflower cultivar germplasm (Sujatha and Lakshminarayana, 2007).

2.4.1.1. Damage:

The larvae feed on the tender leaves (Plate 2.9), shoots, bracts and petals. Later, the larvae spread in the field causing defoliation. The larvae also feed on the developing seeds in capitulum.

2.4.1.2. Management

- Handpick the *Spodoptera* larvae and destroy.
- Spray any of the following insecticides: Dichlorvos 76EC 500 mL/ha or Phosalone 35 EC 1000 mL/ha.

2.5. SUGARCANE, *Saccharum officinarum*

2.5.1. Woolly Aphid, *Ceratovacuna lanigera* (Aphididae: Hemiptera)

The sugarcane woolly aphid, *Ceratovacuna lanigera* was first detected in India in West Bengal in 1958, but it quickly disappeared in 1960. The aphid again reappeared in epidemic form in 2002 in Sangli Province of Maharashtra. A sudden infestation of sugarcane woolly aphid was recorded in January 2006 in the mid country hills of Badulla, Sri Lanka after its first outbreak in 1905. Subsequent outbreaks were recorded during November and December of the same year in other areas (Kumarasinghe and Basnayake, 2009).

The pest breeds on plants of the family Poaceae, but has been also observed on members of Bixaceae, Theaceae and Combretaceae. It has been recorded on ten species of plants in India. It reproduces parthenogenetically and has an anholocyclic (absence of sexually producing generation) life cycle.

Invasion of sugarcane woolly aphid, *C. lanigera* in Maharashtra during 2002 is another example of pest's reaction to climate change and getting mostly naturally regulated. It spread to other parts of Maharashtra covering an area of 1.43 lakh hectares by March 2003 and caused 30 per cent losses in sugarcane yields. Later it migrated throughout the state except Konkan region and gradually spread to parts of Karnataka, Tamil Nadu and Andhra Pradesh (Bade and Ghorpade, 2009). In Karnataka, it was first detected in August 2002 in Athani, Belgaum district. However, within a short span of next 12 months, it quickly spread to Bagalkot, Gulbarga, Bidar, Bijapur, Haveri, Dharwad, Davangere and down south in Shimoga districts (Bambawale *et al.*, 2009).

2.5.1.1. Damage:

Initial aphid infestation was seen on the under surface of leaves along the midrib and then over the entire under surface, covering it with flocculent, waxy secretion (Plate 2.10). Copious honeydew excretion often covers the entire upper surface of the leaves, leading to growth of sooty mould (Ghosh, 1988). Due to continuous sap-sucking, the crop becomes stunted (Lopez, 1931), and continuous infestation leads to reduction in the length, circumference, weight and sugar content of the stalk, and in susceptible varieties in Vietnam (Anon 1963), loss in tonnage as well as sugar recovery (Tripathi, 1995). Gupta and Goswami (1995) assessed the effect of 25 and 100% aphid-infested leaves on yield and quality parameters of sugarcane and found that 100% infestation had detrimental effects on the length (11.6% reduction), girth (3.5% reduction), weight (16.6% reduction), length of internode (18.4% reduction) and width of leaf (4.9% reduction). Juice quality parameters also exhibited considerable reduction. The per cent reduction in sucrose, brix, glucose, purity and commercial cane sugar (CSS) was 53.3, 32.3, 25.3, 31.7 and 64.0, respectively.

2.5.1.2. Management:

The general but important recommendations by Patil *et al.* (2002) include proper sanitation, avoidance of transport of infested leaves from one area to another, removal of other alternate hosts in the vicinity, removal of infested leaves by employing labourers, avoiding lodging of canes, burning of trash after harvest and avoiding use of canes from infested fields for planting.

Thirty-eight natural enemies have been recorded on the aphid from different parts of the world. Recent surveys in the pest-affected areas of Maharashtra (western India) and Karnataka (southern India) have yielded several indigenous predators which include coccinellids, neuropterans, syrphids and a pyralid.

Deng *et al.* (1987) and Rabindra *et al.* (2002) advocated biological methods by augmentation of the coccinellids, *Synonycha grandis* and *Anisolemnia dilatata*. The proposed methods are logical, as indiscriminate use of pesticides may lead to outbreak of secondary sucking pests. Successful suppression of the woolly aphid by field release of *S. grandis* was demonstrated by Deng *et al.* (1987). A total of 44,660 beetles were released into sugarcane fields, resulting in 90 to 96.6% reduction in aphid population.

Sprays of the entomopathogenic fungus, *Metarhizium anisopliae* are being recommended and are also providing effective aphid control (Lingappa, 2003; Nirmala, 2003). These recommendations make sense as many pathogenic fungi have performed well in the sugarcane ecosystem (David and Easwaramoorthy, 1986), being an annual crop with high humidity within the crop canopy, facilitating propagation and rapid spread of fungi. *M. anisopliae* can be easily multiplied on artificial media.

Integrated pest management involving mainly mechanical and biological control appears to be the best option. There is ample scope for more directed work on this important emerging pest, especially in the areas of pest ecology and distribution. The role of resistant varieties, biological, cultural and mechanical control in managing the pest needs to be carefully evaluated. IPM practices involving habitat management, spacing, irrigation, nitrogen fertilization, resistant varieties, traps, natural enemies, sanitation (removal of older leaves and burning of trash), use of pathogens and safe chemicals need to be developed.

2.6. RICE, *Oryza sativa*

Recently, brown plant hopper (BPH), *Nilaparvata lugens* became serious pest and its first serious epidemic occurred in 1973 over half a million ha resulting in 10-70 per cent loss in grain yield under Indian conditions. Series of BPH outbreaks since then were recorded in Karnataka (1975 and 1985), Andhra Pradesh (1976-1983), Telangana (1980), Madhya Pradesh and Orissa (1976), and Tamil Nadu (1977 and 1982). Large scale cultivation of gall midge resistant varieties gave alarming reports of development of new biotypes damaging resistant varieties have been received from Andhra Pradesh and Maharashtra during the last decade. It was also reported to have caused heavy losses in Jawa and Indonesia. The widespread epidemics of rice hispa were reported in Orissa, Himachal Pradesh, West Bengal and Tripura during 60s and 70s. Yellow stem borer is a key pest of rice, causing losses up to 95 per cent until 1960s. The pest declined in severity during 1970s due to large scale cultivation of dwarf rice varieties (Krishnaiah, 1993). A serious problem of rice hispa occurred in Andhra Pradesh and Tamil Nadu, particularly in areas cropped with BPH-resistant varieties. The pest reduces yield by 40-57 % (Kataki *et al.*, 2001).

2.6.1. Brown Plant Hopper, *Nilaparvata lugens* (Hemiptera: Delphacidae)

2.6.1.1. Key Facts

- Emerged in severe form in Northern India and resulted in failure of more than 333,000 ha crop during 2008-09.
- In Delhi, 40-50% hopper burn damage was recorded.
- Reasons for outbreaks:
    - High humidity (>90%) with temperature between 25-32°C.
    - Excessive use of nitrogenous fertilizers.
    - Closer crop canopy (spacing 15 x 10 cm).
    - Indiscriminate use of pesticides.
    - Tolerance to neonicotinoids.
    - Mortality of natural enemies.
- Temperatures of 26.4 to 30.0°C, RH 55-99% and few showers of rain are conducive for BPH population build up in Punjab.

2.6.1.2. Effect of Elevated CO_2 on BPH population on Pusa Basmati 1

- Females laid more eggs (324.3±112.3 eggs/female) on the rice plants exposed to elevated CO_2 of 570±25 ppm than at 380 ppm ambient CO_2 (231.7 ± 31.8 eggs).
- More than two-fold increase in population (435.4±62.0 hoppers/hill) at peak incidence compared to ambient CO_2 (121.4 ± 36.8 hoppers/hill) during kharif 2010.
- Crop under elevated CO_2 suffered higher yield loss (26.5%) due to higher BPH population and sucking rate compared to ambient CO_2 (12.4%).

2.6.1.3. Damage:

BPH infest the rice crop at all stages of plant growth. As a result of feeding by both nymphs and adults at the base of the tillers, plants turn yellow and dry up rapidly. At early infestation, round, yellow patches appear, which soon turn brownish due to the drying up of the plants. This condition is called 'hopper burn' (Plate 2.11). Temperature is a critical factor in the life activities of the insect. The hatchability and survival rate are the highest around 25 °C. Eggs are very sensitive to desiccation and soon shrivel when the host plant starts wilting. The population growth of brown plant hopper is maximum at a temperature range of 28 to 30 °C.

Excessive use of urea as nitrogenous fertilizer and insecticides can lead to outbreaks by increasing the fecundity of the brown plant hopper, and by reducing populations of natural enemies (Preap *et al.*, 2001, 2006). In 2011, the Thai government announced an initiative to address brown plant hoppers that included restricting the use of outbreak-causing insecticides including abamectin and cypermethrin - the decision was supported by the International Rice Research Institute (IRRI), Philippines.

2.6.1.4. Biology:

The brown plant hopper is dimorphic, with fully winged 'macropterous' and truncate-winged 'brachypterous' forms (Plate 2.12). The macropterous forms are potentially migrants and are responsible for colonizing new fields. After settling down on the rice plants, they produce the next generation in which most of the female insects develop as brachypters and males as macropters. Adults usually mate on the day of emergence, and the females start laying eggs from the day following mating. Brachypterous females lay 300 to 350 eggs, while macropterous females lay fewer eggs. The eggs are thrust in a straight line generally along the mid-region of the leaf sheath. Eggs hatch in about six to nine days. The newly hatched nymphs are cottony white, and turn purple brown within an hour; they feed on plant sap. They undergo five instars to become adults.

2.6.1.5. Management:

Rice varieties with resistance to BPH, e.g. IR 64, are important for preventing outbreaks (Alam and Cohen, 1998; Sangha *et al.*, 2013). However, in areas with low insecticide use, high levels of BPH resistance are not usually necessary (Cohen *et al.*, 1997). Chemical mutagenesis can significantly increase or decrease BPH resistance levels of rice (Sangha *et al.*, 2008). Some chemical insecticides, e.g. imidacloprid, can affect the gene expression of rice and thereby increase susceptibility to BPH (Yao *et al.*, 2012).

In an attempt to make BPH control more species-specific, researchers are trying to develop methods of turning off specific BPH genes for digestion-, defense- and xenobiotic metabolism. Many novel genes for these functions have been detected in tissue from BPH intestines (Yan-Yuan *et al.*, 2012).

Some plant lectins are antifeedants to BPH and if properly formulated may have the potential to protect rice from BPH (Rao *et al.*, 1998).

Predators of this insect include the spiders, *Pardosa pseudoannulata* and *Araneus inustus* (Preap *et al.*, 2001). In some cases, BPHs lay eggs in the rice seed beds (also known as rice nurseries) shortly before transplanting, so enter the field in this manner (Preap *et al.*, 2002).

Differential mortality of predators and hoppers does not appear to be the primary factor for insecticide-induced resurgence (Chelliah and Heinrichs, 1980). Some insecticides evidently increase the protein content of BPH male accessory glands, and thereby increase plant hopper fecundity (Wang *et al.*, 2010). Some insecticides increase the amount of amino acids and sucrose available in the phlocm of rice plants, and thereby increase BPH survival (Wu *et al.*, 2001).

2.6.2. Swarming Caterpillar, *Spodoptera mauritia* (Lepidoptera: Noctuidae)

2.6.2.1. Key Facts

- In eastern India, it has emerged as a regular pest in Odisha, Jharkhand, Bihar and Chhatisgarh.
- During 2008, a severe outbreak was observed in Cuttack and Sonepur districts, Odisha where 6-8 larvae/hill were recorded in the initial crop stage.
- During 2009, severe outbreak of this pest was observed in about 1.25 lakh ha of *kharif* paddy in 13 districts of western Odisha recording about 80-90% damage (Anon, 2010).

Rice swarming caterpillar or armyworm, *Spodoptera mauritia* during the last few years has emerged as a serious pest in eastern India and caused severe losses to wet season rice production. It has become a regular pest in western Odisha and has also been recorded in coastal districts of the state. In Odisha, the swarming caterpillars have swarmed over thousands of hectares of land in 2009 completely damaged the paddy plants in the fields that look as if grazed by cattle especially in Sambalpur district where complete loss had been reported. Sundergarh district in Odisha was also severely affected by this pest and complete loss occurred in some places. This pest also shares the same status in Jharkhand, Bihar and Chhatisgarh. During *kharif* 2007, a severe outbreak was observed in Tangi Block of Cuttack district. In India it is found in all the rice growing areas especially along the west coast and delta in Kerala and Tamil Nadu (David and Ananthakrishanan, 2004).

2.6.2.2. Damage:

The swarming caterpillars damage paddy crop by cutting off leaf tips, leaf margins, leaves and even the plants at the base (Plate 2.13). It is more severe on the seedlings in nursery and direct seeded crops and at early tillering stage in transplanted rice. Newly hatched larvae cause the plants to look sick with withered tips and cut leaves.

They appear suddenly in masses and move like an army from field to field so that the seeds or direct seeded fields look as if grazed by cattle, warranting for re-sowing or replanting. Generally a transplanted crop is not seriously affected. Rice plants older than 6-7 weeks are usually not attacked by this pest. The swarming caterpillar occasionally caused yield loss up to 20% (Dale, 1994). Attacked plants are recorded to be damaged from 10-100% depending upon the density of the pest larvae. Under severe infestation a complete damage of the plants has been reported which warrants re-sowing or replanting the crop. During *kharif* 2008, outbreak was reported from 13 blocks of Sonepur district in western Odisha where 6-8 larvae/hill were recorded in the initial stage. During kharif 2009, severe outbreak of this pest was observed in about 1.25 lakh hectares of paddy in 13 districts of western Odisha recording about 80-90% damage (Anon, 2010).

2.6.2.3. Host Range:

Swarming caterpillar attacks a wide range of graminaceous plant species. Rice is the main host of S. *mauritia.* This pest, in general, breeds on other hosts like different kinds of grasses and occasionally on maize, oat and sorghum. The nut grass, *Cyperus rotundus* is a common weed in paddy fields in western Odisha and is found to be an alternate host of this pest during the off-season.

2.6.2.4. Factors Favouring Population Build Up:

Prolonged dry conditions followed by heavy rainfall favours its outbreak. Wind and rain storm helps in migration of moths to long distances. Pest occurs throughout the year on alternate hosts and move to paddy in *kharif* season. Heavy rainfall leads to high mortality of larval population. Pest migrates from older rice to grassy areas for off season survival.

2.6.2.5. Management

i) Cultural Methods

- Crop rotation in endemic pockets helps.
- Deep ploughing the field in summer exposes the larvae and pupae for predation by birds.
- Remove excess nurseries and weeds from the field and bunds.
- Flooding the nurseries and small fields brings out the larvae to the surface, which get predated by the birds. Ducks if let into field, will feed on the caterpillars. A herd of ducks can easily destroy these caterpillars if let into the fields.
- Use of bamboo perches facilitates predation by birds.
- In case of severe infestation, small plots can be isolated and the movement of the caterpillars can be prevented by digging a trench around the infested field wherever possible.

ii) Mechanical Methods

- The inability of *S. mauritia* larvae to swim in water is a weakness and in flooded fields they are forced to stay on the plants which they defoliate. Therefore, kerosene oil may be poured into the stagnant water in the bunded fields (kerosene at 2 L per hectare). With the use of a long rope stretched across the field (two persons walk through the field) the paddy plants are shaken rigorously. The larvae fall into the kerosenized water and ultimately die.

- Application of malathion (2% dust applied at 25 kg ha^{-1}) or chlorpyriphos dust (1.5%) or endosulfan dust (5%) along the bunds of the field kills the pest and also prevents the migration of larvae to other rice fields. The technique was widely adopted because it proved to be a cheap and effective method for managing the *S. mauritia* within a short time.
- Before the paddy season, grasses near the fields earmarked for paddy, may be destroyed mechanically which provide shelter and act as reservoir for migration of larvae.
- As the moths of *S. mauritia* are attracted towards light, therefore, light traps can also be utilized for mass trapping of the moths.

iii) Biological Control/Natural Enemies

a) **Parasitoids:** Under natural conditions, the egg parasitoids have been found effective in containing the population of *S. mauritia.* In Hawaii, under natural conditions, the parasitization of *S. mauritia* egg masses ranged from 80-90 per cent by *Telenomus nawai* (a scelionid egg parasitc). The parasitoid was also released in Island Guam from Hawaii for controlling rice swarming caterpillar. In Hawaii, a braconid, *Apanteles marginiventris*, a solitary internal larval parasite has also been recorded. This parasite preferred to oviposit on the minute first instar caterpillars although it could also parasitize second and third instar larvae.

b) **Predators:** The vertebrate predators such as the common crow, *Corvus splendens*; the jungle crow, *Corvus macrorhynchos*; the cattle erget, *Bubulcus coromondus*; the paddy berd, *Ardeola grayi*; the white breasted water hen, *Amaurovius phoenicocurus*; and the mynah, *Acridotheres tristis* have been found to predate on the larvae of this pest.

c) **Pathogens:** In Hawaii, a nuclear polyhedrosis virus (NPV) of this insect was discovered (Bianchi, 1957). It is believed that the NPV might have entered Hawaii together with its host. Cross infectivity test confirmed that the exposed SWC larvae succumbing to the virus showed the typical symptoms of NPV infection i.e., their skins become fragile and their internal contents a fluid mass, a typical "wilt" condition. Larvae which become infected in their early instars and die before the fourth instar usually begin to turn pale 2 to 4 days prior to death and, at death, have a whitish or creamy appearance. After death the larvae rapidly darken in colour. Older infected larvae which die in the fifth or older instars gradually turn slightly pale with a pinkish tinge several days before death, but otherwise remain nearly as dark brown as the healthy larvae. Thus, in older larvae, change in colouration is not as marked as in the younger larvae.

iv) Chemical Methods

- Spraying of chlorpyriphos 20 EC at 2.5 litres/ha or quinalphos 25 EC at 2.0 litres/ha or triazophos 40 EC at 1.0 litres/ha or dichlorvos 76 SL 600 ml/ha on the paddy crop during evening hours.
- Dusting of chlorpyriphos 1.5 % D or endosulphan 5% D at 30 kg/ha controls this pest.
- Spraying at early stage is very effective.

v) **Integrated Pest Management:** Important components of IPM are regular pest monitoring, cultural and mechanical means of pest management, and augmentation and conservation of natural enemies and intelligent usage of insecticides.

2.7. SORGHUM, *Sorghum bicolor* and PEARL MILLET, *Pennisetum glaucum*

The maize stem borer (*Chillo partellus*) is a traditional destructive pest of maize and sorghum causing 20-87 per cent loss in yield under varied agro-climatic conditions. The shoot fly (*Atherigona* spp.) was a minor pest of maize throughout the country except in parts of Southern India. Heavy incidence of *Pyrilla* was recorded on grain and fodder sorghums in the entire northern belt of the country (Kishore, 2005).

2.7.1. Grey Weevil, *Myllocerus* spp.

The grey weevil, *Myllocerus* spp. and sugarcane leaf hopper, *Pyrilla perpusilla* have assumed pest status on sorghum and pearl millet.

2.7.1.1. Damage:

Cotton grey weevil, *Myllocerus undecimpunctulus maculosus* can cause serious damage to sorghum seedlings, particularly in high yielding hybrids and cultivars. The adults feed on the foliage, starting from the leaf margin (Plate 2.14). During severe infestations, the entire leaf blade is eaten up, leaving only the midribs. In seedlings, the weevils even feed on the tender stems, which often results in total plant loss (Kishore and Srivastava, 1976).

2.7.1.2. Management:

Shinde *et al.* (1970) reported that mevinphos, dicrotophos, and fenitrothion were more effective against *Myllocerus* spp., while Verma *et al.* (1969) found ethyl-parathion, malathion, endrin, mevinphos, telodrin, and formothion to be more toxic.

2.7.2. Leaf Hopper, *Pyrilla perpusilla*

2.7.2.1. Key Facts

- The sugarcane leaf hopper has emerged as a major pest on sorghum and pearl millet.
- Heavy incidence of *Pyrilla* was recorded on grain and fodder sorghums in the entire northern belt of the country.
- Sorghum genotypes MH 410, MH 285, MP 207, MP 234, MP 246 and MP 241 exhibited 4.1 to 6.0 egg masses per plant.
- The colour of leaves of these genotypes was pale green and comparable with leaf colour of sugarcane (Kishore, 2005).

The sugarcane leaf hopper, *Pyrilla perpusilla* have assumed pest status on sorghum and pearl millet. Heavy incidence or *Pyrilla* was recorded on grain and fodder sorghums in the entire northern belt of the country. A perfect hopper burn was also observed on some sorghum varieties at the farm of Indian Agricultural Research Institute, New Delhi. High populations of *Pyrilla* have also been recorded on pearl millet (Kishore, 2005).

2.7.2.2. Damage:

Sugarcane leaf hopper, *Pyrilla perpusilla* (Plate 2.15) is the most destructive sap-sucking pest of sugarcane but it has also been recorded as a pest of other crops such as wheat, maize and millet. This pest occurs in epidemics in sub-tropical areas every 5 - 6 years. It sucks sap from the lower surface of the leaves leading to withering, which can adversely affect the grain yields (Kushwaha *et al.* 1980). *Pyrilla perpusilla* sucks phloem sap from leaves and excretes honeydew onto foliage, leading to fungal diseases. This direct and indirect damage affects sugar yield and quality.

2.7.2.3. Management:

Jotwani and Butani (1978) recommended malathion and fenitrothion for the management of sugarcane leaf hopper, *P. perpusilla.* Foliar sprays of *Metarhizium anisopliae* at 10^7 spores/ml/ha or releases of *Pyrilla* adults seeded with *M. anisopliae* at 250 adults/ha were also found effective.

2.7.3. Leaf Roller, *Cnaphalocrosis trapezalis* (Lepidoptera: Crambidae)

2.7.3.1. Key Facts

- It was a minor pest of sorghum.
- Outbreak in Karnataka in 2013.
- Severe incidence of *C. trapezalis* was noted, mostly in the early stages (4-5 leaf stage).
- The overall incidence ranged from 35.0 to 58.7% with an average of 46.85%.
- Attributed to climate change.

In the year 2013 there was severe outbreak of maize leaf roller, *Cnaphalocrosis trapezalis* on sorghum particularly in Yadgir District of Karnataka. Though it was listed as a minor pest on sorghum earlier by Nair (1970), it appeared in severe form causing greater damage to sorghum. This could be attributed to climatic changes in the recent years. The incidence of *C. trapezalis* was noticed at different stages of growth, mostly in the early stages of the crop particularly when it is four to five leaved and gradually declines as the crop growth advances. The overall incidence during cropping period ranged from 35 to 58.70% with an average of 46.85 % (Shankara Murthy and Nagaraj, 2014).

2.7.3.2. Damage:

The eggs are laid on the upper surface of leaves and the larvae feed on the inside surface of a folded leaf (Plate 2.16). The larva rolls the two edges of leaves by pasting its saliva and feeds within the rolled leaves by scraping the green matter resulting in drying of leaves. As a result of larval feeding, the leaves start drying up from the tips. Under severe incidence, the entire plant dries up. In each leaf, a greenish yellow larva with setae over its body was found. The larva pupated inside the rolled leaves. Broad leaved and succulent varieties are highly susceptible.

2.7.3.3. Biology:

The eggs are laid on the upper surface of leaves and the larvae feed on the inside surface of a folded leaf. The larva is slender, pale yellowish green in colour, and about 20 mm in length, it has small oval spiny patches scattered over the body, from which stout bristly hairs arise. The larvae are fully grown in 11-20 days. Pupation occurs within a leaf fold, and is completed in 6-8 days.

2.7.3.4. Management:

Low usage of nitrogen fertilizer decreases the percentage of leaf folder infestation. Early planting is recommended to escape a high degree of defoliation. Hand pick rolled leaves and spray carbaryl 50 WP at1 kg/ha.

2.8. PULSE CROPS

2.8.1. Changing Pest Scenario

- Changes in cropping patterns, climate change has resulted in emergence of serious pests such as spotted pod borer, *Maruca vitrata*, pod sucking bug, *Clavigralla* spp. (Sharma 2005).
- Heavy rains during October-November often result in outbreaks of *Helicoverpa armigera* and *M. vitrata* in southern India, while early warming of weather in North India (3-5° C higher than the normal in March) result in heavy *H. armigera* damage in pigeon pea and chickpea (Sharma, 2014).

2.8.2. Gram Caterpillar, *Helicoverpa armigera*

Helicoverpa armigera had become a menace in pulse growing regions and started causing considerable damage to chickpea and pigeon pea. The epidemic of *H. armigera* on these crops was reported in 2001. Further, 2008 crop season has also witnessed intense comeback of Helicoverpa on chickpea and other pulses. However, after the introduction of *Bt* cotton in 2002 and its subsequent rapid adoption, its infestation in these crops has significantly declined in the cotton-based cropping systems. *H. armigera* subsequently moved from cotton to pigeon pea and then to chickpea. As *H. armigera* is not able to survive on *Bt* cotton, its cycle gets disrupted and there is no significant movement of the pest from cotton-to-pigeon pea-to-chickpea (Bambawale *et al.*, 2009).

2.8.2.1. Management:

Timely sowing to exploit host avoidance phenomenon, intercropping with sorghum, use of trap crops like *Vicia sativa* and African giant marigold, use of *Ha*NPV at 250 LE/ ha, spraying NSKE at 125 L/ha, need based use of insecticides like 0.07% endosulfan or 0.04% fenvelarate are recommended.

i) **Integrated Pest Management:** Efficacy of intercropping with bio- and chemo-intensive modules against *H. armigera* for four consecutive years showed that chickpea + mustard (4:2) recorded least pod damage (8.75%) with chickpea equivalent yield of 1672 kg/ha as against 13.60% pod damage and 1402 kg/ha yield in sole crop. Among control modules, bio-intensive IPM proved most effective recording 7.92% pod damage and 1748 kg/ha yield as against 15.34% pod damage and 1260 kg/ha yield in control. Chickpea intercropped with mustard in combination with BIPM resulted in highest CBR (Table 2.4) (Ali *et al.*, 2009).

Table 2.4. Effect of IPM modules and intercrops on pod damage and yield in chickpea

Treatment	Pod damage (%)	Chickpea equivalent yield (kg/ha)
Chickpea + mustard 4: 2 row ratio	8.75	1672
Chickpea + linseed 2: 1 row ratio	10.35	1573
Chickpea sole	13.60	1402
Bio-intensive IPM (BIPM*)	7.92	1748
Chemo- intensive IPM (CIPM**)	9.42	1684
Untreated control	15.34	1260

* BIPM: Seed treatment with *Trichoderma harzianum* + Vitavax (4g + 1g/kg seed), first spray *Ha*NPV at 250 LE/ha with adjuvant at ETL; Second spray of NSKE 5%

** CIPM: Seed treatment with Thiram + Carbendazim (2g + 1g/kg seed); First spray of Endosulfan 0.07% at ETL followed by second spray with *Ha*NPV at 250 LE/ha

2.8.3. Spotted Pod Borer, *Maruca vitrata* (Pyralidae: Lepidoptera)

2.8.3.1. Key Facts

- *M. vitrata* is becoming predominant insect pest in recent years in all pigeon pea and cowpea growing areas of India.
- Wide host range and coincidence of high humidity and moderate temperature with the flowering of the crop in India.
- This pest not only damage the pods of the plant but also feeds on flowers, buds and sometimes stem of the plants are infested.
- Incidence was high in late sown conditions and also in varieties having clustering type of branching habit (Halder and Srinivasan, 2007).
- Up to 42% damage has been recorded in cowpea during *rabi* season in Andhra Pradesh (Halder and Srinivasan, 2011).

During recent years due to introduction of short duration pigeon pea cultivars, *M. vitrata* is becoming predominant insect pest in all pigeon pea growing areas of India. *Maruca* has emerged as one of the major constraint because of the coincidence of high humidity and moderate temperature in September October coinciding with the flowering in short duration pigeon pea cultivars in India (Sharma *et al.*, 1999).

2.8.3.2. Economic Importance:

In pigeon pea, losses due to *M. vitrata* have been estimated to be US$ 30 million annually (ICRISAT, 1992). Vishakantaiah and Jagadeesh Babu (1980) observed the infestation of *Maruca* on pigeon pea varying between 9 and 51% at Bangalore, Karnataka. Singh (1997) reported 70 -80 % yield loss in pigeon pea, whereas it was 17-53% in cowpea (Liao and Lin, 2000) and 100% in black gram (Giraddi *et al.*, 2000).

2.8.3.3. Damage:

M. vitrara larvae feed on flower buds and pods by webbing them. This typical feeding habit protects the larvae from natural enemies and other adverse factors including insecticides. Larvae move from one flower to another and each may consume 4-6 flowers before larval development is completed. Third instar larvae are capable of boring into the pods, and occasionally into peduncles and stems (Plate 2.17) (Taylor 1967).

In the field experiments undertaken during 2005 and 2009 at Agricultural Research Station, Gulbarga, Karnataka, India; the results indicated that the incidence of spotted pod borer was high in early (140-150 days) and late maturing (190-200 days) varieties, moderate in medium duration (170-180 days) varieties. It is noteworthy to mention that the incidence was high under late sown conditions and also in varieties having clustering type of branching habit.

2.8.3.4. Management:

M. vitrata was controlled primarily through use of chemical insecticides (Booker, 1965; Dina, 1988). But dependence on only chemicals may lead to the problems such as development of resistance, outbreak of secondary pests and pesticide residues in agricultural produce. To avoid this situation, it is very essential to follow the concept of integrated pest management (IPM). The primary components of IPM are the use of resistant cultivars, conservation of natural enemies and safe use of chemicals. Use of pest resistant cultivars is an effective, cheap and environmentally safe component of IPM programme.

Several natural enemies have been recorded on *M. vitrata*, and pathogens such as *Bacillus thuringiensis, Nosema* sp., and *Aspergillus* sp. play an important role in regulating its populations under field conditions. Cultural practices such as intercropping, time of sowing, density of sowing, and weeding reduce the pod borer damage. Several insecticides have been found to be effective for controlling this insect pest.

2.8.4. Pod Wasp, *Tanaostigmodes cajaninae*

A pod wasp, *Tanaostigmodes cajaninae* which was earlier considered as a minor pest is on the increase in recent years emerging as a major pest on pigeon pea. This pest was first reported from India by Lateef (1977) on pigeon pea.

2.8.4.1. Damage:

Feeding by the grubs produces malformed, atrophied pods that do not grow at all. Locules or individual seed compartments that escape invasion may develop into normal seeds, while locules containing pod wasp remain undeveloped giving an abnormal appearance to the pod (Plate 2.18).

In a survey made during 2006-08 in Tamil Nadu on the occurrence of pest, its seasonal incidence and their natural enemies, it was observed that the pest is on the increasing trend over the years.

2.8.4.2. Management:

The pest can be effectively tackled by (Durairaj *et al.*, 2000):

- Maintaining field sanitation by eradicating alternative hosts and weeds.
- Need based application of insecticides like monocrotophos 0.04 per cent, endosulfan 0.07 per cent, chlorpyriphos 0.05 per cent and botanicals like Neem oil 3 per cent or Neem seed kernel extract 5 per cent may be helpful.

- This species is probably a secondary pest resulting from the increasing use of pesticides on pigeon pea which have destroyed its natural enemies. Under natural conditions, the wasp is attacked by a number of parasitoids. Hence the use of insecticides should be strictly need-based and judicious.

2.9. JUTE, *Corchorus* spp.

2.9.1. Mealy Bug, *Phenacoccus solenopsis*

2.9.1.1. Key Facts

- A new pest in jute crop.
- First recorded on cotton during the year 2005.
- Outbreak: During, 2013 attained the status of serious pest in jute in West Bengal.
- Reasons:
  - Warm and dry conditions during the seedling stage of the crop.
  - Surge in the maximum and minimum temperatures.
  - Decrease in rainfall and number of rainy days.

A wide spread infestation of mealy bug, *Phenacoccus solenopsis* was observed on cultivated jute, *Corchorus olitorius* (Malvaceae) in many parts of Southern West Bengal, India for the first time in May 2009 (Plate 2.19). The infested fields were surveyed for the pest, extent and nature of damage and its alternate hosts. On the basis of morphology and damage symptoms on the host, it was identified as *Phenacoccus solenopsis* (Sternorrhyncha: Coccoidea: Pseudococcidae). Subsequently, in 2011 and 2014 again similar kind of damage caused by *P. solenopsis* on jute was observed (Satpathy *et al.*, 2016).

2.10. SUBABUL, *Leucaena leucocephala*

2.10.1. Leucaena psyllid, *Hereopsylla cubana* (Order: Homoptera, Family: Psyllidae)

2.10.1.1. Key Facts

- It is native to Central and South America.
- Entered India in 1988.
- Infestation coincides with planting of subabul in South India.
- Indigenous Cocinellid predator, *Cheilomenes sexmaclata* is not effective.
- Coccinellid predator, *Curinus coeruleus* introduced from Mexico successfully suppress *H. cubana.*
- Outbreak is not noticed in recent years. In 1988 the insect was detected in southern India and in 1991 in western India.

Veeresh (1990) reported that the cultivation of Leucaena in India began in 1972 and now occupies an area of 10,000 ha in Karnataka State alone. Within six months of its entry into India, the psyllid invaded the entire area under Leucaena. The Karnataka Plantation Company had planned to extend the area of Leucaena to over 4000 ha in 1988 but abandoned the idea for fear of losing the plantations and instead, destroyed their nursery of over 200,000 seedlings.

2.10.1.2. Damage:

Both nymphs and adults feed on Leucaena terminal shoots by sucking plant sap (Plate 2.20). Young flowers are also infested but less frequently. The older leaves are in most cases not attacked. Feeding results in terminal shoots which may become desiccated and consequently, growth is stunted. High psyllid populations produce copious amounts of honeydew which gives rise to the development of sticky black sooty mould coating terminal shoots and leaflets. In the case of heavy infestations, complete defoliation is not uncommon. Although plants can recover later in most cases, death of damaged plants has been known to occur.

Infestation of young terminal shoot tips begins with a few gravid females laying scattered eggs between leaf folds. With increasing numbers of females, heavy egg laying on the shoot tips could turn them yellow in colour due to the large numbers of eggs deposited on young shoot tips or leaflets. After the infestation tapers off or declines, the new shoots and leaves become somewhat distorted, deformed and necrotic. There is no data available to determine if this is merely a reaction to feeding injury or if the Leucaena psyllid is the vector of a plant pathogen.

In many areas, Leucaena psyllid exhibits seasonal abundance with populations peaking during the cooler part of the year. In areas north of the equator, this occurs between late September and early March. During warmer periods, numbers of psyllids are often at very low levels and difficult to detect. Exceptions to this pattern are the existence of "ecological pockets or refugees" where populations remain at higher numbers throughout the year and can be detected at any time.

2.10.1.3. Management:

The coccinellid predator, *Curinus coeruleus* has since become the most abundant species of the family Coccinellidae in Hawaii and is an important predator of the Leucaena psyllid. *Olla v-nigrum* population was found to increase significantly and actively feeding on psyllids. Both coccinellids have played an important role as "indigenous" naturalb enemies in suppressing the populations of *H. cubana* significantly during the early years of its presence in Hawaii.

An encyrtid nymphal parasitoid, *Psyllaephagus yaseeni* was screened, field tested and finally approved for field release in June 1987 to become firmly and widely established in Hawaii (Funaski *et al.*, 1987).

2.11. CONCLUSIONS

In recent years the pest scenario in field crops is changing, and there were outbreaks of sap feeder like aphids, jassids, mirids and mealy bugs in cotton; tobacco caterpillar, *Spodoptera litura* on soybean and sunflower; woolly aphid, *Ceratovacuna lanigera* on sugarcane; brown plant hopper, *Nilaparvata lugens* and swarming caterpillar/armyworm, *Spodoptera mauritia* on rice; grey weevil, *Myllocerus* spp. and sugarcane leaf hopper, *Pyrilla perpusilla* on sorghum and pearl millet; maize leaf roller, *Cnaphalocrosis trapezalis* on sorghum; *Helicoverpa armigera* on chickpea and pigeon pea; spotted pod borer, *Maruca vitrata* on pigeon pea and cowpea; pod wasp, *Tanaostigmodes cajaninae* on pigeon pea; mealy bug, *Phenacoccus solenopsis* on jute; and Leucaena psyllid, *Hereopsylla cubana* on subabul. Hence, there is an urgent need for rapid and accurate detection, diagnosis, and development of management recommendations for these new pests.

3

Emerging Insect Pest Problems in Fruit Crops

3.1. INTRODUCTION

Various species of mealy bugs have started appearing in serious proportions on fruit crops such as guava, papaya, banana, citrus, avocado, mango, pomegranate, custard apple, and mulberry. They have intensified their severity of occurrence in different parts of the country and have become indicator insects for the current ecosystems due to slow changes in climate in the last one decade. Among these, the papaya mealy bug, *Paracoccus marginatus,* has become quite alarming in Tamil Nadu, challenging the pesticides or other IPM measures. *Maconellicoccus hirsutus* causes extensive damage to citrus, grape, guava, and mulberry (Tanwar *et al.*, 2007; Rajendran, 2009). The exotic spiralling whitefly, *Aleurodicus dispersus* is causing serious damage to several fruit crops including guava, papaya, banana, etc. in Peninsular India. The sapodilla seed borer, *Trymalitis marginatus* probably introduced accidentally from Sri Lanka into India through infested fruits of sapodilla, *Manilkara achras*, is causing enormous direct fruit losses. It is an introduced pest in Konkan region of Maharashtra. The crop suffered badly during 1999-2000 to the extent of 40-90 per cent (Puri and Mote, 2003). In a span of a decade this insect has dispersed to south India (Gujarat, Karnataka and Tamil Nadu) causing up to 35% fruit damage. The fruitfly, *Bactrocera caryeae* is one of the most serious emerging pests of mango in India which has created problem in the export of fresh fruits. Several mites have also emerged as major pests of fruit crops in northern India (Table 3.1) (Singh and Raghuraman, 2011).

Table 3.1. Emerging scenario of mite pests in fruits in north India

Emerging mite pests	Family	Hosts
Citrus brown mite, *Eutetranychus orientalis*	Tetranychidae	Apple, ber, citrus
Red tea mite, *Oligonychus coffeae*	Tetranychidae	Mango
Red vegetable mite, *Tetranychus neocaledonicus*	Tetranychidae	Mango
Litchi mite, *Aceria litchii*	Eriophyidae	Litchi
Mango bud mite, *Aceria mangiferae*	Eriophyidae	Mango

Source: Singh and Raghuraman (2011)

3.2. GAUVA, *Psidium guajava*

3.2.1. Spiraling Whitefly, *Aleurodicus dispersus* (Homoptera: Aleyrodidae)

3.2.1.1. Key Facts

- Native to West Indies and Central America.
- The exotic whitefly is causing serious damage to several fruit crops including guava, papaya, banana, etc. in Peninsular India.
- Polyphagous, feed on nearly 500 species of plants from 90 families.
- Introduced in India in 1994 (David and Regu, 1995).
- 53% yield loss in tapioca and heavy yield losses have been observed in groundnut, banana, papaya, guava and chilli in India.
- Emerged as a new pest of mulberry in 2010 and causing huge economic damage to silk worm rearing (Mani, 2010).
- Effective parasitoids include *Encarsia haitiensis* and *Encarsia guadeloupe.*
- Native predators feed on immature stages.

3.2.1.2. Damage:

Spiralling whitefly is not actually a fly, but a sap sucking bug, and derives its name from the characteristic egg spirals that the adult whitefly deposits on foliage and fruit. Spiralling whiteflies predominately occur as a winged adult or whitefly stage and a sedentary nymph stage. Whiteflies feed on the undersides of foliage (Plate 3.1). Heavily infested plants with very high whitefly populations soon develop a black sooty appearance from mould growing on the sugary secretions that the whiteflies and their nymphs produce. This sooty mould, in combination with leaf damage, reduces the plant's ability for photosynthesis (Kumashiro *et al.*, 1983). It also weakens, or in some cases kills the plant. When its natural biological agents are not present, spiralling whiteflies multiply at a rapid rate, producing thousands of individuals on a single plant and can rapidly assume major pest status.

3.2.1.3. Biology:

Eggs (0.3 mm long) are diamond shaped, microscopic, and embedded in the silken spirals produced by the females. Each egg hatches into a tiny active crawler, roughly the same size as the egg. This crawler moves out over the foliage of the host plant and then transforms into an inert, sedentary (nymph) stage that attaches to the underside of the leaves where it sucks nutrients from the leaves. The nymph stage (0.5 mm - 1 mm long) has no visible legs and grows progressively through a series of moults (instars), each instar producing more 'wax' and sugar secretions. The final instar acts as a pupa, out of which the adult whitefly emerges. The time from egg to adult can be less than three weeks in summer, longer in cooler conditions. The female whitefly (which is identical to the male) can lay large numbers of eggs.

3.2.1.4. Host Range:

Spiralling whitefly infests a broad range of horticultural plants including banana, citrus, avocado, papaya, mango, custard apple, guava, mulberry, tomato, bell pepper, eggplant, cucurbits, tapioca, coconut, and many ornamental species (dahlia, gerbera, gladiolus), shade trees and weeds.

3.2.1.5. Management

i) **Biological Methods:** The predators, *Cryptolaemus montrouzieri, Axinoscymnus puttarudriahi* and *Mallada astur* are commonly associated with *Aleurodicus dispersus*. Two aphelinid parasitoids, *Encarsia haitiensis* and *E. guadeloupae* have appeared recently on the spiralling whitefly in Peninsular India (Table 3.2). The buildup and dispersal of these parasitoids would bring down the population of spiralling whitefly (Mani *et al.*, 2000a; Mani and Krishnamoorthy, 2002). The major parasitoids of the spiralling whitefly are presented in Table.

Table 3.2. Important parasitoids of the spiralling whitefly, *Aleurodicus dispersus* in India

Taxonomic group	Species	Reference(s)
Hymenoptera, Aphelinidae	*Encarsia guadeloupae*	Ramani 2000; Mani et *al.* 2000b; Beevi and Lyla 2001
	Encarsia sp. nr. *haitiensis*	Ramani 2000; Mani et *al.* 2000b; Beevi and Lyla 2001
	Fncarsia sp. nr. *meritoria*	Beevi et *al.* 1999; Geetha and Swamiappan 2001
	Encarsia sp. (closely related to *E. haitiensis* and *E. meritoria*)	Srinivasa et *al.* 1999

ii) **Chemical Methods:** Controlling the pest using insecticides is not recommended. Overseas experience indicates that spraying with insecticides has little long-term impact on the pest and usually exacerbates the problem by destroying the biological control agents and increasing insecticide resistance in whitefly populations.

iii) **Cultural Methods:** Use of clean planting material delays the appearance of the whitefly population. Pruning the heavily infested trees and shrubs is recommended to minimize the spiralling whitefly incidence. Subsequent to the pruning the population rapidly increased within 4-5 months on guava.

3.3. PAPAYA, *Carica papaya*

3.3.1. Mealy Bug, *Paracoccus marginatus*

3.3.1.1. Key Facts

- First recorded in Mexico in July 2008.
- This pest is accidentally introduced into India through Sri Lanka.
- In India, the pest was reported for the first time in Coimbatore, Tamil Nadu in 2008 on papaya (Muniappan *et al.*, 2008). Later it moved to Kerala and a few places in Karnataka, Pune area of Maharashtra, West Bengal and Andhra Pradesh (Tanwar *et al.*, 2010; Seni and Sahoo, 2013).
- In 2009 it caused severe damage in Coimbatore, Erode, Tirupur and Salem districts of Tamil Nadu.
- Outbreak in 2009, standing mulberry crop over 2,500 hectares in Tirupur was destroyed by the pest.
- Dhobi *et al.* (2014) reported it on *Bt* cotton (BGII) leaves, stem and bolls in cluster forms at Anand, Gujarat.

- The pest attack has been reported in a large number of cultivated plants, including cotton, mulberry, tapioca, papaya, mango, jackfruit, brinjal, guava, amla, tomato, plantations, custard apple, okra, teak, jatropha and colocasia (Seni and Sahoo 2013).
- Crop losses to the tune of 10 to 15 per cent are expected. The warm and sunny days are conducive for the spread of the pest.
- Mechanism of dispersal through man (pickers), wind, animals and ants.
- Heavy mealy bug infestation not only reduces plant yield but affects fruit quality due to presence of mealy bug and sooty mould.
- The heavy infestation kills the plants in a short period.
- A number of effective parasitoid species such as *Acerophagus papayae, Anagyrus loecki, Anagrus californicus, Aseudaphycus* spp. have been reported.

The incidence and severity of papaya mealy bug, *Paracoccus marginatus* on cotton with its expanding host range across crops of industrial importance *viz.,* cotton, mulberry, tapioca, papaya and jatropha was found in Tamil Nadu. The Papaya mealy bug has caused havoc in agricultural and horticultural crops, ever since its first report from Coimbatore in 2007. In 2009, it caused severe damage to economically important crops and huge losses to farmers in Coimbatore, Erode, Tirupur and Salem districts of Tamil Nadu. In the same year, standing mulberry crop over 1,500 hectares in Tirupur was destroyed by the pest. Recently the pest was noticed in Karnataka, certain parts of Andhra and Mallapuram and Thrissur districts of Tamil Nadu. Now it has spread to Pune area of Maharashtra and also likely to be reported from other parts of the country as well. The pest is reported to cause serious damage to tropical fruits, especially papaya.

3.3.1.2. Damage:

The papaya mealy bug feeds on the sap of plants by inserting its stylet into the epidermis of the leaf, as well as into the fruit and stem (Plate 3.2). In doing so, it injects a toxic substance into the leaves. The result is chlorosis, stunting of plants, deformation of leaves, early leaf and fruit drop, a heavy build up of honeydew, and death of plants. Heavy infestations are capable of rendering fruit inedible due to the build up of thick white wax. Papaya mealy bug has only been recorded feeding on areas of the host plant that are above ground, namely the leaves and fruit.

3.3.1.3. Biology:

In general, mealy bugs have piercing-sucking mouth parts and feed by inserting their mouth parts into plant tissue and sucking out sap. Mealy bugs are most active in warm, dry weather. Females have no wings, and move by crawling short distances or by being blown in air currents. Females usually lay 100 to 600 eggs in an ovisac, although some species of mealy bugs give birth to live young ones. Egg-laying usually occurs over the period of one to two weeks. Egg hatch occurs in about 10 days, and nymphs, or crawlers, begin to actively search for feeding sites. Female crawlers have four instars, with a generation taking approximately one month to complete, depending on the temperature. Males have five instars, the fourth of which is produced in a cocoon and referred to as the pupa. The fifth instar of the male is the only winged form of the species capable of flight. Adult females attract the males with sex pheromones. Under greenhouse conditions, reproduction occurs throughout the year, and in certain species may occur without fertilization.

3.3.1.4. Host Plants:

Economically important host plants of the papaya mealy bug include papaya, hibiscus, avocado, citrus, cotton, tomato, eggplant, peppers, beans, peas, sweet potato, mango, cherry, mulberry and pomegranate.

3.3.1.5. Management

i) **Ant Control:** Mealy bug-control often involves the control of attendant ants that are important for the proper development of mealy bugs. Without the ants, mealy bug populations are small and slow to invade new areas and the field would be free of a serious infestation. Therefore, management of mealy bug often includes the control of ant species.

ii) **Chemical Methods:** Pesticide formulations used for the management of mealy bugs include acephate, carbaryl, chlorpyrifos, diazinon, dimethoate, malathion, and white mineral oils. Typically, twice the normal dose is applied when treating for mealy bugs because mealy bugs are protected by thick waxy, cottony sacs, and often are concealed inside damaged leaves and buds. Thus, chemical controls are only partially effective and require multiple applications. Furthermore, problems with insecticide resistance and non-target effects on natural enemies make chemical control a less desirable control option to combat the papaya mealy bug.

iii) **Biological Methods:** Natural enemies of the papaya mealy bug include the commercially available mealy bug destroyer (*Cryptolaemus montrouzieri*), lady beetles, lacewings, and hover flies, all of which are generalist predators that have a potential impact on mealy bug populations. In addition to predators, several parasitoids may attack papaya mealy bug.

Four genera of encyrtid endoparasitoid wasps specific to mealy bugs have been considered as potential biological control agents: *Acerophagus papayae*, *Anagyrus loecki*, *A. californicus*, and *Pseudaphycus* sp. (Meyerdirk and Kauffman, 2001). The release of the four genera of parasitoid wasps has brought a 99.7% reduction in the density of mealy bug populations at research sites in the Dominican Republic, and a 97% reduction at research sites in Puerto Rico, with parasitism levels between 35.5% and 58.3% (Kauffman *et al.*, 2001, Meyerdirk and Kauffman, 2001). All four species of parasitoids have been observed attacking second and third instars of *P. marginatus*. However, *Acerophagus* sp. emerged as the dominant parasitoid species in both Puerto Rico and the Dominican Republic (Meyerdirk and Kauffman, 2001).

The parasitoids, *Anagyrus loecki* and *A. papayae* appear to be promising biological control agents. The reduction of the papaya mealy bug population density levels below detectable levels was observed in a six-month period following the introduction of these exotic parasitoids (Muniappan *et al.*, 2006).

3.3.1.6. Successful Containment of Papaya Mealy Bug in India:

Through quick action on the part of USDA, USAID, and the US Government's research and corporate partners around the world, an infestation of an invasive crop pest that appeared in Tamil Nadu, India, was successfully controlled in just five months, yielding a successful model for political and technical food security stakeholders to follow for future threats to smallholder farmers due to crop pests.

From 2008-2010, the papaya mealy bug—an invasive species of insect native to Mexico—ravaged crops in the state of Tamil Nadu, wiping out an estimated US$ 93 million worth of papaya, mulberry and tapioca crops, eventually spreading to other Indian states and crops. By 2010, nearly US$ 319 million in crop losses could be attributed to the papaya mealy bug infestation, and the pest threatened to spread even farther throughout India.

When the outbreak was reported to USAID's Integrated Pest Management Collaborative Research Support Program in Tamil Nadu, a comprehensive plan was set in motion to contain and control the pest infestation. Together with the Government of India, state universities, and corporate partners, USDA's Animal and Plant Health Inspection Services (APHIS) worked with the Puerto Rican Department of Agriculture to breed and ship three kinds of parasitoids (parasite-like organisms that kill their hosts and prevent reproduction) to India to control the papaya mealy bug. Approximately 800,000 of these parasitoids were released across Tamil Nadu over a period of eight months. Within five months of the first release, the infestation had effectively been controlled.

Not only was the speed of the containment process impressive, but the innovative solution of using parasitoids was particularly important for smallholder farmers, who would otherwise have needed to use high concentrations of pesticides to control the outbreak. Tamil Nadu Agricultural University, Coimbatore, which invested US$ 200,000 to breed the parasitoids, estimates that farmers saved US$ 35 million that would otherwise have gone toward expensive pesticides.

While invasive pest species remain a perennial threat to crops around the world, the papaya mealy bug outbreak and the international cooperation that successfully controlled it demonstrate the importance and value of collaborative partnership across countries and sectors to promote healthy crops and global food security.

3.4. SAPOTA, *Manilkara zapota*

3.4.1. Seed Borer, *Trymalitis margarias* (Lepidoptera: Tortricidae)

3.4.1.1. Key Facts

- Invasive species probably introduced accidentally from Sri Lanka into India through infested fruits of sapodilla, and is causing enormous direct fruit loss.
- Sapota seed borer is introduced pest in Konkan region of Maharashtra (Puri and Mote, 2003). First noticed at 'Kankradi' area of Dahnu Thane, the crop suffered badly during 1999-2000 to the extent of 40-90 % during September-March.
- The area of 225 ha was found to be affected by this pest comprising of 19 villages in Dahanu tehsil during February 2000.
- In a span of a decade this insect has dispersed to south India (Gujarat, Karnataka and Tamil Nadu) causing up to 35% fruit damage.
- Heavy infestation of seed borer was one of the reasons in many regions attributed to declining sapodilla yields.
- Fruit damage:
    - 40 -90 % in Dahanu areas of Maharashtra (2000)

- 21 and 40 % in Thane and Ganevi (2003-04).
- 25 % incidence in Bangalore during 2006-07.

- Monophagous pest attacking immature fruits of sapota.
- The pest has 'jump' dispersed to much larger distances down the south viz., Karnataka and Tamil Nadu causing 35% annual losses (Kamala Jayanthi and Verghese, 2010).

3.4.1.2. Damage

- Exclusively feed on endosperm of seed (Plate 3.3).
- Full grown larvae excavate a tunnel to come out from fruit for pupation.
- Through the exit hole, fungi as well as ants enter inside fruits and make them unfit for consumption.
- As high as 40 to 80 per cent damage to mature fruits has been noticed.
- Due to infestation of this pest, quality of fruit deteriorates and hence market price goes down and farmers also lose their goodwill in market.

3.4.1.3. Biology:

The biology of *T. margarias* (Plate 3.4) indicated that the female moth laid 161.8 ± 77.236 eggs. The average pre-oviposition, oviposition and post-oviposition period lasted for 1.83 ± 0.185, 1.44 ± 0.373 and 1.61 ± 0.443 days, respectively. The duration of egg, larval and pupal stages lasted for an average of 11.16 ± 0.715, 12.33 ± 0.832 and 13.15 ± 1.663 days, respectively. The adult male and female survived for a period of 4.125 ± 0.690 and 4.865 ± 0.971 days, respectively. The total life cycle from egg laying to adult emergence complete in 38.50 ± 3.117 days.

3.4.1.4. Management

- Sanitation is to be maintained for eliminating the sources of seed borer infestation.
- Collection and destruction of the off season stray mature fruits after main harvest till November will bring down the pest incidence.
- Erecting light traps in the field during cropping season will attract adult moths and will minimize the pest incidence.
- Effective management of seed borer relies primarily on the stage of spray intervention. The time of application is crucial in the management of seed borer.
  - First spray intervention should be made when the fruits are of small lime size and thereafter the sprayings should be repeated at fortnightly intervals during main fruiting season.
  - Alternating the sprays of Deltamethrin 2.8 EC at 1 ml/l and *Bt* at 1 ml/l at fortnightly interval will bring down the seed borer infestation.
  - Critical stage for IPM intervention was determined as marble to lime size of fruits.
  - Decamethrin at 0.028% was best treatment for control.

3.5. BANANA, *Musa* spp.

3.5.1. Root Mealy Bug, *Geococcus citrinus*

3.5.1.1. Key Facts

- Root mealy bug of banana is a pest of recent origin in Kerala.
- First reported from the districts of Palghat, Trichur and Ernakulam districts of Kerala.
- There are two species of mealy bugs, namely, *Geococcus citrinus* and *G. coffeae.*
- Studies indicate Nendran is most susceptible variety.

3.5.1.2. Damage:

Adults and nymphs suck sap from the lateral roots of banana colonizing at the junction of laterals with main root resulting in drying up of such roots. As a result, the roots turn brown at the site of colonization and later dry up (Plate 3.5). Such roots cannot absorb water and nutrients from the soil. Yellowing and narrowing of leaves, general weakening of the plant, reduction in bunch weight etc. are the symptoms observed. In severe case, the plants will topple down by wind due to destruction of anchoring roots. The bunch size is reduced and filling of fingers is adversely affected. Usually the infestation goes unnoticed as it occurs below the soil surface.

3.5.1.3. Host Range:

This mealy bug infestation was also observed in the roots of weeds seen in the banana field. *Cyperus kyllinga, C. pangorei, Ludwigia parviflora, Centella asiatica, Axonopus compressus, Commelina bengalensis,* and *Colocasia antiquorum* are recorded as collateral hosts for this mealy bug (Smitha *et al.*, 2005).

3.5.1.4. Management:

Among the soil ameliorants, application of sodium silicate and calcium oxide at the time of planting effectively reduced the population of root mealy bug. Drenching with 3 % neem seed kernel extract (NSKE) at monthly intervals was superior to neem cake, neem oil and pongamia oil. The entomopathogenic fungus, *Cephalosporium lecanii* was the best among the three fungi screened. At five months after planting, *C. lecanii* recorded 1.95 colonies per sample followed by *Hirsutella* sp. (2.25 colonies per sample) and the control recorded 5.70 colonies per sample. Among the synthetic chemicals, drenching chlorpyriphos (0.05%) at monthly intervals at 2.5 ml l^{-1} effectively reduced the root mealy bug population and was found to be the best treatment. Field experiment conducted with the best treatment in each of the category and their combinations revealed that, almost all treatment combinations containing chlorpyriphos caused maximum reduction of root mealy bug. Among the combinations without synthetic insecticides, sodium silicate alone and its combination with NSKE and *C. lecanii* were effective in reducing the mealy bug population at sixth and seventh month of the crop. Application of chlorpyriphos gave the highest benefit cost ratio of 2.46 followed by sodium silicate (2.30) (Smitha and Mathew, 2010a).

The following recommendations have been suggested for the management of root mealy bug:

- Immersion of seedlings in boiling water for 10 to 30 seconds prior to planting.
- Entomopathogenic fungus, *Beauveria bassiana* PPRC-56 caused 54% mortality.
- Drenching a seed-water suspension of 10% *Millettia ferruginea* into root-zone.
- Application of Chlorpyrifos 48% EC and Diazinon 60% EC at 1.7 L of solution (after dilution of insecticide in 1:5 L of water) per field-grown plant and poured on root collar area, resulted in high levels of root mealy bug mortality.
- Irrespective of the seasons, two cultivars viz., Palayankodan and Kodappanillakunnan were completely free from root mealy bug (Smitha and Mathew, 2010b).

3.6. MANGO, *Mangifera indica*

3.6.1. Fruitfly, *Bactrocera caryeae*

This is one of the most serious emerging pests of mango in India which has created problem in the export of fresh fruits. Its infestation is more in southern states than in northern states.

3.6.1.1. Distribution:

Surveys carried out by the scientists of the Indian Institute of Horticultural Research, Bengaluru, Karnataka, India, during 2001-04 revealed that *B. caryeae* was restricted to only limited area in South India (Table 3.3). The major mango belts especially in North India were free from this pest. This opened Japanese market for mango export from India in 2006.

Table 3.3. Distribution and incidence of *Bactrocera caryeae* in India

Incidence of Bactrocera *caryeae* (%)	Distribution
0.00	**Karnataka:** Hassan, Dharwad, Raichur, Hunsur (Mysore), Ramanagaram, Bangalore **Tamil Nadu:** Thiruvallur, Vellore, Krishnagiri **Andhra Pradesh:** Hyderabad, Mehaboobnagar (Rangareddy), Krishna, Chittoor
52.58	**Karnataka:** Chethalli (Kodagu district)
16.96	**Goa**
8.87	**Kerala:** Thiruvanthapuram

3.6.1.2. Damage:

The maggots feed on pulp of mango fruits and the infested fruits start rotting and fall down. As a result the brown patch appears around the place of oviposition and the infested fruits start rotting. The maggots come out of affected fruits to pupate in the soil.

3.6.1.3. Management:

The following management practices are recommended:

- Affected fruits should be collected and destroyed.
- Rake up the soil below the tree and drench with chlorpyriphos 20 EC at 2.5 ml/ L.
- Setting up of methyl eugenol traps to lure the males in the orchard at 10/ha.
- Spraying malathion at 2 ml + Jaggery at 10 g/ L or Carbaryl at 4 g + Jaggery at 10 g/ L at fruit ripening stage.

i) **Integrated Pest Management:** The following pre- and post-harvest IPM combining sanitation + intercultural ploughing + 3 insecticidal sprays at 15 days interval prior to harvest gives 95 -99% control (Verghese *et al.*, 2000).

a) **Pre-harvest Management:** 45 days before harvest the following precautions need to be taken:

- Destroy all fallen fruits at weekly intervals.
- Install at least ten methyl eugenol bottle traps (0.1 %) per ha.
- Plough/rake the soil at the tree basin at frequent intervals (1 or 2 times between flowering and harvest) to expose pupae to sun.
- Avoid delay in harvesting.

Three weeks before the harvest, spray decamethrin 2.8 EC at 0.5 ml/L + azadirachtin (0.3%) at 2 ml/L and take up timely harvest.

b) **Post-harvest Management:** hot water treatment of fruits within 24 hr after harvest at 48°C (maintained by thermostat) for 1 hr gave 100% fruit fly infestation-free fruits in cvs. Banganapalli and Totapuri. Hot water treatment in 5% salt solution at 55°C for 30 min (without thermostat) gave 100% control in cvs. Banganapalli, Totapuri and Alphanso.

3.6.2. Red Banded Fruit Borer, *Deanolis sublimbalis* (Lepidoptera: Pyralidae)

3.6.2.1. Key Facts

- Earlier limited to the east coast of Odisha and Andhra Pradesh.
- Became a major problem in recent years due to movement of fruits (cv. Banganpalli) from the east coast to the different parts of the country.
- The banded caterpillar damage more than one mango in its larval stage.
- Mango belts of Maharashtra, Gujarat, Andhra Pradesh, Uttar Pradesh, Bihar, Karnataka, West Bengal, etc. are in danger.

Red banded caterpillar, *Deanolis sublimbalis* (Syn. *Deanolis albizonalis, Noorda albizonalis, Autocharis albizonalis*) was the minor pest of mango but very recently emerged as a serious pest all over the world. Sahoo and Jha (2009) reported that fruit borer is a predominant pest which has recently emerged as a serious pest of mango. The damage due to this pest showed 10-52% damage from pin head stage to full maturity in Malda, Murshidabad, Nadia and Hooghly districts of West Bengal, India.

3.6.2.2. Damage:

Mango fruit borer is an important pest where the young caterpillars attack fruit from marble size onwards at the distal end. The caterpillars feed within the fruit and may make series of tunnels in the young kernel. Infested fruits fall. Fully grown caterpillars pupate inside the fruits (sometimes pupation takes place outside) and the moths emerge through the exit holes. The characteristic symptom of damage is appearance of small holes (which also serves as exit holes) with dark pale brown ring encircling it (Plate 3.6).

3.6.2.3. Biology:

According to Sujatha and Zeheruddin (2002) the pest has a short life cycle of less than one month with 3 to 4 generations during the fruiting season. Fecundity ranged from 4 to 38 with pre-ovipositional period of 2 days. The larvae passed through 3 larval instars within 12.4 days and they pupate in the dead wood, cracks and crevices of the bark for a period of 13 to 43 days after pre-pupal period of 2 to 7 days. They also noted the pest has got pre-pupal diapauses during off season.

3.6.2.4. Management

- Collection and destruction of all fallen fruits.
- Sticky bands are fixed around the tree trunk which acts as a barrier for pupation of fruit borer larvae. The bands would also prevent the weaver ant (*Oecophylla smaragdina*) which could predate on natural enemies of fruit borer.
- Bagging fruit with paper bags.
- An effective pheromone lure has been identified by Hortresearch in New Zealand which trapped hundreds of male moths and was significantly more attractive than virgin female moths, catching six times more moths than caged virgin females. This lure can now be used for routine trapping of male moths and has the potential to be used for mating disruption. Initial testing suggested the lure was effective for at least four weeks in the field under tropical conditions (Gibb *et al.*, 2006).
- Thiacloprid (Calypso) (37.5 ml/L) was identified as the only effective chemical control option, reducing fruit borer damage to almost nil. Apply a total of 4 sprays at 14 day intervals commencing at egg laying of the first generation.
- Spraying endosulfan 35 EC at 2 ml/L at marble size onwards and repeating with decamethrin 2.8 EC at 1 ml/L after two weeks in case of heavy infestation.
- No spray should be given a fortnight before harvest.
- Two species of egg parasitoids, *Trichogramma chilonis* and *T. chilotraeae* (Trichogrammatidae), were recorded as egg parasites of red banded caterpillar in the Philippines by Golez (1991).

3.6.3. Leaf Weevil, *Rhynchaenus mangiferae* (Coleoptera: Curclionide)

3.6.3.1. Key Facts

- Restricted to tropical and subtropical fruits native to South China and Taiwan.
- It infests mango and litchi in Punjab.

- Reported from Ludhiana, Patiala, Amritsar, Moga, Firozpur, Jalandhar, Hoshiarpur, Kapurthala and PAU Campus.

3.6.3.2. Damage:

Rhynchaenus mangiferae is recorded as damaging mango leaves in South India. The larvae mine in the tender leaves and cause crinkling of leaves. In severe outbreaks, there may be 20-30 weevils on every leaf. The feeding by adults on the leaves (Plate 3.7), as well as the mining by larvae causes the leaves to dry out or become distorted, and the photosynthetic area can be severely reduced.

3.6.3.3. Biology:

The female weevils lay eggs in the mid-rib of leaves. The total life cycle from egg to adult takes about 12 days.

3.6.3.4. Management:

The synthetic pyrethroids permethrin and fenvalerate, both at 0.02%, and monocrotophos at 0.05% were the most effective, resulting in 100% mortality of the pest for up to 15 days after application. The systemic insecticides phosphamidon, dimethoate and methyl demeton (demeton-S-methyl), all at 0.05%, were moderately effective, causing 85-90% mortality for up to 15 days. Effective control of the pest could be obtained with alternate sprays of either permethrin or fenvalerate and monocrotophos during emergence of new growth (Shukla and Prasad, 1985).

3.7. AMLA, *Embelica officinalis*

3.7.1. Anar butterfly, *Deudorix isocrates* (Lepidoptera: Lycaenidae)

In recent years, anar butterfly has emerged as major pest of amla in Punjab.

3.7.1.1. Damage:

Affected fruits are generally deformed at the point of entry of larvae. Frass may be seen exuding out of the borer hole (Plate 3.8). Such fruits weaken, rot and fall down before maturation. In case of severe attack, it may cause considerable loss. Attack of this insect occurs during September-October, coinciding well with the fruiting season. The violet brown female butterfly lays shining white eggs, singly on young fruits. The larva of this pest bores the fruit and feeds on seeds, making this portion hollow from inside.

3.7.1.2. Management

- Cultivation of pomegranate and guava should be discouraged close to amla plantation as these are major host plants of this pest.
- Infested fruits should be identified, collected and destroyed to prevent further spread of infestation.
- Spray spinosad 0.25 ml/l or carbaryl (2g/l) at pea size stage of amla fruits. The spray may be repeated after two weeks, depending upon the intensity of attack.
- Release of *Trichogramma chilonis* at 2.5 lakhs/ha four times at 10 days interval.
- Amla cvs. Chakayia and Kanchan are reported to be tolerant (Table 3.4).

Table 3.4. Reaction of amla cultivars to anar butterfly

Cultivars	Fruit Infestation (%)	Fruit Drop (%)
Seedling Amla	20.42	51.34
Banarsi	14.36	42.94
Francis	12.18	36.26
Kanchan	7.76	22.86
Chakayia	6.82	9.35

3.8. LITCHI, *Litchi chinensis*

3.8.1. Key Facts

- The trees having highly damaged canopies (>50% foliage) by red weevil, *Apoderus blandus* and fruit borer, *Conopomorpha cramerella* represented as much as 40.0% while partially damaged (10-30% foliage) plants represented up to 20.8%.
- Indiscriminate use of pesticides to control fruit borer complex in litchi, particularly synthetic pyrethroids by the farmers seem to be responsible for high incidence of these pests.

Field surveys carried out in litchi growing areas of Bihar during 2009-10 revealed the occurrence of two insect species in severe form which were hither to either unnoticed or of minor importance. They were *Apoderous blandus*, a weeveil and *Conopomorpha cramerella*, a borer pest on fruits (Kumar *et al.*, 2011). The weevil was brownish red in colour, about 5-7 mm in length and with a strong rostrum, causing heavy damage to new flush. Young growing trees of less than five years of age were devastated.

3.8.2. Red Weeveil, *Apoderous blandus* (Coleoptera: Curculionidae)

3.8.2.1. Damage:

It was found feeding on young foliage during November 2009 and in subsequent months during 2010. It was a solitary feeder but extensively damaged young foliage and shoots. The insect fed on leaves by pricking the surface of leaf (Plate 3.9). As a result of feeding, leaves dried up from the pricked parts giving blighted appearance to the twigs. Young growing trees of less than five years of age were devastated. Infestation of the weevil was observed throughout the year but was less during December and January due to extremely low temperatures. Considering the extent of leaf damage it causes, it can be regarded as second only to fruit borer in economic importance.

3.8.2.2. Management

- The grubs of these weevils feed on organic matter in the soil below the canopy, hence, ploughing and exposing these grubs reduces the problem.
- Hand picking of the adult weevils reduces their problem to some extent.
- When severe damage is seen, spraying of Carbaryl at 2ml/l may be undertaken.

3.8.3. Fruit Borer, *Conopomorpha cramerella* (Lepidoptera: Gracillariidae)

Due to the changing dynamics of pests, *Conopomorpha cramerella* which was earlier regarded as minor pest became very serious, affecting litchi cultivation not only in Bihar but also in other states like Uttar Pradesh.

3.8.3.1. Damage:

Generally, two generation of this insect occurs during litchi season. Earlier the fruits infestation that occurred during May (Plate 3.10) was important but now the generation which starts in March, when the litchi fruit (18-20 mm length) is still in stage I of development (growth of pericarp and seed coat) have become important. A large number of fruits drop due to early infestation of this pest. Overlapping generations during the fruiting period (April to June) have been observed. Indiscriminate use of pesticides to control fruit borer complex in litchi, particularly synthetic pyrethroids by the farmers seem to be responsible for high incidence of *C. cramerella* in litchi orchards.

3.8.3.2. Management:

The management of fruit borer complex warrants the integration of alternative methods such as use of pheromone trap, biocontrol agents (*Trichogramma* sp.), removal and destruction of dropped fruits and wild hosts such as *Eugenia jambolana* and *Cassia tora* from orchards, prophylactic spray of neem (*Azadiracta indica*) based insecticides and need based application of chemical insecticides. Minimal use of pesticides in litchi is more relevant because of greater hazards of pesticide residues in the fruits.

The following practices have been recommended for the management of litchi fruit borer:

- Moths can be excluded by enclosing the fruit panicles in nylon mesh bags, but is uneconomic in areas with high labour costs.
- National Research Centre on Litchi, Muzaffarpur, Bihar, recommended the release of *Trichogramma chilonis* at 50,000 eggs/ha and use of pheromone trap. However, the pheromone is not giving consistent results in all the situations.
- Fallen fruits should be removed to reduce the build-up of moths and ploughing may be undertaken after fruit harvesting.
- Bearing trees should be inspected during early flush development and sprayed if necessary. The leaf flush before flower initiation is very important as it supplies the carbohydrates needed for fruit development. Young, non-bearing trees do not need to be sprayed. This also allows the parasitoids to build up in the orchard.
- Fruits may be inspected weekly from fruit set to detect eggs of borers, which are very small and almost invisible to the naked eye. Infested fruit should be picked and destroyed at infestation levels of 1 to 2 per cent. When the pest becomes more active, spraying 0.05 per cent fenitrothion or dicholorvos or carbaryl at 2g/l may be undertaken.
- Permethrin is applied at weekly intervals, until two weeks before harvest. Spraying with cypermethrin, deltamethrin or fenthion during early fruit set is recommended to prevent damage later in the season.

- Affected shoots may be removed and all agronomic efforts such as ploughing and nutrient management should be done so that flushing takes place before the month of September.
- Neem based products may be applied at the time of new shoot emergence to avoid heavy population of the pest.
- Monocrotophos at 0.05 per cent may be applied in case of severe shoot damage.

3.9. CONCLUSIONS

Emergence of crop pests has been affecting crop production from time immemorial. There were outbreaks of mealy bug, *Paracoccus marginatus* on papaya, avocado, citrus, mango, cherry, mulberry and pomegranate; root mealy bug, *Geococcus citrinus* on banana; *Aleurodicus dispersus* on several fruit crops including guava, papaya, banana; seed borer, *Trymalitis marginatus* on sapodilla; fruitfly, *Bactrocera caryeae*, red banded fruit borer, *Deanolis sublimbalis*, and leaf weevil, *Rhynchaenus mangiferae* on mango; anar butterfly, *Deudorix isocrates* on amla; and red weevil, *Apoderus blandus* and fruit borer, *Conopomorpha cramerella* on litchi. The emerging and future pest problems have to be tackled with IPM approach as a part of a sustainable crop production technology.

4

Emerging Insect Pest Problems in Vegetable Crops

4.1. INTRODUCTION

Insect pests are the major biotic constraints to vegetable crop production in India. Apart from causing direct damage many of them also act as vectors for several viral diseases. The crop losses to the tune of 30-40 per cent have been reported in vegetable crops. Insect pests of much significance to vegetable crops and yield loss are given in Table 4.1.

Table 4.1. Yield losses due to major insect pests in vegetable crops in India

Crop	Pest	Yield loss (%)
Tomato	Fruit borer, *Helicoverpa armigera*	24-73
Brinjal	Fruit and shoot borer, *Leucinodes orbonalis*	11-93
Chillies	Mites, *Polyphagotarsonemus latus*	34
	Thrips, *Scirtothrips dorsalis*	12-90
Okra	Fruit borer, *H. armigera*	22
	Leaf hopper, *Amrasca biguttula biguttula*	54-66
	Whitefly, *Bemisia tabaci*	54
	Shoot and fruit borer, *Earias vittella*	23-54
Cabbage	Diamondback moth, *Plutella xylostella*	17-99
	Cabbage caterpillar, *Pieris brassicae*	69
	Cabbage borer, *Hellula undalis*	30-58
Cucurbits	Fruit fly, *Bactrocera cucurbitae* - Bitter gourd	60-80
	Cucumber	20-39
	Ivy gourd	63
	Musk melon	76-100
	Snake gourd	63
	Sponge gourd	50

Sources: Shivalingaswamy *et al.* (2002); Dhillon *et al.* (2005); Satpathy *et al.* (2005); Raju *et al.* (2007); Singh *et al.* (2007); Ghosal *et al.* (2012).

In recent years, with changes in the cropping systems and climate, and introduction of highly input intensive high yielding varieties/hybrids, a shift in pest status has been observed. Most of the plant protection recommendations in vegetables so far indicated the calendar based application of insecticides. This has become a common practice over the years by most of the farmers growing vegetables in the country. Many pests have adapted to new hosts, developed resistance to pesticides and often there are secondary outbreaks.

Over the past 100 years, annual rise in temperature with an average of 0.56°C, along with warming during post-monsoon and winter season has led to exacerbation of pest problems like:

- *Helicoverpa armigera* : Vegetables, pulses and seed crops
- *Spilosoma obliqua* : Oil seeds and vegetable crops
- *Pieris brassicae* : Crucifers
- *Liriomyza trifolii* and several sucking pests : Vegetable crops

Emerging scenario of mite pests in vegetable crops in North India is presented in Table 4.2.

Table 4.2. Emerging scenario of mite pests on vegetables in North India

Emerging mite pests	Family	Hosts
Citrus brown mite, *Eutetranychus orientalis*	Tetranychidae	Cucurbits
Two spotted spider mite, *Tetranychus urticae*	Tetranychidae	Bean, brinjal, cucurbits, okra
Red vegetable mite, *Tetranychus neocaledonicus*	Tetranychidae	Vegetables

Source: Singh and Raghuraman (2011)

4.2. RECENT CHANGES IN THE PEST SCENARIO

Some of the insect pests of vegetable crops have become major and are gradually attaining the major pest status in different regions of the country due to changes in the ecosystem and habitats. Fruit borer, *Helicoverpa armigera* on tomato; whitefly, *Bemisia tabaci* and serpentine leaf miner, *Liriomyza trifolii* on tomato and cucurbitaceous crops; fruit fly on fruits and vegetables; mealy bugs on several horticultural crops; gall midge on brinjal; stem fly on okra; leafhopper on bitter gourd; red spider mite on okra, brinjal, cowpea, Indian bean; and nematodes on several vegetable crops are some of the examples. In recent times, there were outbreaks of gall midge (*Asphondylia capparis*), which was known to be a minor pest is gradually becoming a regular problem in chilli, bell pepper and brinjal in the states of Andhra Pradesh and Karnataka and on brinjal in Chattisgarh. Various species of mealy bugs in cotton, vegetables and papaya have intensified their severity of occurrence in different parts of the country and have become indicator insects for the current ecosystems due to slow changes in climate in the last one decade. Several national and international polyphagous pests like termites, white grubs, hairy caterpillars, gram pod borers and *Spodoptera litura* are also becoming severe and expanding their host-horizon (Table 4.3). This envisages that there is a need to relook into the cropping systems for reorientation and vegetables should be given importance with adequate insect pest management measures. In this context, integrated pest management (IPM) system was developed out of the need for sustainable crop protection strategy against the backdrop of increasing pesticide use and deleterious effect of residues in the environment.

Table 4.3. Changing pest scenario in vegetable crops in India

Insect Pest	Major host	Other hosts infested	Reference
Serpentine leaf miner, *Liriomyza trifolii*	Tomato	Brinjal, cowpea, French bean, squash, leafy vegetables, cucurbits	Srinivasan *et al.*, 1995
Spiralling whitefly, *Aleurodicus dispersus*	Guava, citrus, tapioca	Okra, bell pepper, brinjal, tomato	Puri *et al.*, 2000; Mani, 2010
Mealy bug, *Phenacoccus solenopsis*	Cotton	Brinjal, tomato, chilli, okra, cucumber, pumpkin	Halder *et al.*, 2013
Hadda beetle, *Henosepilachna vigintioctopunctata*	Brinjal	Bitter gourd, cowpea	—
Fruit borer, *Helicoverpa armigera*	Gram, cotton, tomato, cabbage	Peas, chilli, brinjal, okra	Puri *et al.*, 2000
Gall midge, *Asphondylia capparis*	Brinjal	Chilli, bell pepper	Nagaraju, 2000
Cabbage butterfly, *Pieris brassicae*	Cabbage, cauliflower, mustard	Knol-khol, radish	Puri *et al.*, 2000
Stem fly, *Ophiomyia phaseoli*	Beans	Okra	—
Red spider mite, *Tetranychus urticae*	Okra, brinjal	Cucumber, cowpea, Indian bean	Mahto and Yadav, 2009
Leaf hopper, *Empoasca motti*	—	Bitter gourd	Puri *et al.*, 2000
Plume moth, *Sphenarches caffer*	Field bean	Bottle gourd	Halder et *al.*, 2014

4.3. EMERGING VEGETABLE PESTS

4.3.1. Solenopsis Mealy Bug, *Phenacoccus solenopsis* (Pseudococcidae: Homoptera)

Polyphagous, soft bodied Solenopsis mealy bug, *Phenacoccus solenopsis* previously known as a pest of cotton but now poses a new threat to most of the cultivated crop plants. Out of a total record of 84 host plants across 28 families recorded at Central cotton growing zone of India up to 2009, 60 plant species from 22 families belonged to weeds (Vennila *et al.*, 2010). Presently, they feed on the host plants covering cereals, pulses, oil seeds, fruits, vegetables, ornamental crops as well as many weeds including *Parthenium*. Amongst vegetable crops, they found to attack on variety of plants belonging to Malvaceae (okra), Solanaceae (tomato, brinjal, potato, chilli), Leguminoceae (filed bean), and Cucurbitaceae (pointed gourd, cucumber, melons and gourds) (Halder *et al.*, 2013).

4.3.1.1. Damage:

Besides sucking the sap, they also secret the copious amounts of honey dew which is deposited on the plants and create black sooty mould and thereby reducing the photosynthetic activity of the plants.

4.3.1.2. Management:

Removal of alternate hosts and weeds like *Parthenium, Vernonia* from the field will help to reduce the pest incidence. Ants also help in spreading the mealy bugs besides giving protection to mealy bugs against their natural enemies. Hence, destruction of the ant colonies during land preparation is advisable to manage mealy bugs. Uprooting and burning the affected plants reduce the pest's load from the field. Spraying of fish oil resin soap (FORS) at 20 g/L of water (Kumar *et al.*, 2011) or entomopathogenic fungi *Verticillium lecanii* (2 x 10^8 cfu/mL) at 5 g/lit of water gives better control (Halder *et al.*, 2013). In case of severe infestation, spraying of buprofezin 25% SC at 1 mL/L or acephate 75 SP at 1gm/L or Malathion 50 EC at 2mL/L of water is recommended (Rai *et al.* 2014b). In another study Arif *et al.* (2008) documented that profenofos is the best insecticides against mealy bug causing maximum mortality after 24 hours. Recently, Patel *et al.* (2010) also reported that buprofezin at 625 g a.i./ha is effective in controlling this pest.

4.3.1.3. Challenges and Issues in Management

- Protective waxy coating over the body.
- Insecticides are not much effective and they also cause residue problem.
- Honey dew attracts the ants which give them protection against natural enemies.
- Ants also held responsible for physical movement of this pest.
- Biological control with Australian lady bird beetle, *Cryptolaemus montrouzeiri* and biopesticides like white halo fungus, *Verticillium lecanii* may be encouraged.

4.3.2. Serpentine Leaf Miner, *Liriomyza trifolii* (Agromyzidae: Lepidoptera)

4.3.2.1. Key Facts

- Economically important pest.
- Introduced to the Indian sub-continent in 1991
- Severe during summer months.
- Cotyledons stage is the critical crop stage.
- Parasitoid, *Hemiptarsenus* sp. is effective against the pest.

The American serpentine leaf miner, *Liriomyza trifolii* which is a native of Southern United States of America and Central America, has spread to other countries in seventies along with Chrysanthemum flowers (Puri and Mote, 2003). It is suspected that this species has entered Karnataka along with plant materials during 1990-91 (Viraktamath *et al.,* 1993) and then spread to many other parts of the country like Andhra Pradesh, Maharashtra, Gujarat, Tamil Nadu, Uttar Pradesh, West Bengal, Delhi, Haryana and Madhya Pradesh. Srinivasan *et al.* (1995) reported 78 host plants covering pulses, fibre, vegetables, ornamental and flowers, green manure crops, fodder crops, narcotics and weeds belonging to 16 families. In vegetables, three species of this pest were found infesting tomato, French bean, cowpea, cluster bean, summer squash, cucumber, melons etc. (Plate 4.1). The losses to the extent of 15-70 % in French bean, 41% in cucumber and 35% in tomato have been reported by Krishna Kumar (1998) from Karnataka.

4.3.2.2. Management:

It has been observed that higher dose of nitrogen favours the population build up of this pest. Judicious application of nitrogenous fertilizer reduces pest build up in endemic areas. Periodically infested leaves should be removed and destroyed. The pest can also be managed by using yellow sticky traps in the field for monitoring their presence (Lopez, 2010) and mass trapping of adults and exploiting the natural enemies. Recently, a new parasitoid, *Neochrysocharis farmosa* (Eulophidae: Hymenoptera) has been reported from this leaf miner. During February-March, the per cent parasitization was 37.5 at Varanasi region, Uttar Pradesh, India. In Hawaii, Guyana and Senegal this pest was controlled by inundative release of parasitoids *viz.*, *Hemiptarsenus, Chrysocharis* and *Chrysonomyia* spp. (Puri and Mote, 2004). In another study, it was also revealed that marigold (*Tagetes erecta*) plant serve as an attractant for this leaf miner besides it also attracts large number of the parasitoid, *N. farmosa.* Application of NSKE 4% with a sticker was found to be effective to deter the leaf miner. In case of severe infestation, spraying of imidacloprid 17.8SL at 0.3 mL/L of water during early stages of crop growth before flowering is effective (Nadagouda *et al.*, 2010). Kumar *et al.* (2010) reported that soil application of phorate 10G at 1 kg a.i./ha or seed treatment with imidacloprid 70WS at 3g/kg seeds or foliar applications of neem seed extract (5% NSKE) 10 days after germination gave better result in controlling the serpentine leaf miner. Rai *et al.* (2014b) reported that spraying of imidacloprid 17.8 SL at 0.35 mL/L of water during the early stage of the crops before flowering and application of dichlorovos 76EC at 0.5 mL/L of water in severe infestation during reproductive phase of the crop is beneficial.

4.3.2.3. Challenges and Issues in Management

- Wide host range makes them difficult to control; it can survive even on weeds also.
- Being an internal feeder it is difficult to control with insecticides.
- Periodical monitoring using yellow sticky traps should be ensured.
- More emphasis should be given on marigold as a trap crop.

4.3.3. Phytophagous Mites, *Tetranychus* spp., *Polyphagotarsonemus latus*

The predominant phytophagous mite species associated with vegetables are *Tetranychus neocaledonicus*, *T. urticae*, *T. cinnabarinus* (Tetranychidae: Acarina) and *Polyphagotarsonemus latus* (Tarsonemidae: Acarina). *T. urticae* also called as two spotted mite or red spider mite is one of the most common and destructive species of mite which is cosmopolitan in distribution and is highly polyphagous in nature. It is commonly associated with plants like okra, brinjal, beans, cowpea and cucurbits (Singh, 2004). *T. cinnabarinus* though mainly feeds on cucurbits like pointed gourd, cucumber, melons, gourds but also found on a number of other crops including brinjal, beans, onion, peas, Cole crops, tomato, sweet potato and various ornamental crops (Butani, 1984). The activity of these tetranychid mites has been observed during the post-monsoon period and dry period. The activity declines with the drop of temperature (Puri and Mote, 2004). *Polyphagotarsonemus latus* or yellow mite has also been identified as serious pest on chilli and beans. Due to infestation of this mite in chilli the underside of the leaves turns reddish and plants become rossetted. The *'Murda'* disease of chilli is due to infestation of this mite. During recent years, the major bottleneck of chilli growers in Varanasi region, Uttar Pradesh is heavy infestation of this mite which needs immediate attention (Singh, 2004). Nicotina and Cioffi (2000) referred to the dangerous diffusion of this broad mite on vegetables and floral crops especially in greenhouses. At cellular level, the typical symptoms of mite can be recognized by thickening of the lower epidermal cells, elongation of the palisade cell layer, shrinkage and deformation of parenchymatous tissues (Rai *et al.*, 2007).

4.3.3.1. Management:

Spraying of chemical acaricides like dicofol 18.5% EC at 2.0-2.5 mL/L or fenazaquin 10% EC at 2.5 mL/L is able to control the pest. During initial stage of broad mite infestation, destruction of severely attacked plant parts mechanically reduces mite population. Conservation and release of native predatory mites if available also can minimize the mite populations from the field. *Amblyseius* sp. is an efficient predatory mite, which regulates the pest under field conditions. Botanicals like *Kochea* and *Calotropis* leaf extract showed strong acaricidal action under field conditions (Rai *et al.*, 2014b). In case of severe infestation, spraying of dicofol at 2.5 mL/L or propargite at 3 mL/L (Singh and Singh, 2004) or fenazaquin at 1.5 mL/L or spiromesifen at 0.75 mL/L or chlorfenapyr at 1.25 mL/L of water should be done at 10 days interval at the initiation of symptoms. Recently, Rai *et al.* (2014) described that integrated module comprising seedling dip with imidacloprid at 1 ml/ lit, spray of buprofezin 1 ml/ lit at 25 days after transplanting (DAT), fipronil at 0.2 g/lit at 35 DAT, *Verticillium lecanii* at 5 g/lit at 45 DAT, chlorfenapyr at 1 ml/lit at 55 DAT, neem oil at 1% at 65 DAT gave yellow mite reduction over control in both chilli cvs. Kashi Anmol (75.48%) and Kashi Gaurav (72.90%).

4.3.3.2. Challenges and Issues in Management

- Tiny and microscopic creature, difficult to observe.
- Their symptoms often resemble with physiological disorders.
- Favoured by the continuous cultivation of vegetables throughout the years.
- Excessive use of insecticides with higher doses.
- Red spider mite can spread from one place to another by means of air current *i.e.*, ballooning.

4.3.4. Fruit Flies, *Bactrocera* spp.

It has been estimated that in India, fruit flies alone cause losses up to Rs. 26,902 million when control measures were not provided (Stonehouse, 2001) and in addition 35 to 40% losses in gourds. Kapoor and Agarwal (1980) and Kapoor (1993) reported 159 host plants of various species of fruit flies from India. Among these, *Bactrocera dorsalis, B. cucurbitae, B. zonata* (Tephritidae: Diptera) were recorded on different vegetable crops. If uncontrolled, these pests pose a serious limiting factor in the production of gourds, cucumber, melon and other cucurbits to the extent that their growing may become highly unprofitable. Besides these, this pest is very important from export point of view. Therefore, successive cultivation and export are highly relie on sound pest management system.

4.3.4.1. Management:

Infected and damaged fruits often serve as source of infestation. So periodically all the infested fruits should be collected and destroyed. Earthing up of soil around the vine helps in exposing the pupae for desiccation and predation by predatory birds. Ploughing the infested field after the crop is harvested can help in killing the pupae. Apply the bait spray containing 20 mL malathion 50 EC + 10 g protein hydrolysate + 500 g of molasses/jaggery in 20 litres of water per acre. When the attack is serious, it should be repeated at weekly intervals. Follow bait annihilation techniques with spray fluid of protein hydrolysate (0.3%) and malathion (0.015%) followed by male annihilation technique (MAT) by adult flies through plastic bottle trap with ethanol, any insecticide (carbaryl/malathion), cue lure (6:1:2) coated in wooden block. Installation of such traps at 25-30/ha is recommended (Rai *et al.*, 2014b).

4.3.4.2. Challenges and Issues in Management

- Pupation takes place in soil, so difficult to control until soil treatment is done.
- Internal feeders are difficult to identify at early stage of infestation.
- More emphasis should be given on behavioural management.
- Lack of promising natural enemies.

4.3.5. Whiteflies, *Bemisia tabaci*

Amongst 1200 species of whiteflies, *Bemisia tabaci* (Aleyrodidae: Hemiptera) has emerged as an important pest in vegetables and caused extensive economic damage. It is a cosmopolitan, highly polyphagous with remarkable ability to adapt to new hosts (Gerling and Kravchenko, 1995). It has been reported to attack more than 600 host plant species (Secker *et al.*, 1998) belonging to 77 families (Basu, 1995). The economic damage may result from direct feeding leading to de-vitalization of the plants, yellowing and finally death of the plant. Beside these, they also secrete honey dew which produces sooty mould and thereby restricting the photosynthetic activity of the plant. They also serve as vectors for many dreaded viral diseases. Earlier this pest was considered as a minor pest of vegetables. For last 4-5 years (2008-2013) this pest is occurring regularly and attained the major pest status and currently observed infesting vegetable crops like okra, brinjal, chillies, cowpea, field bean, tomato, cucumber etc. and cause significant loss both in field and greenhouse conditions. This pest damages the crops by direct feeding through sucking the cell sap. Both nymphs and adults feed on lower surface of the leaves causing deformation of young leaves. In recent years, severe incidence of leaf curl virus in tomato and chilli, yellow vein mosaic virus (YVMV) disease in okra, little leaf in brinjal, and apical leaf curl virus in potato and many viral diseases in cucurbits, beans and cowpea transmitted and spread by whitefly, has evoked widespread concern in northern India. Similarly, the detection of biotype-B of this pest in Karnataka on tomato causing severe incidence of leaf curl has sent a scare among the tomato growers in the country. Efforts are, however, on to evolve suitable management strategy for B-biotype of whitefly as a pest and vector is the need of the hour (Puri and Mote, 2004).

4.3.5.1. Management:

It is recommended to use yellow sticky traps for monitoring this pest at early crop stage. Many weed plants harbour whiteflies, removal of weed hosts was found to reduce both the incidence of whiteflies and associated viral diseases. Conservation of natural enemies like *Encarsia brevivena*, *Eretmocerus corni*, *E. mundus* (are some of the parasitoids), *Chrysoperla zastrowi sillemi*, *Mallada boninensis*, *Coccinella septempunctata* (are the predators), and *Beauveria bassiana* and *Paecilomyces farinosus* (are the entomopathogens) recorded from *B. tabaci* were found promising for the management of this pest. Seed treatment with imidacloprid 70WS at 2.5 g per kg of seed provides protection for 25-30 days. Protection of seedlings in nursery by nylon net (200 mesh) covering for 25-30 days (Shivalingaswamy *et al.*, 2006) followed by need based application of systemic insecticides after transplanting is able to reduce the whitefly incidence. Spraying of systemic insecticides like imidacloprid at 1 ml/3 lit of water or thiomethoxam at 1 g/3 lit of water or dimethoate at 2 ml/lit of water is recommended. Ameta and Kumar (2005) reported that Spiromesifen 240 SC at 500 ml/ha showed promising result in controlling whitefly in cotton.

4.3.5.2. Challenges and Issues in Management

- Wide host range and polyphagous nature made it difficult to control.
- Indiscriminate use of insecticides leads their resurgence.
- Yellow sticky traps should be used to monitor the early population.
- Need for conservation of natural parasitoids, *Eretmocerus* spp. and *Tetrastichus* spp.

4.3.6. Giant African Snail, *Achatina (Lissachatina) fulica*

Giant African snail, *Achatina* (*Lissachatina*) *fulica* (Stylommatophora: Achatinidae) is a polyphagous and one of the world's most damaging land snails (Plate 4.2). It is originated in east Africa and now spread all over the world (Raut and Barker, 2002) mainly to the countries that have tropical climates with warm, mild year round temperatures and high humidity. In India, the occurrence of this snail has been reported in different states in 1914. It was first reported from Aurangabad district during 1973 and in Konkan region during 1989. Being polyphagy, it is reported to feed on at least 500 different types of plant species (Capinera, 2011) and it voraciously feed on many vegetables like colocasia, elephant foot yam, cucumber, cowpea, field bean, pea, okra, tomato etc. Its sporadic outbreak was noticed in about 300 ha in Sapur during 1998 and in 150 ha area of Nashik of Maharashtra and causing havoc damage to the crops like bitter gourd, beans, bottle gourd, chilli, tomato and cauliflower (Puri and Mote, 2004).

4.3.6.1. Management:

Adult snails are large enough and can be seen easily. They can be handpicked form their resting sites in the evening and early morning and can easily be destroyed. Clean cultivation and establishment of physical barriers like wrapping of tree trunks with copper foils around the trees etc. that restrict the movement of snails has long been practiced. Ravikumara *et al.* (2007) recorded highest number of snails attracted to papaya stem waste, followed by vegetable waste, fish meal waste, areca leaf waste, banana sheath waste, sugarcane bagasse, leaf litter and farmyard manure. Sajeev

(2011) adopted a control method of initially baiting the snail using cabbage leaves followed by spray with tobacco decoction, made with copper sulphate mixture (TDCS mixture). Beside these, metaldehyde bait (2.5%) is long being practiced to control this malacofaunal pest (Jayashankar *et al.*, 2013). According to Javaregowda (2006), metaldehyde (2.5%) was most effective and registered highest mortality after one day of application, followed by monocrotophos bait. Basavaraju *et al.* (2001) observed 100% mortality under laboratory conditions using carbofuran (600 g carbofuran 3G + 60 kg rice bran + 6 kg jaggery/ha) bait against the snail.

4.3.6.2. Challenges and Issues in Management

- It thrives in high humid areas.
- It spreads through egg or juvenile stages.
- Management required community based approach.
- Very few known natural enemies.

4.4. TOMATO, *Solanum lycopersicum*

4.4.1. Silver Leaf Whitefly, *Bemisia argentifolii*

4.4.1.1. Key Facts

- The silver leaf whitefly or 'B' biotype of *B. tabaci* was reported for the first time in 1999 from Kolar district, Karnataka, India.
- This whitefly has wider host range, rapid reproduction ability inflicts more damage and is more resistant to insecticides.
- Recently, high incidence of this pest caused failure of the tomato crop in Bangalore, through direct damage to plants, as well from vectoring tomato leaf curl virus.
- The silver leaf whitefly could soon spread to affect a wider range of crops, including many vegetables, tobacco and cotton.
- Main impact of this new species is its nature of insecticide resistance, ability to transmit new diseases.
- Since, *Bemisia tabaci* is a highly invasive species, we need to know.
  - How biotypes are formed.
  - Basis for broadening of host range.
  - The reproductive mechanisms between and within biotypes.

4.4.1.2. Damage:

Silver leaf whiteflies (Plate 4.3) have piercing and sucking mouth parts and both nymphs and adults feed on lower surface of the leaves causing deformation of young leaves. Severe cases leads to chlorotic leaves accompanied by wilting. Whiteflies also excrete honeydew, causing sooty mould.

4.4.1.3. Management

- Use virus resistant hybrids.
- Raise nurseries in nylon nets/polyhouses.
- Dip the roots in systemic insecticides like imidacloprid (0.3 mL/L) or thiomethoxam (0.3 g/L) for 5 minutes before planting. If portrays are used

for raising nursery, drench the protrays with the chemicals one day before transplanting.

- Fifteen days after planting, spray with the systemic insecticides used for root dipping.
- Rogue out the virus affected plants as soon as the symptoms are observed.

4.4.2. Pinworm, *Tuta absoluta* (Lepidoptera: Gelechiidae)

4.4.2.1. Key Facts

- Tomato pinworm, *Tuta absoluta* was observed for the first time infesting tomato and potato crop in Pune, Ahmadnagar, Dhule, Jalgaon, Nasik and Satara regions of Maharashtra, and six districts of Karnataka, India.
- This pest has been classified as the most serious threat for tomato production worldwide.
- Severe infestation (50-87% plants affected) was observed in several tomato fields.
- If *T. absoluta* invades rest of the world, the tomato pest management cost may raise up to $ 500 million per year.

Tuta absoluta, a lepidopteran tomato leaf miner also called as pin borer, is considered as one of the most devastating tomato pests in the countries it has invaded so far. It has originated from Peru (South America) and then invaded many other countries in South America, Europe, Africa and Asia. *T. absoluta* larvae can completely destroy the tomato leaves by mining leaves, stems and buds and burrowing tunnels in the fruits, affecting the marketability of fresh tomatoes and yield losses up to 100%. The pest was detected and identified in October 2014 from Pune, Maharashtra in India by the Scientists of Indian Council of Agricultural Research, which now poses most serious threat to tomato cultivation in the country. The damage of this pest on tomato crop has been reported from Gujarat, Maharashtra, Telangana, Andhra Pradesh, Karnataka and Tamil Nadu.

4.4.2.2. Diagnostic Features of *T. absoluta* Damage on Tomato:

Feeding damage is caused by all the larval instars. Larva attacks all the parts of tomato except roots throughout the crop stage i.e., from seedling to harvest and can cause up to 100% crop damage mainly on the leaves and green fruits. *T. absoluta* can be easily recognized on tomato plants by the presence of large blotch or mine on the leaves with dark frass inside and pinhead size holes on the developing fruits (Plate 4.4). On the leaves, the larvae feed on the mesophyll tissue, forming large mines or extensive galleries. In case of serious infestation, leaves die completely. The larval entry and exit holes on the fruits are small and pin-head sized. Sometimes the entry and exit holes are used by secondary pathogens, leading to fruit rot. Multiple holes and presence of many larvae in a fruit can be noticed (Table 4.4). The other potential host plants for *T. absoluta* in India are brinjal and potato apart from other Solanaceous weed hosts.

Table 4.4. Extent of damage on tomato crop by *T. absoluta* in different states during March 2015 to February 2016

State/District	Village	% leaf blotch	% Fruit damage
Tamil Nadu: Dharmapuri	Palakodu	5.7-55.5%	3.5-43.4 %
Krishnagiri	Rayakottai, Kelamangalam, Ulimaranapalli, Binnamangalam, Hosur, Soolagiri, Bagalur	11.8-46.2%	7.5-52.4%
Karnataka: Chintamanai	Doddaulluru, Nandagudi, Kadagaskanahalli, Kuruhatti	1.5-41.7%	5.0-32.5%
Kolar	Malur	3.5-64.3%	0.0-41.8%
Bengaluru Rural	Devanahalli, Doddaballapur, Hessarghatta, Rajanukunte, Hoskote	5.5-33.5%	11-28.5%
Bengaluru Urban	Anekal, Jigini, Halehalli	4.5-45.5%	2-32.7.0%
Raichur	Kadagammadoddi, Kapur, Katlekur, Singandoddi, Hosur	10.2 – 60.3%	2.6-35.8%
Gujarat: Surendranagar	Sayala, Limbudi	1-4.5%	Nil
Rajkot	Dhoraji, Gondal, Jetpur	2-3%	Nil
Junagadh	Vadal, Choki	5.5-17.0%	5-12.5%
Jamnagar	Jalansar	1-4%	Nil
Anand	Anand	>2%	Nil
Andhra Pradesh	Madanapalli, Kuppam	13.0-36.7%	0.0-21.8%
Telangana	Warangal, Adilabad	0.0-2.8%	>1.0%

4.4.2.3. Management

i) **Biological Methods:** Three species of *Trichogramma* could successfully parasitize *T. absoluta*. Parasitism by *Trichogramma achaeae* was 28.8% followed by *T. pretiosum* (*thelytokous*) (22.7%) and *Trichogrammatoidea bactrae* (12.5%). No parasitism was recorded in the cages where *T. chilonis* was released.

Anthocorid predators, *Amphiareus constrictus* and *Blaptostethus pallescens* were observed to be efficient predators of *Tuta absoluta* eggs, feeding on 90 to 100% of the eggs when released in a ratio of 1 predator: 10 eggs.

Four *Bt* isolates from National Bureau of Agricultural Insect Resources, Bangalore, Karnataka, India, along with standard MTCC-8997 expressing the coleopteran specific proteins were tested against early second instar larvae of *Tuta absoluta* by tomato leaf dip methodology. The most toxic was NBAIR-4 isolate with LC_{50} of 301.3 ppm, followed by NBAIR-1, which showed LC_{50} as 373.7 ppm. The formulations are being tested for field efficacy.

ii) **Rapid Action Management Plan Advocated to Farmers**

- Destruction of infested tomato plants and fruits by burying deep inside the soil or by burning.
- Crop rotation with non-Solanaceous crops.

- Covering of nursery with pest proof net and use of pest free seedlings for transplantation.
- Preservation / augmentation of natural enemies like *Nesidiocoris tenuis, Necremnus* sp., *Orius* sp., and *Trichogramma* spp.
- Installation of *T. absoluta* pheromone baits for monitoring and mass trapping male moths both in nursery and main field (40 traps /ha).
- Use insecticide both in nursery and main field, if moth catches in pheromone trap exceeds 20-30 moths/trap/week.
- Recommended the following insecticides (as per the decision taken at the 355th meeting of CIB Registration Committee held on 29.04.2015) for managing the pest on tomato:
  - Chlorantriniliprole (Rynaxypyr) 10.26% OD at 0.3 ml/lit
  - Cynatriniliprole (Cynaxypyr) 18.5% SC at 0.3 ml/lit
  - Flubendiamide 20% WG at 0.3 ml/lit
  - Indoxacarb 14.5% SC at 0.5 ml/lit
  - Imidacloprid 17.8% SL at 0.3 ml/lit
  - Neem formulations (Azadirachtin content of either 1% or 5%) at 2-3 ml/lit

4.4.3. Western Flower Thrips, *Frankliniella occidentalis* (Thripidae: Thysanoptera)

4.4.3.1. Key Facts

- Western flower thrips (WFT) was first reported from Kolkata in 2015.
- Highly polyphagous and the most destructive pest species in the order Thysanoptera.
- WFT transmits tomato spotted wilt virus (TSWV).
- Body pale yellow.
- Brown patches on abdominal tergites.
- Considering its economic importance, it is a serious pest and vector of tospoviruses.
- Occurrence in other parts of India need systematic monitoring.

4.4.3.2. Damage:

Thrips cause both direct and indirect injury to crops. Direct damage occurs when the thrips cause injury by feeding or oviposition (Plate 4.5) wounds in tomatoes. Indirect damage refers primarily to the transmission of viruses by thrips. Examples of plant viruses transmitted by *Frankliniella occidentalis* include *Tomato spotted wilt virus* (TSWV) and *Tomato chlorotic spot virus* (TSCV). Rugman-Jones *et al.* (2010) has reported that WFT transmitted TSWV alone caused > $1 billion economic loss worldwide in early 1990s.

4.4.3.3. Biology:

The developmental time from egg to adult on the tomato leaf was 12.91 ± 0.04 days. Survival of immatures on tomato was high (75–80%).

4.4.3.4. Management:

Cultural control techniques may be used to manage *Frankliniella occidentalis*. In northern Florida, UV-reflective mulch is used to reduce the number of thrips in tomato. UV-reflective mulch interferes with the host-finding behaviour of thrips (Stavisky *et al.*, 2002; Demirozer *et al.*, 2012) (Plate 4.6). A reduction in nitrogen fertilization may result in a decrease in the number of *F. occidentalis* (Stavisky *et al.*, 2002).

Among the most important natural enemies of this species are minute pirate bugs, *Orius* spp. (Plate 4.7). Suppression occurs at a ratio of approximately one *Orius insidiosus* to 180 thrips, and control occurs at a ratio of approximately one *Orius insidiosus* to 50 thrips. Other thrips predators include the predatory mites *Amblyseius swirskii* and *Neoseiulus cucumeris.*

4.4.4. Serpentine Leaf Miner, *Liriomyza trifolii* (Agromyzidae: Lepidoptera)

4.4.4.1. Key Facts

- Serpentine leaf miner is an introduced pest (1991) from USA.
- Spread to other countries along with Chrysanthemum flowers.
- Damage occurs due to larval mining on leaves and ovipositional injury by adults.
- Has wide host range of 55 - 79 plant species (Srinivasan *et al.*, 1995; Puri and Mote, 2004).
- In vegetables, three species of this pest reported and found infesting tomato, French bean, cowpea, cluster bean, summer squash, cucumber, melons, etc. The losses to the extent of 15-70 % in French bean, 41% in cucumber and 35% in tomato have been reported (Krishna Kumar, 1998).

4.4.4.2. Damage:

The major form of damage is the mining of leaves by larvae, which results in destruction of leaf mesophyll. The mine becomes noticeable about three to four days after oviposition by adults and becomes larger in size as the larva matures (Plate 4.8). The pattern of mining is irregular. Both leaf mining and stippling can greatly depress the level of photosynthesis in the plant. Extensive mining also causes premature leaf drop, which can result in lack of shading and sun scalding of fruit. Wounding of the foliage also allows entry of bacterial and fungal diseases.

Although leaf mining can reduce plant growth, crops such as tomato are quite resilient, and capable of withstanding considerable leaf damage. It is often necessary to have an average of one to three mines per tomato leaf before yield reductions occur.

4.4.4.3. Biology:

The adult is a tiny fly with transparent wings. Adult females lay about 200 eggs. The eggs are deposited singly by piercing the lower surface of leaf, below the epidermis.

The eggs are oval and creamy in colour. The maggots are legless, pale yellow in colour with a tapering anterior body. Pupation is in the soil and the entire life cycle is completed in 15 days.

4.4.4.4. Management

i) **Biological Control:** Because parasitoids often provide effective suppression of leaf miners in the field when disruptive insecticides are not used, there has been interest in release of parasitoids into crops. Species of Eulophidae such as *Diglyphus begini*, *D. intermedius*, *D. pulchripes*, and *Chrysocharis parksi* are generally found to be most important. Several species of parasitic wasps, particularly *Chrysocharis parksi* and *Diglyphus begini,* attack leaf miner larvae; left undisturbed, parasites often keep leaf miners numbers below economic injury levels. Parrella *et al.* (1989) published information on mass rearing *Diglyphus begini* for inoculative release.

ii) **Cultural Practices:** Because broadleaf weeds and senescent crops may serve as sources of inoculum, destruction of weeds and deep ploughing of crop residues are recommended. Adults experience difficulty in emerging if they are buried deeply in soil.

   Cultural practices such as mulching and staking of vegetables may influence both leaf miners and their natural enemies. The leaf miner numbers were higher when tomatoes were grown with plastic mulch or tied to stakes. At least part of the reason seems to be due to lower parasitoid activity in plots where tomatoes were staked.

iii) **Chemical Control:** Spraying Triazovip at 1 ml/l along with neem formulations (Achook at 5ml/l), at 35-45 days after transplanting will be beneficial.

4.5. BRINJAL/EGG PLANT

4.5.1. Mealy Bug, *Phenacoccus solenopsis* (Pseudococcidae: Hemiptera)

Mealy bug, *Phenacoccus solenopsis* which was hitherto considered as a minor pest, has attained the status of major pest of brinjal in West Bengal. This pest has appeared in a serious proportion for last 4-5 years in West Bengal. Brinjal is an important vegetable crop in West Bengal.

4.5.1.1. Damage:

The mealy bugs attack isolated old plants which become covered with white mass (Plate 4.9). Insects suck sap from plants making them pale and weak.

4.5.1.2. Biology:

The female nymphs moulted three times for becoming adults while the male took four moults. The female and male nymphs completed development in 14.45 ± 0.96 days and 19.53 ± 1 days, respectively at 20-36.2°C temperature. The ratio of the male: female were 1: 1.75 and 1: 1.87. The duration from nymphs to adult female was 28 days. The pre-oviposition period was 6.37 ± 1.29 days. The average oviposition period was 5.89 ± 0.89 days. Fecundity rate was 205-278 eggs/female (Alam *et al.*, 2011).

4.5.1.3. Management:

Spraying 0.15% malathion or 0.4% monocrotophos gave effective control of mealy bugs. It was also effectively controlled by release of the predator, *Cryptolaemus motrouzieri* (Krishnamoorthy and Mani, 1996).

Recommendations for the management of *P. solenopsis* include:

- Large number of incidental hosts that have low population of *P. solenopsis* found within fields, field borders and roadside during offseason should be removed and disposed by burial or burning.
- Management of *P. solenopsis* on weed hosts by roadside and field borders should be a priority in all zones to prevent spatial spread and limit severity on brinjal crop.
- The extent of offseason management determines the magnitude of incidence and severity of *P. solenopsis*.
- Brinjal season cultural practices should focus on field sanitations and proper weed management.

4.5.2. Gall Midge, *Asphondylia capparis*

4.5.2.1. Damage:

In parts of South India, gall midge, *Asphondylia capparis* is a regular pest on flowers and fruits of brinjal and sweet pepper (Tewari *et al.*, 1987; Krishna Kumar *et al.*, 1997). Maggots of this pest feed on developing ovary/fruits. When young flowers are attacked, stamens and pistils degenerate and fruits fail to form. Following the emergence of the fly, the fruit cracks along the inner side of the curve and is rendered unfit for marketing (Plate 4.10).

4.5.2.2. Management

- Spraying of 4% NSKE is effective in controlling this pest.
- The gall midge was found parasitized by *Eurytoma* sp. to the tune of 8-15% (Tewari and Krishnamoorthy, 1986).

4.6. CHILLI, *Capsicum annuum*

4.6.1. Gall Midge, *Asphondylia capparis*

Gall midge, *Asphondylia capparis* (Cecidomyiidae: Diptera) is an emerging pest of chilli in parts of Tamil Nadu and Warangal district of Andhra Pradesh. Occurrence of gall midge in chilli crop is being recorded in eastern India especially in Odisha for the last few years. Chilli gall midge incidence commences from October and continues till January end with peak incidence during November and December. The infestation declines in summer months. Earlier infestation by this gall midge resulted in 42.10% damage to floral parts in Tamil Nadu (Rangarajan and Mahadevan, 1974).

4.6.1.1. Damage:

A. capparis is considered as major pest that causes paprika floral galls (Nagaraju, 2000) as the maggots feed on the ovary of the flower bud, flower and tender fruits. Infestation by the gall midge can easily be identified as the affected bud remains unopened, while the affected flower dries and drops, severe flower and fruit drops, reduction in yield, fruit size, seed number, and production of malformed fruits (Plate 4.11). The damage is more severe in areas where chilli is grown mostly under irrigated conditions. Maryana *et al.* (2006) observed flower buds or very young pods of chilli transformed into galls and opined that when young pods were attacked, they do not grow normally and remain smaller than normal pods. Sertkaya *et al.* (2006) noticed damaged fruits by *A. capparis* in the last week of July in both 2004 and 2005 and of the damaged fruits ranged from 1.41 to 69.71% in Antakya province of Turkey.

4.6.1.2. Management:

Two hymenopteran parasitoids were recorded during the survey on chilli gall midge *viz.*, *Eurytoma* sp. (Eurytomidae), a larval-pupal parasitoid and *Bracon* sp. (Braconidae), a larval parasitoid. Highest parasitization of 31.15% by *Eurytoma* sp. was observed in the month of November and the combined effect of parasitoids *Eurytoma* sp. and *Bracon* sp. ranged from 6.67- 24.58% with an average of 15.27 ± 6.69% in Karnataka (Basavraj *et al.* 2011). Tomar *et al.* (1997) from Madhya Pradesh reported that *Dinarmus* sp. (Hymenoptera: Pteromalidae) as a new larval parasitoid of chilli gall midge, *A. capparis.*

4.6.1.3. Challenges and Issues in Management

- Can be recognized only after symptoms are visualized.
- Attack only the reproductive parts (flower bud, fruits) of the plants.
- Damage is more severe under well irrigated conditions.

4.7. POTATO, *Solanum tuberosum*

4.7.1. Changing Pest Complex

- In Punjab, during 2010-12, severity of damage by thrips, mites, aphids and butterfly ranged between 7.54 to 15.08 % in early sown seed crop and from 0.01 to 0.62 % in the normally planted crop.
- *Myzus persicae* and *Aphis gossypii* have shown deviation from earlier trends and the persistence of these insects throughout the season has given an alert to initiate research on IPM to manage these insect pests. Aphids of themselves can cause wilting damage by sucking out nutrients from foliage and stem tissues. But, the real problem caused by aphids in commercial fields is their ability to carry pathogenic viruses. In potato fields, the common viruses are potato leaf roll virus (PLRV), the mosaic viruses (e.g., PVA, PVY) and alfalfa mosaic virus (AMV, "calico virus"). Persistence is the key to the spread of a virus by aphids in the potato field. For this reason, the most important of the aphids in potato is the Green Peach Aphid (GPA) (*Myzus persicae*) which will persistently carry PLRV, usually the major viral problem on potato. Other aphids that will carry viruses non-persistently are potato aphids which may carry PLRV and others as the GPA, and alfalfa aphid which carries the non-persistent AMV/calico.

4.7.2. Aphid, *Myzus persicae*

4.7.2.1. Damage:

Wingless green peach aphids tend to be pear shaped. They may have variable colours but usually are straw-coloured or light-green (Plate 4.12). Hind cornicles slant so they converge. A key characteristic is tubercles at the base of the antennae. Potato aphids are more elongate and larger, about 1/6 to 1/8 inch, than the green peach aphid. They often are bright green, but pinkish forms occur. They are similar in general appearance to the English grain aphid. Both potato and green peach aphids produce winged forms. Winged aphids are most important in spread of viruses to plants since they move readily within and between fields. Wingless green peach aphids tend to occur on lower, older potato leaves. Potato aphids more commonly colonize the upper part of the potato plant. During the summer, aphids reproduce asexually (no males) and give live birth to young. Under optimal conditions, a generation may be completed in 10 days. Green peach aphids overwinter in the egg stage on various *Prunus* species, particularly plum and apricot, in and around towns. It may also be introduced on infested greenhouse plants.

4.7.2.2. Management:

Cropping in aphid-free periods or when their population is low by adjusting the planting dates in the Indo-Gangetic Plains, where about 90% of seed crop is grown i.e. up to 15th October in North Western Plains, up to 5th November in North Eastern Plains. Remove all weed hosts susceptible to viruses and for aphids (especially those having yellow flowers) from within and around the vicinity of seed potato plots. The haulm cutting of seed crop should be done as soon as the aphid number crosses the critical level i.e. 20 aphids/100 leaves. Care must be taken to prevent any regrowth of stumps before harvest.

Monitor flights of winged aphids by using yellow traps (yellow dish pans with water and a bit of bleach). Sweep fields at least weekly to monitor aphids. Make sure to get the sweep net deep in the canopy since a lot of aphids are found on lower leaves. Pull up yellowish plants to check for aphid colonies. Apply pesticides only when necessary. Use of some insecticides can exacerbate aphid problems, because the control has more effect on natural enemies than on aphids. Insecticides can only provide limited control of aphid transmitted virus diseases. The green peach aphid can be highly resistant to insecticides registered on potatoes. Therefore, choose insecticides carefully for effective control. Protect natural predator populations that are important in aphid management. Trap crops around the edges of fields can help aphid control. Direct damage from aphid feeding in a strictly commercial field needs high populations. If aphid levels are approaching 50 aphids per plant, treat the field.

Several predators and parasitoids help to control aphids. Among the more important insects are lady beetles, damsel bugs, green lacewings, minute pirate, and parasitic wasps.

4.8. CAULIFLOWER, *Brassica oleracea* var. *botrytis*

The trends in relative importance of crucifer insect pests in Asia are changing. In recent times, leaf miners (mainly *Liriomyza* spp. and *Chromatomyia* spp.), *Spodoptera exigua,* aphids and *Crocidolomia binotalis* are becoming more important (Table 4.5). Increasing pest status of *Plutella xylostella* in most parts of the world is due to the absence of natural enemies and fast development of insecticide resistance.

Table 4.5. Changing trends in relative importance of crucifer insect pests in Asia

Relative importance	Insect pest/s	Remarks
Becoming more important	Leaf miners (mainly *Liriomyza* spp. and *Chromatomyia* spp.) *Spodoptera exigua*	Believed these are spread around rapidly and widely by increasing movements of trade and people in recent times. Farmers spray heavily against them.
Becoming more important following effective suppression of *Plutella xylostella* by parasites	Aphids *Crocidolomia binotalis*	These have been observed to increase recently in areas with effective IPM programmes against *P. xylostella*. Believed to be a case of pest succession where the niche previously occupied by *P. xylostella* is now taken over by these species.

4.8.1. Diamondback Moth, *Plutella xylostella*

The diamondback moth, *Plutella xylostella*, has consistently remained the most destructive insect pest of crucifer vegetables worldwide (Plate 4.13). The major outbreaks of *P. xylostella* are more likely in the fields that are sprayed frequently and heavily with insecticides. The absence of natural enemies and fast development of insecticide resistance are believed to be the major causes of increasing pest status of *P. xylostella* in most parts of the world. Currently, this insect has become resistant to almost all classes of insecticides used against it in south-east Asian countries (Dhaliwal and Koul, 2010). An outbreak of diamondback moth, *Plutella xylostella* on cauliflower was reported in Aligarh, western part of Uttar Pradesh, during September to first fortnight of October during 2006. The infestation increased gradually from first fortnight of August and led to total loss (100 % yield loss) of the crop (Ahmed *et al.* 2009). Moreover, the current climatic change may lead to increase in severity of this pest in many regions of the world. For instance, *P. xylostella* may have two additional generations per year in Japan (Cherim 2007).

4.8.1.1. Management

i) **Cultural Methods:** Sprays of 4% NSKE/4% pulverized NSPE were found to be highly effective against DBM. Field sprayed with NSKE/NSPE had obtained higher yield, incurred 15% less cost and realized 43% higher net returns. The performance of powder formulation of 'Soluneem' was on par with 4% NSKE in controlling DBM. Sprays of neem soap or pongamia soap both at 1% were highly effective in reducing DBM, aphids (*Brevicoryne brassicae*) and leaf webber (*Crocidolomia binotalis*). Further, the soap sprays significantly increased the yield of cauliflower and were relatively safe to the parasitoid, *Cotesia plutellae* (Srinivasan and Krishna Moorthy, 1993; Krishna Moorthy *et al.*, 1998; Krishna Murthy and Krishna Kumar, 2002).

ii) **Cultivation under Protected Environment (Netted Structures):** In parts of some countries in the Asian region (e.g., China, Philippines, Malaysia, Thailand, Vietnam), particularly in areas where strong winds are not a recurring problem, cultivation of crucifers and other vegetable crops under protected environment is being practised. Essentially, two kinds of netted structures are used; those that are covered completely with a net (ca 256 mesh) and those with nets on the side and a plastic sheet on top. One major reason for such a practice is to exclude insect pests, hence avoiding or minimising the use of chemical insecticides. Other advantages include avoiding damage from direct impact of rain, wash-off of applied chemicals against diseases, soil spillage onto crops, better fertilizer use, and less weed problems.

iii) **IPM using Indian Mustard as a Trap Crop:** IPM using Indian mustard as a trap crop involves planting of paired rows of mustard after every 25 rows of cauliflower and spraying of 4% NSKE at primordial formation. Two more sprays of 4% NSKE may be given at 10-15 day interval after the first spray. The IPM gave 152% more returns than pure cabbage crop. IPM controls diamondback moth, leaf webber (*Crocidolomia binotalis*); stem borer (*Hellula undalis*); aphids (*Brevicorne brassicae, Hyadaphis erysimi*) and bug (*Bagrada cruciferarum*) (Srinivasan and Krishna Moorthy, 1991) (Plate 4.14).

4.9. COWPEA, *Vigna unguiculata*

4.9.1. Spotted Pod Borer, *Maruca vitrata* (Pyralidae: Lepidoptera)

Spotted pod borer, *Maruca vitrata* previously considered as a minor pest of red gram had now became a serious problem in many vegetables like cowpea, field bean, cluster bean and pea. The activity of *M. vitrata* was observed in almost all the states mainly southern, northern and central part of the country during *rabi* and pre-*kharif* season. Overlapping generations, short life cycle, wide host range and protcctive web forming nature of this pest are main reasons for its cosmopolitan distribution.

4.9.1.1. Damage:

The damage due to this pest reached up to 42% in cowpea during *rabi* season in Andhra Pradesh (Halder and Srinivasan, 2011). This pest not only damage the pods of the plant but also feeds flower buds and sometimes stem of the plants are infested (Plate 4.15) (Halder *et al.*, 2006; Halder and Srinivasan, 2007).

4.9.1.2. Host Range:

It is a serious pest of cowpea, pigeon pea, black gram, green gram, beans and soybean in India.

4.9.1.3. Life Cycle:

The duration of total life cycle varied from 31 to 36 days on cowpea (Naveen *et al.*, 2009).

4.9.1.4. Management:

Clean cultivation and removal of the web reduce the pest load from the field. Conservation of bioagents like *Apanteles* sp. is effective. A larval-pupal parasite, *Agathis* sp. accounted for 32.52 to 48.14 per cent mortality of *M. vitrata.* Need based spraying of azadirachtin at 0.005 % or NSKE 4% or *Bacillus thuringiensis* var *kurstaki* at 1 kg/ha is recommended (Rai *et al.*, 2014b). Chandrayudu *et al.* (2008) recorded the efficacy of commercial formulation of Bt at 0.0025% in suppression of pod damage due to spotted pod borer in cowpea. Bhat *et al.* (1988) reported that neem seed extract was the next best treatment to monocrotophos against the pod borers on cowpea. For seed production, application of DDVP at 0.75 ml/lit of water at flowering and early pod formation stage gives good control. Recently, Mittal and

Ujagir (2005) reported that Spinosad at 90 g ai/ha was found to be effective against pod borer complex of pigeon pea including *M. vitrata*. Cowpea cv. HC-270 was reported to be highly tolerant to *M. vitrata* (Table 4.6) (Halder and Srinivasan, 2011).

Table 4.6. The chronological order of various insecticides with green pod yield in kg/ha and incremental cost benefit ratio, respectively based on overall rank for bioefficacy against pest complex plus safety to lady bird beetles

Insecticides	Green pod yield in kg/ha	Benefit: cost ratio
Lambda-cyhalothrin at 0.0025%	3337	31.68
Methyl-o-demeton at 0.025%	3160	29.99
Fenvalerate at 0.01%	2882	33.50
Deltamethrin at 25% tablet	6962	49.53
Alphamethrin at 0.01%	3742	28.53
Deltamethrin at 0.0014%	5729	71.53
Endosulfan at 0.07%	5150	45.50
Quinalphos at 0.05%	4045	29.21
Indoxacarb at 0.00725%	4454	18.45
Methomyl at 0.08%	5364	16.53
Azadirachtin at 0.001%	2498	13.08
Bacillus thurungiensis var. *kurstaki* at 0.75 kg/ha	2188	2.51
NSKE at 5%	1890	12.98

4.9.1.5. Challenges and Issues in Management

- Difficult to control this pest with conventional insecticides due to protective web.
- Concealed in the web and thereby also safe from its natural enemies.
- Over lapping generations and short life cycle.
- Need for development of biointensive integrated pest management (BIPM) strategy.

4.9.2. Hadda beetles, *Henosepilachna vigitioctopunctata* and *Epilachna dodecastigma* (Curculionidae: Coleoptera)

Hadda beetles, *Henosepilachna vigitioctopunctata* (Plate 4.16) and *Epilachna dodecastigma* (Curculionidae: Coleoptera) are well-known pests of eggplant and potato. But recently, its serious incidence was reported in cowpea as foliage feeder in many parts of the country particularly eastern Uttar Pradesh and Bihar. More than 80% leaves were infested by the grubs and adults of this beetle on cowpea during summer season (Halder *et al.*, 2011). Similarly, its serious incidence was also observed from bitter gourd, *Momordica charantia* in Jammu (Jamwal *et al.*, 2013), Allahabad in Uttar Pradesh (Maurice and Ramteke, 2012) wild bitter gourd, *Momordica dioica* in Maharastra (Deshmukh *et al.* 2012) and Ivy gourd, *Coccinia grandis* in Allahabad (Maurice and Ramteke, 2012).

4.9.2.1. Management

i) **Biological Control:** Conservation of larval-pupal endoparasitoid, *Pediobius foveolatus* (Eulophidae: Hymenoptera) which cause up to 36.6% reduction in beetle population in and around Varanasi, Uttar Pradesh (Halder *et al.*, 2011). Amongst the entomopathogenic fungi, under laboratory conditions black muscardine fungus, *Metarrhizium anisopliae* at 5 g/L gave good control (Anon, 2014). In another study, Vishwakarma *et al.* (2011) documented that, significantly maximum reduction in the population of *Epilachna* beetle (74.91%) was achieved in treatment of *Beauveria bassiana*, when used at 3.0 g/L of water.

ii) **Botanicals:** Islam *et al.* (2011) observed that crude aqueous extracts of castor (*Ricinus communis*) leaves had highest larvicidal toxicity (LC_{50}=18.40%) besides significant reduction in both oviposition and egg-hatch, prolonged larval duration ($P<0.001$), and inhibited pupal formation and adult emergence.

iii) **Chemical Control:** Need based application of malathion or carbaryl at 1 kg a.i./ha or cypermethrin at 0.4 ml/l is able to control this polyphagous pest (Rai *et al.*, 2014b). Insecticides such as emamectin benzoate 5 SG at 10 g a.i./ha and thiodicarb 75 WP at 70g a.i./ha proved to be the most effective against Hadda beetles with highest reduction in pest population of 75.3 and 73.9%, respectively under field conditions.

iv) **Integrated Pest Management:** Integration of the biopesticide, *Metarhizium anisopliae* ($1x10^8$ cfu/g) with neem oil provided highest protection over control and indicated compatibility and synergistic activity at 1:1 combination.

4.9.2.2. Challenges and Issues in Management

- Polyphagous in nature and have diverse host range.
- Efforts are to be directed for mass rearing and artificial release of *P. foveolatus*.
- Egg stage is more vulnerable and there is a need to search for suitable ovicide(s).

4.10. FIELD BEAN, *Lab lab purpureus* var. *lignosus*

4.10.1. Plume Moth, *Sphenarches caffer*

White plume moth, *Sphenarches caffer* (Pterophoridae: Lepidoptera), is a serious pest of lablab and beans (Nair, 1995; Sujithra *et al.*, 2010). Recently, it attained pest status as a foliage feeder in bottle gourd as they damaged the leaves and buds of bottle gourd by scrapping the leaf portion thereby reducing the photosynthetic activity of the plants in and around Varanasi region. However, damage is more severe when they feed on the emerging buds resulting in restricted growth of the buds with characteristic black excreta inside it (Halder *et al.* 2014). During the peak summer months of May-June when atmospheric temperature was around 45°C in Varanasi its incidence was also observed and from mid-October onwards when *rabi* season bottle gourd was in its vegetative stage there was no incidence of this plume moth.

Sujithra *et al.* (2010) from Tirupati, Andhra Pradesh reported *S. caffer* as a major pod borer of field bean.

4.10.1.1. Damage

- Tiny caterpillar (Plate 4.17) bore into the pods and feed on developing seeds.
- The pest does not bore completely, but the posterior abdominal portion remains outside.
- Sometimes the caterpillars emerging from eggs laid on flower buds enter into these buds and feed inside.

4.10.1.2. Biology:

Total life cycle of plume moth ranged between 22.5 to 45.3 days on natural diet as compared to 22.5 to 51.0 days when reared on artificial diet at prevailing room temperature of 29 ± 2°C and relative humidity of 70-80% (Harathi *et al.*, 2015).

4.10.1.3. Management:

Conservation of solitary larval endoparasitoid, *Apanteles paludicole* (Braconidae: Hymenoptera) (maximum parasitization 40.91%) (Halder *et al.*, 2014) and chalcid pupal parasitoid, *Tropimeris monodon* are beneficial. Need based application of *Bacillus thuringiensis* at 1 kg/ha or malathion at 2 ml/lit is able to control this pest.

Tamil Nadu Agricultural University cv. Purple Pod consistently proved to be resistant to plume moth, *Sphenarches caffer*, due to operation of certain of mechanism of non-preference. Mere scrapings by the test insect *S. caffer* were observed on the resistant genotype, TNAU purple pod (Harathi, 2008).

4.11. CONCLUSIONS

In recent past, with changes in the cropping pattern, ecosystems and habitat, climate, and introduction of input intensive high yielding varieties/hybrids, a shift in pest status has been realized in time and space. Many pests have expanded their host horizon, developed resistance to pesticides and often there are secondary outbreaks. Incidence of chilli gall midge, *Asphondylia capparis* in parts of Tamil Nadu and Andhra Pradesh; solenopsis mealy bug, *Phenacoccus solenopsis* in brinjal, tomato, okra and cucurbits; Hadda beetles, *Henosepilachna vigitioctopunctata* and *Epilachna dodecastigma* on cowpea and bitter gourd; plume moth, *Sphenaeches caffer* in bottle gourd are some of the examples. The emerging insect pests in vegetable ecosystems, development of suitable control measures and some issues/challenges in their management needs to be undertaken on war footing.

5

Emerging Insect Pest Problems in Plantation and Tuber Crops

5.1. INTRODUCTION

The coffee berry borer, *Hypothenemus hampei* has spread in the major coffee growing areas (3,88,000 ha) in southern states of India. Annually more than Rs. 20 crores is spent towards control measures. Tea mosquito bug, *Helopeltis antonii* is a serious constraint in cashew (west coast-Kerala, Karnataka, and east coast-Tamil Nadu). The pest may spread to new areas due to climate change and states like Andhra Pradesh, West Bengal and Orissa may come under this pest attack (Gupta *et al.*, 2009). The rhinoceros beetle, *Oryctes rhinoceros* has become a problem on oil palm in southern states of India. The coconut leaf beetle is one of the most damaging pests of coconut and other palms. About 80-90% of seedlings were found infested by the pest, damaging 40% of the leaf area in Kasargod, Kerala.

Corm borer, *Aplosonyx chalybaeus* is a serious, univoltine pest of colocasia (taro) in north eastern hill states and hilly regions of West Bengal. It is a regular and endemic pest causing 20–30 % damage to the foliage and 80–90 % to the corms resulting in severe losses to the tribal farmers of the eastern Himalayan region of India.

5.2. PLANTATION CROPS

5.2.1. Coffee, *Coffea canephora / Coffea arabica*

5.2.1.1. Berry Borer, *Hypothenemus hampei* (Order: Coleoptera, Family: Scolytidae)

The coffee berry borer, *Hypothenemus hampei* is native to Central Africa and has now invaded many coffee producing countries worldwide. It was first reported in Gudalur, Nilgiri district of Tamil Nadu during February 1990, and in March 1990 in the adjoining district Wynad of Kerala (Kumar *et al.*, 1990). It is specific to coffee, affects both Arabica and Robusta cultivars, and is of quarantine significance. It was introduced accidentally either through coffee brought by refugees from Sri Lanka or through illegally imported coffee seeds (Singh and Ballal, 1991). The fertilized female bores an entrance hole at the terminal pore and lays bean shaped eggs in coffee seeds. The coffee berry borer has spread in the major coffee growing areas (388,000 ha) in southern states in India.

i) **Economic Importance:** The damage caused by *H. hampei* is mainly a decrease of coffee yield due to abscission of berries, loss of weight, and a decrease on coffee quality and, therefore, coffee price. It has been estimated that there is a weight loss of 55% on coffee grains attacked by *H. hampei* (Montoya, 1999); however, the decrease of weight of the total coffee production is about 18% (Borbón, 1990). Crop loss of up to 96% has been reported to occur in some Eastern African countries (Magina, 2005). Annually more than Rs. 20 crores is spent towards control measures.

ii) **Damage:** *Hypothenemus hampei* is known to feed and reproduce only in the seeds of coffee. The pest attacks both Arabica and Robusta types and is also found on other species of coffee. The female enters the coffee berry by cutting an irregular hole, generally at the tip of the berry (Plate 5.1). Berries in all stages of development are prone to attack. Breeding occurs within berries with hardening or hard endosperm. The damage is mainly due to feeding inside the berries in the field. In general, only one of the two beans in a berry is damaged and hence maximum loss could be 50% of infested berries. Damage from 5-90% of the berries has been reported. Berries of the previous season left on the plant or fallen on the ground serve as pest reservoirs. This serves as a source of infestation for the next season crop.

iii) **Host Range:** Berry borer find its shelter in many leguminous plants like Crotalaria, *Tephrosia* sp., *Phaseolus lunatis*, Centrosema, and seeds of *Hibiscus* sp., berries of *Vitis lanceolaria*, *Liqustrum pubinerve* and *Oxycanthus* sp. (Le pelley, 1968). Though it harbours on other plants feeding and breeding is possible only on coffee berries. All the cultivated varieties of coffee are attacked by the borer without any preference (Anon, 2003).

iv) **Biology:** The adult female after entering the berry tunnels into the bean, excavating irregular galleries while feeding and laying bean shaped eggs. The brownish black female adult measures 1.5-1.9 mm. The males are smaller and generally incapable of flight. The population is usually female dominated (10: 1). The females can lay 30-70 eggs during its life. The eggs hatch in 4-9 days and the larvae feed and make extensive tunnels. The larval period lasts for 10-26 days. The total development from egg to adult lasts for 25-35 days. On emergence, the female remains inside the berry for a few days, during which period mating takes place between siblings. The females then leave the berry to infest fresh berries. Adults are long lived, a female can live for 282 days with an average of 156 days. Females can survive up to 2 months in buried beans and mobile females can survive by feeding on immature berries. Three to four generations a year have been reported.

v) **Ecology:** The optimum temperature and RH requirements for reproduction and development was found to be around 25-26°C and 90-95%, respectively. An elevation of 500-1000 metres has been found to be optimum for infestation. Intermittent rain favours emergence and development. The incidence was found to be more under dense levels of shade as compared to open or less dense shade.

vi) **Management**

a) **Cultural Methods:** Cultural and phytosanitary measures that aim at disrupting the breeding are especially useful in places like India where there is a distinct gap between harvest and availability of suitable fruits in the new crop. Timely and clean harvest (removing fallen berries) helps in reducing the infestation.

Ungleaned estates recorded an average infestation of 13% as compared to 1.5% in gleaned estate. The use of picking mats while harvesting also helps in reducing the gleanings (fallen berries on the ground) to an extent of 73% and in reducing the infestation to the extent of 72% in the next crop as compared to areas harvested without picking mats.

Coffee dried to 11.5% moisture helps in killing all pest stages as also dipping infested fruits in boiling water for 2 min or cold water for 2 days.

The beetles migrating drying yards can be trapped by using oil coated polythene sheets and also in the berries on plants surrounding the drying yard by using them as trap plants, which can be clean harvested and destroyed.

Thinning of shade trees and training and pruning of unwieldy bushes to avoid excessive shade help in reducing the infestation.

b) **Chemical Methods:** Spraying 0.07% endosulfan was found to be the most effective among all insecticides tested. It is important to spray during the period when the females are entering the berry i.e. during September-October. Spot application in infested patches also proved useful to prevent the spread of infestation. Spraying of chlorpyriphos, quinalphos, methomyl and malathion is also effective.

Mass trapping using chemical attractants (1: 1 mixture of ethanol and methanol is best) could be used for the management of the berry borer. A trap density of about 70 traps/ha was found optimal. Large number of beetles can be trapped during the harvest and post-harvest periods.

c) **Biological Methods:** Among indigenous natural enemies, *Beauveria bassiana* was the most abundant and caused epizootics under high humidity conditions. High volume sprays gave better results than low volume sprays. The fungus is most effective at a temperature of 23-28°C and RH of more than 90%. A spore load of 1 x 10^{10} has been found effective. Trials have shown that treatment during September with spray solution containing 10^8 conidia/ml resulted in 51-60% mortality of the berry borer.

Among parasitoids, only 4 are considered promising. They are *Prorops nasuta, Cephalonomia stephanoderis, Phymastichus coffea* and *Heterospilus coffeicola,* which are all native of Africa. Of these 4 promising parasitoids, three have been introduced into India. *Prorops nasuta* and *Cephalonomia stephanoderis* were imported from Mexico in 1995 and *Phymastichus coffea* and *P. nasuta* from Columbia in 1999. The parasitoid, *C. stephanoderis* has established in several areas of Kodagu district, Karnataka; Wynad district, Kerala and Lower Palanis, Tamil Nadu.

d) **Integrated Methods:** The pest could be effectively tackled by adopting the following measures:

- Maintenance of optimum shade and good drainage.
- Thorough and clean harvesting of the crop in time.
- Use gunny bags/picking mats/polythene sheets while harvesting.

- Collecting the gleanings and dipping the infested berries in boiling water for 2 minutes or in cold water for 48 hours for killing all the stages of the borer inside the berry.
- Spreading the polythene sheet smeared with castor oil over the harvested coffee heap during the day for trapping all the beetles.
- Maintenance of trap plants around drying yard followed by clean harvesting from the trap plants and treating the berries in boiling/cold water.
- Timely spraying of entomopathogenic fungus, *Beauvaria bassiana.*
- Mass trapping of beetles using lures.
- Fumigating the storage material and transport structures effectively to check the pest.

5.2.2. Coconut, *Cocus nucifera*

5.2.2.1. Mite, *Aceria guerreronis* (Order: Acari, Family: Eriophyidae)

i) Key Facts

- A national pest causing extensive damage to coconut.
- Sathiamma *et al.*, (1998) first reported the incidence of *A. guerreronis* in Amballoor Panchayat of Eranakulum district of Kerala (India). The infestation had spread throughout peninsular India and in parts of Pondicherry and Lakshadweep.
- Prasad and Ranganath (2000) also found the presence of perianth mites' infestation in Andaman.
- Rao *et al.* (2001) have reported the incidence of eriophyid mite *A. guerreronis* on coconut for the first time in costal Odisha.
- It has spread to Tamil Nadu, Karnataka and Andhra Pradesh (Ramaraju *et al.*, 2002; Begum and Babu, 2013; Reddy and Naik, 2000; Nair, 2000).
- In the recent past, the pest has spread rapidly to all coconut growing areas of India (Muthiah, 2007).
- Copra production dropped from 18–20 kg to 10–12 kg per 100 nuts after the coconut mite upsurge at the end of the 1990s. Weight loss of copra was also noticed.
- Annually more than Rs. 100 crores is being spent towards management of this pest (Sreejith, 2011).
- Bagde and Ashte (2014) reported 64.16 to 89.42% nut infestation in Thane district of Maharashtra
- Yield reduction of 67.2% due to intense early and late nut fall has been recorded in Kerala (Sangeetha Kaimal, 2015).

ii) Damage:

Recently coconut gardens in some parts of Kerala, Tamil Nadu, Andhra Pradesh and Karnataka were seriously affected by a non-insect pest called eriophyid mite. These mites infest by sucking sap from the soft tissues of buttons. In the initial stages, symptoms are seen as triangular patches close to perianth. Later, because of the continuous desapping by various stages of mites present beneath the inner bracts of perianth, brown coloured patches are formed (Plate 5.2). As the nuts grow in size, the injured patches become warts and then develop into longitudinal splits on the surface of nuts. The liquid oozing from these patches dries and as a result dried decayed matter is noticed. The damage thus caused affects the quality of husk and dehusking becomes difficult.

iii) Management:

Being a micro pest inhabiting under the periyanth, management of the pest at field level is not easy. However, by adopting the following integrated plant protection measurers the mite population could be reduced considerably:

- Adoption of phytosanitary measures in coconut gardens such as cleaning the crown of the palm, keeping the plantation clean and burning of all immature nuts fallen due to mite infestation.
- Spraying biopesticides on the bunches
  - 2% neem oil - garlic emulsion (20ml neem oil + 20g garlic + 5g bar soap in 1 litre water). Emulsion has to be prepared on the same day of application itself.
  - Other neem based pesticides at 0.004% Azadirachtin can also be used. If the pesticide formulation contains 1% Azadirachtin, 4 ml has to be used in 1 litre of water. Wherever spraying is difficult root feeding may be resorted to with Azadirachtin 5% formulation (7.5 ml + 7.5 ml water) or Azadirachtin 1% formulation (10 ml + 10 ml water).
- Spraying has to be done 3 times a year - December-February, April-June and September-October. While spraying, ensure that the spray fluid falls over the perianth region especially on button and tender nuts. On an average 1-1.5 litre spray fluid is required per palm. Care should be taken to harvest mature bunches before spraying.
- The fungus, *Hirsutella thompsonii* has been found to be an effective microbial agent for annihilating this pest (Plate 5.3).
- Following palm health care practices also may be adopted.
  - Recycling of biomass generated within the coconut system by vermi-compost method or by using lignin degrading fungus.
  - Raising green manure crops in the coconut basins (like sun hemp, cowpea, calapagonium etc.) and incorporation into the soil in the basin itself. These act as mulch during summer and slowly decompose and provide nutrients when incorporated into the soil.
  - Application of recommended dosages of fertilizers, in two split doses, as per the package of practices prevailing in the respective states.
  - Recommended level of irrigation during summer months i.e. 250-500 litres of water per tree per week (based on evapotranspiration in the given area).
  - Soil moisture conservation by following methods.
    i. Burial of coconut husk in the basin.
    ii. Mulching the basins (2 m radius) with coconut leaves/green manure/ green leaf manure.
    iii. Mulching with coir pith wherever available (2 m radius).

- Neem/garlic oil/soap emulsion has emerged as one of the potent measures for the control of mites.
- Dissolve 1 kg biocatch in 200 litres of water and keep it overnight. Filter the solution using a muslin cloth. To the filtrate add 500 ml of Nimbecidine. Spray the resultant solution on the frond and newly developing nuts at 15 days interval.

5.2.2.2. Leaf Beetle, *Brontispa longissima* (Coleoptera: Chrysomelidae)

i) Key Facts

- The coconut leaf beetle is one of the most damaging pests of coconut and other palms.
- Native to Indonesia.
- Extensive loss in recent years in Maldives, Myanmar and Indonesia.
- 80-90% of seedlings were found infested by the pest, damaging 40% of the leaf area in Kasargod, Kerala.
- 'Red alert' quarantine has been issued and thorough screening of baggage for ornamental palms at entry points into the country.

Scientists at research institutions across the State are constantly on the alert against the coconut leaf beetle, an invasive pest that has ravaged the crop in large tracts across several neighbouring countries. With improved air and sea connectivity with the Maldives, a hotspot of the pest, Kerala is considered a frontline State in the battle against the coconut leaf beetle (*Brontispa longissima*). The Indian Council of Agricultural Research (ICAR) has classified it as a very destructive pest posing an immediate threat for States like Kerala.

Once the pest establishes a beachhead in Kerala, it could blight the coconut crop, triggering a crisis worse than the dreaded Eriophyid mite that had destroyed tens of thousands of palms across the State. It would be difficult to eradicate it.

Countries such as India, Bangladesh and Sri Lanka, which are the major coconut growers, are at high risk because neighbouring countries including Myanmar and Maldives are already infested with *Brontispa longissima.*

ii) Distribution:

The leaf beetle is a native of Indonesia and Papua New Guinea, including the Bismarck Archipelago. It is currently distributed in Australia, many Pacific Islands, Malaysia, Singapore, Cambodia, Laos, Thailand, Vietnam, the Maldives, Philippines, Myanmar and China.

iii) Damage:

The larvae and adults of the beetle feed on the soft tissues of the young leaves (Plate 5.4), resulting in stunted growth of the tree and reduced nut production. Affected leaves dry up, resulting in stunting of the palm and reduced nut production. Prolonged attacks on young palms can lead to their death. Millions of palms have been damaged and substantial yield loss has been reported from the affected countries.

iv) Host Range:

The beetle attacks more than 20 palm species with coconut (*Cocos nucifera*) being the most favoured host. Other hosts include Royal palm (*Roystonea* sp.), Alexandra palm

(*Archontophoenix alexandrae*), Sago palm (*Metroxylon sagu*), California fan palm (*Washingtonia filifera*), Mexican fan palm (*W. robusta*), Bottle palm (*Hyophorbe lagenicaulis*), Chinese fan palm (*Livistonia chinensis*), Madagascar palm (*Chrysalidocarpus lutescens*) and Areca nut palm (*Areca catechu*).

v) Biology:

The eggs are brown and flat (1.4 mm long and 0.5 mm wide), commonly laid in longitudinal rows (surrounded by debris and excrement) in the unopened leaflets of both young and mature palms. They hatch in 3 - 7 days to form larvae that are white with two pincer-like spines at the rear end of the body. Older larvae have an average length of 8 - 10 mm. Larvae undergo four to six larval instars in a period of 30 - 50 days. The pupae are yellowish-white and measure 9 - 10 × 2 mm. The pupal period is four to six days. The whole cycle from egg to adult takes 5-9 weeks (Plate 5.5). The adult leaf beetle is fully mature two weeks after emergence from the pupa and lives for two to three months.

vi) Spread:

The coconut leaf beetle spreads mostly through the movement of infested palms. Its natural spread is very slow since the beetles cannot fly long distances. Shipments of ornamental palms from infested countries have been the main source of spread within the Asia-Pacific region.

vii) Management

a) **Mechanical Methods:** Blockading and cutting of coconut palms up to three kilometres from the infestation spot are done to prevent the beetle from spreading. The pest is also controlled by pruning, clean culture and proper disposal of infested coconut palms and parts thereof.

Since the spread of the beetle is mainly through the movement of beetle-affected palms, all palms meant for transportation from known areas of infestation should be checked to make sure they are beetle-free. A simple inspection of the young leaves in the throat of the palm will be sufficient. To avoid further spread, non-infested countries in the region should adopt strict quarantine measures to control the import of plant materials, soil and any organic materials from infested countries.

b) **Chemical Methods:** Several insecticides including imidacloprid, dieldrin, aldrin, phosdrin, aldicarb, dichlorvos, fenthion, monocrotophos, chlorfenvinfos, idiofenphos, trichlorophon, quinalphos, deltamethrin, dimethoate, cypermethrin, diazinon, azinophos, methidathion and chlordane are being used to control the coconut leaf beetle. Some insecticides are also injected into the trunk of infested palms. However, the effect of these treatments lasts only for 3-4 months. Repeated applications may be impractical and uneconomic and cannot be used as a long-term control measure. In China, hanging insecticide bags on infested palms has been attempted successfully to check the spread of the beetle.

c) **Biological Methods:** Two parasitoids of coconut leaf beetle viz., *Tetrastichus brontispae* and *Asecodes hispinarum* (Plate 5.6), have been successfully used in several countries to control the beetle. Use of the entomopathogenic fungus, *Metarrhizium anisopliae* is also promising (Fig. 5.6).

5.2.3. Oil Palm, *Elaeis guineensis*

5.2.3.1. Rhinoceros Beetle, *Oryctes rhinoceros*

There are indications of shift of insect pests of plantation crops to new crops and new areas. The rhinoceros beetle, *Oryctes rhinoceros* is an established pest of coconut palms in India. Quite recently, it has become a problem on oil palm in southern states of India. The incidence of the pest in oil palm plantations closer to natural forest areas has been noticed at many places.

i) **Damage:** The adult beetle feeds on the softer tissue of the rachis, resulting in snapping off of the fronds and spears at the feeding sites. *O. rhinoceros* adults feed in the crown region of oil palm (Plate 5.7). On oil palms they bore through petiole bases into the central unopened leaves. This causes tissue maceration and the presence of a fibrous frass inside and at the entrance to the feeding hole is an indication of its activity within (Catley, 1969). A single attack may be followed by others on the same palm (Barlow and Chew, 1970; Young, 1975). These attacks subsequently produce fronds which have wedge-shaped gaps or the characteristic V-shaped cuts to fronds (Wood, 1968; Sadakathulla and Ramachandran, 1990).

ii) **Biology:** *O. rhinoceros* eggs are laid in rotting palm materials including trunks, tree stumps, compost heaps, dung hills, saw dust or garbage dumps. The eggs hatch after 8-12 days and the entire larval stage is spent inside the breeding medium. Development of the larva takes 80-200 days: first instar takes 10-21 days, second instar takes 12-21 days and third instar takes 60-125 days. The beetle then enters the prepupal stage of about 8-13 days, subsequently pupating within a pupal chamber made from the food substrate. The pupal stage lasts 17-30 days. Adults may live up to 6 months or more (Bedford, 1980; Khoo *et al.*, 1991).

iii) **Management:** A combination of cultural, chemical and biological methods is now being used in an integrated manner to successfully manage outbreaks of the pest. Field sanitation and elimination of breeding sites are essential components of the pest management operation. This pest can be suppressed by using pheromone traps (Plate 5.8) and the virus, *Baculovirus oryctes*. The entomopathogenic virus, *Oryctes rhinoceros nudivirus* (OrNV), which was originally discovered in Malaysia (Huger, 1966), attacks both the larval and adult stages of *O. rhinoceros* (Fig. 5.8). Oryctes larvae also succumb to attack by the green entomopathogenic fungus, *Metarrhizium anisopliae* (Sundara Babu *et al.*, 1983). Bacterial pathogens which have the potential to control larval stages of *Oryctes* were identified as *Acinetobacter calcoaceticus* (Kannan *et al.*, 1980), *Pseudomonas alcaligenes* (Gopal and Gupta, 2002), *Bacillus thuringiensis* and *B. popilliae* (Sundara Babu *et al.*, 1983; Norman *et al.*, 2005).

5.2.4. Cashew, *Anacardium occidentale*

5.2.4.1. Tea Mosquito Bug, *Helopeltis antonii*

Tea mosquito bug, *Helopeltis antonii* is a serious constraint in cashew (west coast-Kerala, Karnataka, and east coast-Tamil Nadu). Cashew tracts of Andhra Pradesh, West Bengal and Odisha are free from this pest. The hot spot areas of the bug across the cashew tract of the whole country have been demarcated, taking into consideration the optimum temperature during flushing and following stages. The pest may spread to new areas under current scenario of climate change and states like Andhra Pradesh, West Bengal and Orissa may come under this pest attack in the changed situation (Gupta *et al.*, 2009).

i) **Damage:** Tea mosquito bugs, also known as *Helopeltis* bugs or mirid bugs, are the most important and serious pests of cashew. These bugs are slender, delicate insects, about 7-10 mm long with long legs and antennae, the antenna being nearly twice as long as the body. The females are red and the males are brown to yellowish red. They lay eggs inserted into the soft tissue near the tips of flowering or vegetative shoots. Nymphs (immature bugs) are yellowish in colour. Both adults and nymphs feed on young leaves, young vegetative and flowering shoots, and developing fruits.

Feeding on the stalks of the tender shoots causes elongated green lesions, sometimes accompanied by exudation of gum. Severely damaged shoots show die-back due to the effect of bug saliva in combination with fungi, which enter the plant tissue through the feeding lesions (Plate 5.9); the subsequent development of numerous auxiliary buds causes a bunched terminal growth known as 'witches' broom. In case of serious infestations, the trees may appear as if scorched by fire.

Attacked leaves are deformed and show angular lesions, particularly along the veins, which may drop off, so that the leaves appear as if attacked by biting insects. Bug feeding on developing apples and nuts causes brown sunken spots (Plate 5.10). The growth of trees is seriously retarded and fruit formation of attacked flowering shoots is reduced. The production loss from the tea mosquito bug is estimated to be about 30 per cent.

ii) **Management:** Intensive removal of stalks during plucking is recommended to reduce the incidence of this pest. Weeds offer excellent hiding places and serve as alternate hosts for *Helopeltis*. Growth of weeds and wild host plants in and around cashew fields may be controlled and this will help to reduce the incidence of pest.

The following practices have been recommended for the management of tea mosquito bug:

- Monitor the crop regularly. *Helopeltis* attack occurs very suddenly and great vigilance is very important to control this pest, particularly during the rainy season or when water is available leading to flushing (production of young shoots) when *Helopeltis* populations normally build up.
- Conserve natural enemies. Weaver ants build nests on cashew trees providing good protection against this and other bug pests (Plate 5.11).
- Do not interplant cashew with crops that are host for *Helopeltis* bugs, such as cotton, tea, sweet potato, guava and mango.
- The pest can be easily controlled by spraying 0.05 % endosulfan (15 ml in 10 1itres of water). Three sprays should be given for complete control of the pest. The first spray of endosulfan should be given in October-November when the new flushes emerge, second of systemic insecticides like monocrotophos and dimethoate in December-January during emergence of inflorescence, and followed by third spray of endosulfan in February at the time of fruit set.

5.3. TUBER CROPS

5.3.1. Taro, *Colocasia esculenta*

5.3.1.1. Corm Borer, *Aplosonyx chalybaeus*

Corm borer, *Aplosonyx chalybaeus* is a serious, univoltine pest of colocasia (taro) in north eastern hill states and hilly regions of West Bengal. Infestations of *A. chalybaeus* were recorded in Darjeeling, Kalimpong in West Bengal.

Corm borer is a regular and endemic pest causing 20–30 % damage to the foliage and 80–90 % to the corms resulting in severe losses to the tribal farmers of the eastern Himalayan region of India. The pest is found up to 3,000 m MSL.

i) **Damage:** Soon after hatching, the white tiny grubs (5-7 mm) bored into the shoot and consumed the developing corm resulting in the death of the plant. The grubs overwinter in the corms as the temperatures during winter in the region ranges between 0°C to 3°C. The damaged plants wither, wilt and became yellow. The dead plants emit foul smell. The adults consume the leaves by damaging up to 20-30%. The adults make circular holes (trenches) of different sizes ranging between 2-3 cm on the leaf (Plate 5.12) and the damaged leaf exuded milky and waxy latex. Number of holes was recorded more on the edges of the leaf than inside.

   Beetles emerged with the start of the first summer showers in the region during middle of May. However, peak adult activity has been noticed from the month of June till August with peak populations (5-6/plant) during July first week. Damaged plants withered, wilted and became yellow. The local cultivars with purple or pink colouration on the leaves were less preferred by *A. chalybaeus*. A maximum of 70-80 grubs were found per plant. One generation per year was recorded. Several wild species of taro *viz.*, *Colocasia esculenta* var. *antiquorum*, *C. esculenta* var. *esculenta*, *C. esculenta* var. *sylvestris*, *C. affinis*, *C. fallax* in Sikkim, up to 3000 meters altitude, were infested by *A. chalybaeus*.

   Adults are of bright metallic in colour and emerge during June–July. They feed on leaves making circular trenches of 2.5 cm size (Fig. 5.12). They hide in the leaf sheath and each plant harbours 3–15 beetles. Crop losses are up to 50–60 % during severe infestation.

ii) **Biology:** The adult female lays eggs in groups of 20–100 in shoot leaf sheath above the ground. Soon after hatching, the young ones bore into the shoot and consume the developing corm resulting in the death of the plant. A maximum of 75 grubs were found per plant.

iii) **Host Range:** Wild relatives of taro are found in the eastern Himalayan region beginning from lower to higher altitudes. Successful establishment of the wild taro species cumulated the pest problem of corm borer.

iv) **Management:** The adults are picked and destroyed by farmers and sometimes eaten by some of the tribes in the region. The corm borer incidence is reduced when taro is intercropped with ginger, sweet potato, yam, and maize than the monocrop. Preliminary experiments suggested that entomopathogenic nematodes and fungi (*Beauveria bassiana*) cause significant increase in mortality of both adults and grubs (Rajasekhara Rao 2013). Earwigs are also found feeding on eggs of corm borer (Plate 5.13).

5.4. CONCLUSIONS

The severity of pest problems has been changing with the developments in agricultural technology and modifications of farming practices. There were outbreaks of berry borer, *Hypothenemus hampei* on coffee; tea mosquito bug, eriophyid mite, *Aceria guerreronis* and leaf beetle, *Brontispa longissima* on coconut; *Helopeltis antonii* on cashew; the rhinoceros beetle, *Oryctes rhinoceros* on oil palm in plantation crops. Among tuber crops, corm borer, *Aplosonyx chalybaeus* on taro has emerged as a serious pest. There is an immediate need to redefine management strategies for the above pests in plantation and tuber crops.

6

Emerging Insect Pest Problems of Forest Crops

6.1. INTRODUCTION

Eucalyptus gall wasp (EGW), *Leptocybe invasa* was first reported in 2001 in Mandya District of Karnataka and Pondicherry in 2002. It was first noticed in southern districts of Kerala, including Thiruvananthapuram district in 2005 and by 2006 spread to all districts of Kerala and Karnataka and also recorded from few districts in Maharashtra (Iasaac and Pillai, 2006). Nearly 60 per cent gall wasp damage on *Erythrina* plants was observed in Wynad District of Kerala in 2006 (Jacob and Devasahayam, 2008). The Erythrina gall wasp was first reported damaging *Erythrina* sp. in Mauritius and Singapore in 2003 (Kim *et al.*, 2004). First reported in Hawaii (July 2005) and quickly spread to all of the Hawaiian Islands. It is a major invasive pest on *Erythrina* spp. in black pepper plantations of Kerala and Karnataka, as it was widely used earlier as a live standard for trailing black pepper. It induces galls in the leaflets and petioles.

Eucalyptus gall wasp might have been introduced from Australia through planting material. Accidental transportations of eucalyptus wasp infested seedlings into uninfected areas for plantation purpose increased the spread to new areas. By 2007, the eucalyptus gall wasp has spread to Tamil Nadu, Andhra Pradesh, Kerala, Pondicherry, Gujarat, Madhya Pradesh, Uttar Pradesh, Maharashtra, Goa and Delhi (Mutitu *et al.*, 2008). It is reported from 22 countries. Severe damage was observed in 2007-08, and infested fresh seedlings were destroyed in huge numbers to avoid its spread. Eleven species of Eucalyptus are hosts for this pest. In India, this pest poses threat to 8 million ha of eucalyptus plantation, paper industry and forest areas. Control of gall infestation in nurseries is difficult even after repeated sprays. Classical biological control may be an ideal solution.

6.2. CORAL TREE, *Erythrina spp.*

Erythrina species is native to India, Malaysia, and Indonesia. *Erythrina variegata* and *E. variegata orientalis* are the most common species.

6.2.1. Gall Wasp, *Quadrastichus erythrinae* (Hymenoptera: Eulophidae)

6.2.1.1. Present Distribution:

The pest is distributed in tropical countries in Africa and Asia. In recent years, it is also found in Australia, Hawaii, Florida, Japan, etc.

6.2.1.2. Anticipated Effect of Climate Change:

Climate change is likely to expand its distribution range and cause serious damage to *Erythrina* trees (Fabaceae) by inducing galls on twigs and leaves.

Erythrina, an ornamental plant used as living fence for black pepper and betel vine has been infested with gall wasp, *Quadrastichus erythrinae* which is considered as one of the most swiftly invasive and destructive arthropod pests to sweep across the world, posing imminent threat to native *Erythrina* species across the world. In India, this pest was noted in Maharashtra and Karnataka during 2004 (Kore, 2006) and reported during 2005 in a serious proportion in Kerala (Faizal *et al.*, 2006). This pest infestation led to Erythrina mortality, resulting in the collapse of the entire canopy of black pepper vines (Jacob and Devasahayam, 2010).

Das and Talukdar (2010) reported its infestation from West Bengal in 2008 attacking *Erythrina variegata* in various proportion and subsequently it was found that the insect was progressing its way northwardly. Galls are induced on developing shoot, petiole, rachis, leaflet and flowers. Adult females lay eggs inside young plant tissue. The larvae develop and pupate within the gall tissue. After pupation within the gall, adult wasps escape through the holes cut by them. Obvious swelling and deformation can be seen on infected tissue parts. Severe infestations may lead to death of the tree. Only biological control agents from its native range could be useful to control this pest.

6.2.1.3. Symptoms:

The damage is severe. Trees are defoliated by the gall wasp and often die as cycles of regrowth and attack result in loss of reserves (Plate 6.1). *Erythrina* seedlings are killed by infestations and this threatens stands of the tree. For example, in Hawaii about 95% of the *Erythrina sandwicensis* (*wilwili* tree) and *Erythrina variegata* - important trees in the endangered lowland dry forest have been killed by the gall wasp.

6.2.1.4. Life Cycle:

The wasp lays its eggs in the youngest leaves and stems. As the larvae hatch they cause the galls to develop. The leaves become swollen and deformed, and look unlike the leaves of healthy trees. Repeated attacks cause dieback and death.

The larvae pupate in the swollen tissues, and the adults cut holes in the leaves and stems to emerge. The wasps are 1-1.5 mm in length, with the female being larger than the male. The life cycle is completed in about 20 days.

6.2.1.5. Spread:

Spread over short distances is by flight, or the wind, and perhaps on clothing. Spread over long distances is thought to be on fallen leaves perhaps on boats or planes.

6.2.1.6. Management

i) **Chemical Control:** Drenches of imidacloprid on some small heavily infested *E. variegata* did not work as trees became more intensely galled and died. A large tree may have enough reserve that it can stand one or more defoliations and recover after insecticidal treatments.

Imidacloprid applied systemically as a root drench or injected through trunks was effective against EGW. One of the most effective trunk injection systems evaluated was the Arborjet system (arborjet.com); it performed consistently and allowed for the most volume of liquid to be injected into a trunk through the fewest locations (Plate 6.2). Imidacloprid was very persistent within the leaves and can provide season- or year-long control.

ii) **Biological Control:** After extensive search for parastioids in Africa and testing, scientists in Hawaii selected *Eurytoma erythrinae* to have the potential to bring the gall wasp outbreaks under control. Tests showed that it was a specific parasitoid of the gall wasp. Releases began in 2008, and have shown a substantial decrease in gall wasp populations. Importantly, *Erythrina* trees recovered with the production of healthy leaves and full canopies, although flowers of *Erythrina sandwicensis* still show damage.

iii) **Cultural Control:** It is important not to move *Erythrina* plants between localities where the gall wasp is present and where it has not yet established.

6.3. EUCALYPTUS, *Eucalyptus teriticornis*

6.3.1. Gall wasp, *Leptocybe invasa* (Hymenoptera: Eulophidae)

Leptocybe invasa, commonly called Blue Gum Chalcid, is a newly described gall inducing wasp species currently spreading in many countries and causing damage to young eucalyptus plantations and nurseries. Galls induced by this wasp can cause serious injuries to trees, thereby weakening them. Information on the taxonomy, distribution, biology and economic impacts of the Blue Gum Chalcid are still meager.

6.3.1.1. Present Distribution:

Believed to be a native of Australia, where its distribution is still unknown, the wasp has spread to most eucalyptus-growing countries in Africa, Asia and the Pacific, Europe, North America and the Near East. In the Asia–Pacific region it is a threat in countries like Cambodia, Thailand, Vietnam, India and New Zealand. Outside its native range, the wasp was first recorded in the Middle East in the year 2000; in Uganda, Kenya and India in 2002; in Tanzania in 2005; and in South Africa in 2007.

6.3.1.2. Symptoms:

L. invasa causes bump-shaped galls on the midrib, petioles and stems of young eucalypt trees, young coppice and seedlings (Plate 6.3). Severely affected trees show leaf fall, gnarled appearance, loss of vigour, stunted growth, dieback and stem and petiole gall caused by *L. invasa* on eucalyptus seedlings and eventually tree death. The leaves of intensively growing trees may carry over 50 galls per leaf.

6.3.1.3. Damage:

The adult wasps lay eggs inside tender leaves, especially the epidermis of the upper side of newly developed leaves, on both sides of the midrib, petiole and in the parenchyma of twigs. After hatching, the larvae remain in a cavity formed within the plant tissues and feed on the plant sap and the injury thus caused results in the formation of galls (Plate 6.4). The larvae develop inside the galls; adults then emerge leaving round exit holes. The attack by the wasp takes place within 1-2 weeks of the bud breakout. In Israel, planting of *E. camaldulensis* was stopped because of extensive attacks by the wasp. In India, the wasp poses a threat to an estimated 8 million ha of eucalyptus plantations.

6.3.1.4. Host Range:

The wasp has a relatively narrow host range. It attacks the leaves and stems of young eucalyptus trees and seedlings. Other host species include *Eucalyptus saligna, E. grandis, E. deanei, E. globulus* s sp. *globulus, E. nitens, E. botryoides, E. camaldulensis, E. gunnii, E. robusta, E. bridgesiana, E. viminalis* and *E. tereticornis.*

6.3.1.5. Biology:

The female adult wasp is 1.1 - 1.4 mm long. The body is brownish in colour with a blue to green metallic shine. Fore coxae are yellow; mid and hind coxae are brown. The scape of the antennae is yellow, with the rest of the segments brown. Males are unknown except for one record from Turkey. Females reproduce by parthenogenesis. Larvae are minute, white and legless.

6.3.1.6. Spread:

The movement of eucalypt seedlings is one of the ways by which the wasp spreads. The adult wasp is very small and hence cannot fly long distances.

6.3.1.7. Management:

No effective control measures are currently available. The following interim measures are recommended to manage the wasp problem:

- Periodic monitoring in nurseries and plantations can reveal beginning of Eucalyptus wasp infestations.
- Severely affected plants in nursery to be destroyed by burning or plant parts may be pruned and burnt especially unwanted coppice shoots which serve as breeding base for gall insects.
- Production, supply and planting of susceptible clones to be restricted or stopped till the problem is contained.
- Singling or pruning of coppice to be carried out as soon as the visible symptoms of infestations are noticed.
- Variation in gall incidence warrants screening of available germplasm to identify gall resistant/tolerant candidates for plantation programmes (Jacob and Ramesh Kumar, 2009).
- A "Poly-monoculture" instead of "Pure monoculture" of tree species will decrease exotic pest invasions and loss of forest resources. Periodic monitoring of infestation in nurseries and plantations is important.
- Mechanical removal of affected plant parts and subsequent burning.
- Utmost caution is warranted to prevent transportation of infected seedlings to pest-free areas.
- Avoid production and planting of highly susceptible clones.
- Use of granular insecticides in nursery, sticky traps, to trap the adult than larvae, identification of resistant clones.

- Application of systemic insecticides like Rogor or Metacid 50 (at 2 ml/ litre water) on foliage at fortnightly intervals is effective.
- Imported parasitoids of eucalyptus gall wasp are under quarantine tests at NBAIR, Bangalore.
- In Australia, *Aprostocetus* sp. (Plate 6.5) and *Megastigmus* sp. (Plate 6.6) play a significant role in controlling *L. invasa.*
- Recently, Israel introduced two parasitoids, namely, *Quadrastichus mendeli* (Plate 6.6) and *Selitrichodes kryceri* (Plate 6.6), from Australia which may offer a potential for biological control.

6.4. OAK, *Quercus* spp.

6.4.1. Processionary Moth, *Thaumetopoea processionea*

6.4.1.1. Present Distribution:

Central and southern Europe to Asia Minor.

6.4.1.2. Anticipated Effect of Climate Change:

The range of this species appears to be extending north. *T. processionea* re-appeared at high densities in Noord-Brabant in the southern half of the Netherlands (Stigter and Romeijn, 1992). It was caught for the first time in Denmark in 1996 (Skule and Vilhelmsen, 1997). It was found at several locations in London in 2006. It is present in the provinces of Antwerp and Limburg in Belgium (Vandenbussche, 1997); and is now resident in the Channel Islands (Kimber, 2008).

The oak processionary moth's northward progression was due to improved synchrony of egg hatch and reduction of late frosts.

6.5. PINE, *Pinus* spp.

6.5.1. Processionary Moth, *Thaumetopoea pityocampa*

6.5.1.1. Present Distribution:

Mediterranean countries

6.5.1.2. Anticipated Effect of Climate Change:

The pine processionary moth is expanding its geographical range in Europe, as a consequence of enhanced winter survival under a warmer climate (Buffo *et al.,* 2007). Increasing winter temperatures will also affect the severity of this species on relic pines in upland regions of southern Europe (Hodar *et al.,* 2003).

6.5.2. Mountain Pine Beetle, *Dendroctonus ponderosae*

The massive population build up of mountain pine beetle (*Dendroctonus ponderosae*) and its northward progression in the North American Pacific Northwest has most likely been due to a combination of warmer winter temperatures, reduced episodes of underbark mortality and increased drought which weakened the trees.

6.5.3. Southern Pine Beetle, *Dendroctonus frontalis*

Northward shift predicted based on both shifts in its pine hosts and in limits of lethal winter temperatures (based on <-16°C). Tree defences are also implicated.

6.6. CONCLUSIONS

The incidences of several insect pests on forest crops have shown an increasing trend. There were outbreaks of gall wasp (EGW), *Leptocybe invasa* on eucalyptus trees; Erythrina gall wasp, *Quadrastichus erythrinae* on coral trees; processionary moth, *Thaumetopoea processionea* on oak and pine trees; and mountain beetle, *Dendroctonus ponderosae* and southern beetle, *Dendroctonus frontalis* on pine trees. Crop pests should be overcome to meet future needs of forest products.

Section III
DISEASE PATHOGENS

7

Emerging Disease Problems in Field Crops

7.1. INTRODUCTION

The truly scary possibility is that a new or re-emerging plant disease could decimate one of the few crops — rice, wheat, corn — that the global diet is based on. We have already had a few close calls. Wheat rust, which is caused by a fungus, devastated wheat crops in Africa, and is poised to spread to other major wheat-producing countries. The rice blast fungus, present in scores of countries, has migrated to wheat, and has become a scourge of farmers in Brazil. Since its first report in Brazil in 1985, wheat blast, caused by *Magnaporthe oryzae* (anamorph: *Pyricularia oryzae*), has become increasingly important in South America, where the disease is still spreading, causing yield losses between 40 and 100% (Kohli *et al.*, 2011). The rice blast, *M. grisea*, which was hither too commonly observed on wheat and black oats, is found on one commercial field of white oat cultivar IAC 7 severely attacked in Assis, in the state of São Paulo (Marangoni *et al.*, 2013).

In 1972, Karnal bunt, a fungal disease of wheat, durum, rye and triticale (a hybrid of wheat and rye) caused by *Tilletia (Neovossia) indica*, became of global importance following its discovery in Mexico. It was reported for the first time from the USA in 1996 (Bonde *et al.*, 1997). Recent interest in the disease has surrounded the potential for its use as a biological weapon and exports of wheat from many regions with Karnal bunt have been banned, leading to severe economic loss for affected countries (Gewin, 2003).

Oerke (2006) updated the global loss data due to disease pathogens and virus diseases for major food and cash crops for the period 2001-03 (Table 7.1).

Table 7.1. Current global losses (%) due to disease pathogens and virus diseases in major food and cash crops

Crop	Pathogens	Viruses	Total
Cotton	7.2	0.7	7.9
Maize	5.8	5.2	11.0
Potato	14.5	6.6	21.1
Rice	10.8	1.4	12.2
Soybean	8.9	1.2	10.1
Wheat	10.2	2.4	12.6
Average	**10.0**	**2.5**	**12.5**

Source: Oerke (2006)

7.2. GROUNDNUT, *Arachis hypogea*

7.2.1. Peanut Bud Necrosis Virus

Peanut bud necrosis disease (PBND) is an important disease of groundnut in south and south-east Asia. It was shown to be economically important in parts of Tamil Nadu, Karnataka, Andhra Pradesh, Maharashtra, and Uttar Pradesh in India. Peanut bud necrosis disease is also currently recognized as economically important in parts of China, Nepal, Sri Lanka, and Thailand. It is caused by *Peanut bud necrosis virus* (PBNV) and vectored by *Thrips palmi* in a propagative manner. It was first reported in India in 1949, and it became economically important during the late 1960s when incidences of up to 100% were recorded in many peanut-growing regions of the country. Incidence of PBND has ranged from 5 to 80%, and yield losses of up to 50%, worth more than $89 million in India alone, have been reported (ICRISAT, 1992).

7.2.1.1. Symptoms:

The primary symptom of PBND on peanut is a mild chlorotic mottle or specks on young, quadrifoliate leaves, which develop into chlorotic and necrotic rings and streaks (Plate 7.1). Necrosis from the leaf extends to the petiole and terminal bud and results in necrosis of the terminal bud (Plate 7.1). Secondary symptoms include stunting, auxiliary shoot proliferation, and malformation of leaflets (Plate 7.2). The virus can cause severe crop losses, especially when the plants are infected before they are a month old. Seeds from such plants are small, shrivelled, mottled, and discoloured (Plate 7.2). Plants infected at a later stage produce normal-sized seeds, but testa on such seeds are often mottled and cracked.

7.2.1.2. Causal Virus:

In 1992, the virus causing PBND was identified as a distinct tospovirus and named PBNV. With ELISA as well as Western blots, PBNV was shown to be serologically distinct from Tomato spotted wilt virus (TSWV) and Impatiens necrotic spot virus (INSV) (Reddy *et al.*, 1992).

7.2.1.3. Transmission

i) **Sap Transmission:** Peanut bud necrosis virus can be transmitted by mechanical sap inoculations if care is taken to extract the virus only from young infected leaflets with primary symptoms.

ii) **Thrips Transmission:** Transmission Studies with *Thrips palmi, Frankliniella schultzei* and *Scirtothrips dorsalis* showed that only *T. palmi* (Plate 7.3) transmitted PBNV. *T. palmi* acquired the virus as larvae, requiring a 5-min acquisition period, but only transmitted it as adults. Adults needed an inoculation feeding period of more than 1 hr to transmit, and the majority transmitted the virus throughout their life.

In field situations, a virus-susceptible peanut genotype had higher densities of *T. palmi* than resistant genotypes. In peanuts, *S. dorsalis* was the dominant foliar species, *T. palmi* was on foliage and flowers, and *F. schultzei* was mostly on flowers. Several common weed species in peanut were infected with PBNV and were colonized by *T. palmi.*

7.2.1.4. Management:

Options available for management of PBNV include:

- Vector control
- Host resistance
- Cultural practices
  - Date of planting
  - Crop density
  - Crop rotation
  - Host-free period
  - Rouging – Removal of symptomatic plants within 45 days post-planting

i) Challenges in Thrips Vector Control

- Polyphagous and survive in different habitats and climates.
- Can reproduce quickly and have many generations in a given season.
- Can 'overwinter' on a broad range of plant species.
- Efforts to control vector thrips with insecticides have been mostly unsuccessful.

Several cultural practices such as adjustments to sowing dates, sowing at the recommended rate, adopting measures to maintain plant population, intercropping with fast-growing cereal crops such as maize and pearl millet can reduce the incidence of PBND. These practices have been shown to reduce infestation by *T. palmi.*

Rouging of infected plants, especially during early stages of plant growth should be avoided because this practice creates gaps in the field and can increase PBND incidence.

Excellent progress has been made in the identification of sources of field resistance to PBND. Some of the field resistant genotypes such as ICGV 86388, show resistance to PBNV and less colonization by vector thrips compared with susceptible genotypes (Buiel *et al.*, 1995; Dwivedi *et al.*, 1995). Cultivars such as ICGS 11, Kadiri 3, and ICGS 44 are field resistant to PBND.

ii) Designing IPM packages for PBNV management

- Seed treatment with *Pseudomonas fluorescens* at 10g/ kg and *Trichoderma viride* at 4g/ kg of seeds.
- Soil application of neem cake at 250 kg/ ha.
- Soil application of *P. fluorescens* at 2.5 kg/ ha.
- Selection of healthy seedlings for transplanting.
- Rouging virus infected plants up to 45 days post-transplanting.
- Installation of yellow sticky traps.
- Spraying neem formulations/ neem seed kernel extract (need-based).

Early planting, increased plant density and use of the cultivar Robut 33-1 appear to minimize losses due to bud necrosis in Hyderabad, but need to be investigated in different regions of India before they can be generally recommended. Eliminating primary sources of infection does not appear to be feasible and rouging of infected plants is unlikely to be effective in controlling virus spread, as most infection is from sources outside the crop.

Groundnuts planted with green gram or black gram showed a very high incidence of bud necrosis and these plants are not recommended for inter-cropping because they are seriously affected by PBND.

The most effective method of controlling PBND would be to grow resistant cultivars. However, attempts to locate sources of high resistance in *A. hypogaea* have so far been unsuccessful. The only known source of resistance to PBND appears to be in wild *Arachis* spp. and these are now being tested.

7.2.1.5. Future Research:

Peanut bud necrosis virus and *T. palmi* have extremely wide host ranges. Therefore, the virus is a potential threat to cropping systems which include legumes, vegetables, and ornamentals. Some of these crops are grown under irrigation and protected with insecticides. These conditions are likely to result in a gradual build up of PBNV inoculum, leading to disease epidemics. Therefore, it is essential to closely monitor the incidence of PBNV in various cropping systems, which include highly susceptible hosts of the virus and the vector.

Since the field-resistant groundnut varieties are of medium duration, attempts should be made to breed early-maturing cultivars for environments where they are needed. To achieve this rapidly, transgenic groundnuts expressing PBNV genes could be developed. sRNA of PBNV has been fully sequenced and the coat protein gene located (Satyanarayana *et al.*, 1995) for utilization in the transformation and regeneration of groundnut.

Thrips are known to have several parasites and predators. They have not been tested for their effectiveness to reduce populations of *T. palmi.* It would also be useful to study the effect of a range of synthetic and natural insecticides on the parasites and predators of *T. palmi,* once they are identified.

Peanut bud necrosis virus is currently known to cause economic losses to many commercial crops other than groundnut. These include chilli, potato, tomato, tobacco, and early-maturing legumes such as green gram and black gram. Data generated for the management of PBNV on groundnut are likely to be applicable to these crops. Due to the specialized skills required for the detection of PBNV, the economic importance of PBNV in many other high-value crops has not been realized.

Variation among isolates of PBNV has not yet been fully investigated. This should be carried out because of its implication for the durability of host-plant resistance. Limited tests conducted in the case of groundnut (Buiel *et al.*, 1995) indicate that the resistance is effective in India in different eco-regions. These tests should be extended to PBNV hot spots in other South Asian countries.

Thrips palmi has recently been detected in large populations in many south-eastern states of USA. In these states, currently, TSWV is considered to be economically important. To date, there is no evidence of the occurrence of PBNV in the USA or in other countries where *T. palmi* may be occurring in large populations. The potential threat from PBNV therefore exists in such places, indicating the need to conduct surveys for the occurrence of PBNV.

7.2.2. Peanut Yellow *Spot Tospovirus (Tospovirus: Bunyaviridae)*

First reported on *Arachis hypogaea* from India (Anon, 1980).

7.2.2.1. Geographical distribution:

Reported from India and Thailand.

7.2.2.2. Symptoms:

The symptoms of peanut yellow spot virus (PYSV) include small chlorotic local lesions, which coalesce and then become necrotic (Plate 7.4). Chlorotic spots first appear along the midribs of young leaflets where the thrips, which spread the virus, feeds. The spots increase in size and join together to become irregular in shape as the leaves approach maturity. At this stage the colour of the spots is distinctively yellow (Plate 7.4).

7.2.2.3. Transmission:

The virus is efficiently transmitted in groundnut by an insect vector, *Scirtothrips dorsalis* (Thysanoptera) in a semi-persistent manner. Larvae could acquire the virus in 30 min and the maximum percentage transmission of 43.8% by individual insects resulted following two days AAP. Single adult thrip transmitted the virus after minimum IAP of 30 minutes. The percentage transmission (33.3%) increased linearly with an increase in IAP up to 1.5 days and maximum up to 55 h of IAP (36.1%). PYSV is persistently transmitted by thrips by more than 75% of their life span (Gopal *et al.*, 2010). Virus retained when the vector moults; not transmitted congenitally to the progeny of the vector, or by contact between plants; or by seed. The virus is transmitted by mechanical inoculation by grafting.

7.2.2.4. Host Range

i) Diagnostically Susceptible Host Species and Symptoms

- *Vigna unguiculata* - large chlorotic local lesions becoming necrotic.
- *Petunia hybrida* - necrotic local lesions.
- *Chenopodium quinoa* - chlorotic local lesions.
- *Datura stramonium* - chlorotic local lesions with necrotic rings.

ii) Susceptible Host Species

- *Arachis hypogaea*
- *Chenopodium amaranticolor*
- *C. quinoa*
- *Datura stramonium*
- *Glycine max*
- *Nicotiana benthamiana*
- *Petunia hybrida*
- *Phaseolus vulgaris*
- *Pisum sativum*
- *Vigna mungo*
- *V. radiata*
- *V. unguiculata*

7.2.2.5. Particle Morphology:

Virions unusually shaped - quasi-isometric, enveloped, 70-90 nm in diameter, and without a conspicuous capsomere arrangement.

7.2.2.6. Management:

To prevent spread of the virus, infected plants should be immediately removed away from neighbouring plants. Control of insects, especially thrips, is important to reduce spread of the virus by vectors.

7.3. WHEAT, *Triticum spp.*

7.3.1. Karnal Bunt, *Tilletia indica*

In 1972, Karnal bunt, a fungal disease of wheat, durum, rye and triticale (a hybrid of wheat and rye) caused by *Tilletia (Neovossia) indica*, attained global importance following its discovery in Mexico, and in 1996, it was reported for the first time in the USA (Bonde *et al.*, 1997). Recent interest in the disease has surrounded the potential for its use as a biological weapon and exports of wheat from many regions with Karnal bunt have been banned, leading to severe economic loss for affected countries (Gewin, 2003).

The main effect of extensive Karnal bunt is to reduce yield (Bansal *et al.*, 1984) and impart a fishy odour and taste to wheat flour, thus reducing the quality of the flour (Aujla *et al.*, 1980). Yield and quality losses are considered by many smut pathologists to be minor (Butler, 1990; Warham, 1986). However, since Karnal bunt is the subject of strict quarantines by several wheat-importing countries, *T. indica* can profoundly affect international trade of commercial grain and the movement of wheat germplasm (Warham, 1986). In 1983, the Animal and Plant Health Inspection Service (APHIS) placed restrictions on wheat coming from countries with Karnal bunt, recognizing that establishment of Karnal bunt in the United States could have major economic ramifications on U.S. wheat exports. The spread of Karnal bunt to the United States and its establishment therein could have placed the country at a marked disadvantage in the international wheat market as the first major wheat-exporting country to have Karnal bunt. As shown by responses following the discovery in March 1996 of Karnal bunt in Arizona, and later in California, the disease has generated considerable concern and debate both within and outside the country.

7.3.1.1. Symptoms:

The fungus causes a reduction in the length of ears as well as in the number of spikelets of bunted ears. Infected plants may be dwarfed. In general, *T. indica* rarely infects more than a few spikelets per ear and then the affected grains are not swollen. Oblong or ovoid sori, 1-3 mm diameter, develop, containing dusty, brown to black spore masses. These characteristically smell of decaying fish (trimethylamine) as do those of *T. tritici*, *T. foetida* and *T. contraversa* (OEPP/EPPO, 1996). Feeding studies have revealed no adverse health effects, but consumers can begin to taste and smell the fishiness when 3% or more of grain is affected.

The grain is partially destroyed, the attack starting at the hilum and running along the suture, leaving the endosperm intact and covered by the whole or partly ruptured seed coat (Plate 7.5). In the case of mild infection, only a black point just below the embryo towards the suture is apparent. In advanced attack, tissues along the suture and adjacent endosperm are replaced by spores. The glumes spread apart, exposing the infected grains, and both glumes and grains may fall to the ground.

7.3.1.2. Spread and Dispersal:

Natural spread can be substantial as teliospores can be carried over long distances and at high altitudes by wind and man-made elements like movement of people, agricultural products and machinery (Warham, 1986). Teliospores can pass through the digestive tracts of animals undamaged (Smilanick *et al.*, 1986), thus making it possible that the pathogen is distributed with farm manure. However, the main mode of international spread is on infected or contaminated wheat seeds. Marshall *et al.* (2003) suggest that, for movement from Mexico into USA, this may involve not only consignments, but small lots of wheat moving in automobiles, trucks and railway cars. Grain imports have also been targeted as possible avenues of introduction. Teliospores of the pathogen may also be transported in straw, i.e. dried up wheat plants with a few ears mixed up in it.

7.3.1.3. Management:

In field trials in India, a single protective application of propiconazole controlled the incidence of *T. indica* on wheat by 71.4%; three sprays reduced disease by 100%. Triadimefon and carbendazim controlled the disease by 87 and 84%, respectively (Aujla and Sharma, 1990). Carboxin + thiram, and chlorothalonil have been used as seed treatments in the USA and Mexico.

Beniwal *et al.* (2000) reported that soaking seeds in water could help eradicate teliospores of *T. indica* without any adverse effect on wheat seed germination. Soaking at 35°C for 12 hr was the most effective treatment.

For organic farming, seed treatment with extracts of *Cannabis sativa, Eucalyptus globulus, Thuja sinensis* and *Datura stramonium* are fully effective against *T. indica* in the field (Borgen, 2004).

The disease is controlled by the use of resistant cultivars in infested areas so that high levels of infection are seldom reached at present.

i) **Cultural Methods:** High nitrogen applications and excessive irrigation favour the disease (Warham, 1986). Crop rotation may help to control the pathogen, but its value is questionable because *T. indica* can survive for up to 4 years in the soil. To prevent the spread of *T. indica* into previously unaffected areas, the use of disease-free seed is essential. The movement of farm machinery and soil from contaminated fields may also be restricted.

ii) **Chemical Methods:** Chemical seed treatments have proved to be ineffective in killing the teliospores of *T. indica* on seeds of wheat, with the exception of mercurial compounds (Warham *et al.*, 1989) which are banned in most countries. Bleach, in combination with heat treatment, is effective (Smilanick *et al.*, 1989). Carboxin + thiram, and chlorothalonil have been used as seed treatments in the USA and Mexico.

Foliar sprays of fungicides may be used to control the air-borne inoculum of primary and secondary sporidia. Propiconazole was shown to be effective against natural infection in India (Singh *et al.*, 1999). In Pakistan, propiconazole and bitertanol reduced the disease by 79 and 67%, respectively (Chaudhry and Khan, 1990).

iii) **Host-Plant Resistance:** Intensive research has been carried out to breed resistant cultivars of bread wheat. It has been reported that all commercially available cultivars in India are susceptible to the pathogen (Warham, 1986). However, Gill *et al.* (1986a, b) reported two cultivars to be either tolerant or field resistant. Cultivars of bread wheat, durum wheat and triticale tested by Warham (1988) were all equally susceptible to inoculation at the boot leaf stage. In Mexico, a few lines of bread wheat showed moderate resistance (Singh *et al.*, 1989).

iv) **Integrated Management:** Integration of fungicidal seed treatment with foliar sprays of phytoextracts, bio-control agent and fungicide revealed that seed treatment with Tilt 25 EC (0.1%) and few sprays of Tilt 25 EC (0.1%), one at flag leaf stage followed by another at 50 per cent emergence of spikes and seed treatment of Tilt plus one spray of *Lantana camara* extract and one spray of Tilt 25 EC gave cent per cent disease control under field conditions. Seed treatment with Tilt 25 EC plus one spray of *Trichoderma viride* at flag leaf stage and another spray of *L. camara* at 50 per cent emergence of spikes gave 95.84 per cent disease control (Table 7.2) (Kumar and Singh, 2014).

Table 7.2. Integrated management of Karnal bunt of wheat

Treatment	Dose	Germination (%)	% Karnal bunt incidence	Disease control (%)	Yield (t/ha)
Tilt 25 EC (ST)	1mL/L	85.67 (69.03)	13.14 (3.45)	47.03 (66.47)	37.50
Lantana camara (ST)	250mL/L	95.00 (75.95)	14.11 (3.62)	43.13 (66.29)	36.53
Trichoderma viride (ST)	5g/kg	93.33 (73.10)	17.13 (4.05)	30.95 (63.30)	35.49
Tilt 25 EC (ST) + 1 spray of Tilt 25 EC	1mL/L + 1mL/L	91.00 (73.01)	2.50 (2.86)	89.92 (71.46)	40.62
L. camara (ST) + 1 spray of Tilt 25 EC	250mL/L + 1mL/L	86.67 (69.39)	3.50 (3.39)	85.89 (67.91)	40.23
T. viride (ST) + 1spray of *T. viride*	5g/kg + 5g/L	92.00 (73.56)	7.54 (4.97)	69.76 (56.61)	36.69
Tilt 25 EC (ST) + 2 sprays of Tilt 25 EC	1m/L + 1mL/L	85.67 (68.63)	0.00 (0.00)	100.00 (89.96)	42.25
L. camara (ST) + 2 sprays of *L. camara*	250mL/L + 250mL/L	89.00 (72.32)	4.52 (3.84)	81.85 (64.76)	37.96
T. viride (ST) + 2 sprays of *T. viride*	5g/kg+ 5g/L	91.33 (73.10)	6.50 (4.62)	73.80 (59.18)	36.87
Tilt 25 EC (ST) + 1 spray of *L. camara* + one spray of Tilt 25 EC	1mL/L + 250mL/L + 1mL/L	90.33 (72.76)	0.00 (0.00)	100.00 (89.96)	43.56

Treatment	Dose	Germination (%)	% Karnal bunt incidence	Disease control (%)	Yield (t/ha)
Tilt 25 EC (ST) + 1 spray of *T. viride* + 1 spray of *L. camara*	1mL/L + 5g/L + 250mL/L	86.00 (66.07)	1.03 (1.81)	95.84 (78.42)	41.66
Control	—	90.00 (72.89)	24.81 (9.05)	—	34.84
CD (P=0.05)	—	**4.24**	**0.12**	**0.83**	**2.43**

ST= seed treatment. Figures in parenthesis values are arc sine value

7.3.2. Blast, *Magnaporthe oryzae*

One of the most fearsome and intractable diseases on wheat in recent decades is wheat blast, caused by the Ascomycetous fungus *Magnaporthe oryzae* (anamorph *Pyricularia oryzae*). Until the 1980s, the blast disease was not known to affect wheat, a main staple crop critical to ensuring global food security. In 1985, the disease was first reported on wheat (*Triticum aestivum*) in Paraná State, Brazil (Igarashi *et al.*, 1986). In South America, blast is now a major threat to wheat production (Goulart *et al.*, 2007; Kohli *et al.*, 2011). Currently, wheat blast is widespread in South American wheat fields, affecting as much as 3 million hectares in the early 1990s and seriously limiting the potential for wheat cropping on the region's vast grasslands (Savannas).

The blast disease is the most important fungal disease on wheat (Prabhu *et al.*, 2003). The first epidemic was detected in Paraná State, southern Brazil. Wheat blast recently emerged as a serious disease in Paraguay and Bolivia (Duveiller *et al.*, 2010). It appeared in the USA in 2011. *M. oryzae* also infects barley, rye, and triticale in areas of central-western and southern Brazil (Anjos and Charchar, 2000). It also infects oat and barley. The rice-infecting isolates of *M. oryzae* are genetically distinct from wheat-infecting isolates and generally do not infect wheat (Maciel *et al.*, 2014; Chiapello *et al.*, 2015; Yoshida *et al.*, 2016).

The wheat blast has become widely distributed across most of the wheat-producing areas in Brazil over the last 25 years because of a lack of resistant cultivars and effective fungicides (Anjos *et al.*, 1996). Despite a continuous breeding effort over the past decades (Cruz *et al.*, 2010), the majority of available wheat cultivars are very susceptible to blast, though 'BR 18 Terena', 'BRS 229', and 'MGS 3 Brilhante' are considered tolerant under field conditions (Cruz *et al.*, 2010; Maciel *et al.*, 2008). However, the tolerance of these varieties has been erratic in different regions of Brazil (Urashima *et al.*, 2005; Urashima *et al.*, 1999). Wheat blast is currently considered the main obstacle to the expansion of wheat cropping in the Brazilian Cerrado areas (Duveiller *et al.*, 2010).

7.3.2.1. Distribution:

Wheat blast has remained restricted to South American countries, Brazil, Bolivia, Paraguay, and Uruguay (Kohli *et al.*, 2011). It appeared in the USA in 2011. However, a severe outbreak of blast in key wheat districts of south-western Bangladesh was observed (Callaway, 2016; Malaker *et al.*, 2016). It has been now noticed in 15% of wheat area in Bangladesh. Areas in the districts of Kushtia,

Meherpur, Chuadanga, Jhenaidah, Jessore, Barisal, Bhola and several others were affected by wheat blast (Malaker *et al.*, 2016). There were reports that wheat production was affected in 15000 hectares with resultant fall in production by 20%. Standing wheat crop was burnt in some areas. Based on the molecular characteristics, wheat blast isolate from Bangladesh was found similar to that of Brazil (Malaker *et al.*, 2016). So far wheat blast has not been sighted anywhere in India.

Experts had feared the possible spread of blast from Latin America to regions of Africa and Asia where conditions are similar. The consequences of a wider outbreak in South Asia could be devastating to a region of 300 million undernourished people, whose inhabitants consume over 100 million tons of wheat each year.

7.3.2.2. Economic Importance:

The production losses caused by wheat blast can vary from very low to 100% (Goulart *et al.*, 1990; Goulart *et al.*, 1992). The disease can occur on all above ground parts. Highest loss occurs when fungal infection occurs at the base of the rachis restricting the development of grains and killing of spike (Kohli *et al.*, 2011). First epidemic in Paraguay caused more than 70% losses (Viedma, 2005). Infection in low lands of Santa Cruz region of Bolivia in 1996 resulted in 80% yield reduction in wheat (Barea and Toledo, 1996). Government owned blast infected fields were burnt in Bangladesh (Callaway, 2016).

7.3.2.3. Symptoms:

Wheat blast infects all above ground parts of plant (Plate 7.6a). The most conspicuous symptom is head/spike infection (Plate 7.6b). Head infection can be easily confused with *Fusarium* head blight. Head infection can occur on the glumes, awns and rachis. Infected glumes support elliptical lesions with reddish brown to dark grey margins and white to light brown centre. Blackening of the rachis, lower nodes, shrivelling of grains, low testa weight have also been observed (Malaker *et al.*, 2016). The pathogen is known to produce non-host specific toxin pyricularin (Agrios, 2005). Depending on the place of infection on the head, drying of partial or full ear head takes place. Symptoms on heads can vary from elliptic lesions and bleached centres, spike bleaching, sterility and empty grains depending on time of infection. Since infection on spike blocks the translocation of photosynthates and nutrients to spike, therefore, results in partial or total spike sterility. Grain fills can occur in case of late infections but it may lead to seed-borne inoculum to the next crop. As plants grow older, lesions are less frequent. Lesions with white centre and of reddish brown margin on upper side, dark grey on the underside of the leaf can be observed on both young and old infected leaves. Infection on seedlings can be very damaging under high temperature and humidity and can lead to total plant death (Igarashi, 1990).

7.3.2.4. Disease Cycle:

Wheat and rye grass isolates of *Magnaporthe* exhibit almost same disease cycle as that of rice. The characteristic feature of this disease is the pyriform conidia (Plate 7.6) that gives rise to the disease. Presence of melamin in the appresoria, arms the conidia to build up a very high pressure essential to puncture the outer plant surface and gain access to the host tissues. Presence of a free film of water is essential during the infection process. After entry into host, the fungus colonizes for 4 days without any visible appearance. Thereafter, conidia are released and process is facilitated by water. In other words disease outbreak is hard to predict as by the time first visible symptoms, appear, blast is already established (Anon, 2013). Young and expanding leaves are more susceptible to spores of blast fungus. Conidia have been detected up to 1000 metre (Urashima *et al.*, 2007) away from the fields. Seed is the primary sources of inoculum to new areas (Greer and Webster, 2001). However, possibility of a grass strain jumping to wheat can also cause wheat blast (Callaway, 2016).

7.3.2.5. Host Range:

A wide range of host plants within the family Poaceae can provide a reservoir of primary inoculum for the wheat-infecting *M. oryzae*. Strains of *M. oryzae* isolated from wheat were able to infect Poaceous plants from six tribes and also produced mature perithecia when crossed with strains from *Eleusine coracana* (Chlorideae), *Brachiaria plantaginea* (Paniceae), and *Setaria indica* (Paniceae) (Urashima *et al.*, 1993). Despite the presence of rice blast in Brazil since 1912 (Averna-Sacca, 1912), rice is considered a non-host for *M. oryzae* isolated from wheat. The wheat- and rice-derived populations of *M. oryzae* were described as genetically distinct and host specific based on distinct DNA-fingerprinting profiles, the absence of cross pathogenicity between wheat- and rice-derived strains, and sexual incompatibility between the two host-specialized populations (Bruno and Urashima 2001).

7.3.2.6. Survival and Spread:

The pathogen can be spread by seed, and also survive on crop residues. Currently, most varieties being planted are susceptible and fungicides have not been effective in controlling the disease.

The pathogen can be dispersed efficiently over short and long distances via air-borne inoculum (Urashima *et al.*, 2007) and infected seed (Goulart *et al.*, 1995). *M. oryzae* is a heterothallic Ascomycete requiring two mating types to produce sexual fruiting structures (perithecia, asci, and ascospores) (Valent and Chumley, 1991). Under laboratory conditions, the wheat-infecting strains cross easily (Urashima *et al.*, 1993). However, the relative importance of sexual ascospores and asexual conidia in epidemics of wheat blast remains unknown.

7.3.2.7. Epidemiology:

According to Kohli *et al.* (2011), most severe blast years coincide with wet years (El Nino phenomenon) characterized by several days of continuous rains and average temperatures between 18-20°C during the flowering stage of the crop followed by sunny hot and humid days. Cardoso *et al.* (2008) observed under controlled conditions, highest blast intensity at 30°C which increased with duration of wetting period and lowest at 25°C with a wetting period of less than 10 hr. However, with increasing wetting period of 40 hr at 25°C blast intensity of 85% was observed. In other words, sprinkler irrigation may expose the plants to blast as was observed for *Fusarium* head scab of wheat. Global temperature rise and EI nino may predispose wheat to blast.

7.3.2.8. Source of Inoculum:

Seed has been shown as a primary/initial source of inoculum (Goulart *et al.*, 1990), however, it plays a minor role in epidemiology. Majority of the spike infection comes from the air-borne conidia from various secondary hosts (Prabhu *et al.*, 1992; Urashima *et al.*, 1993). Several grass weeds, *Cenchrus echinatus, Eleusine indica, Digitaria sanguinalis, Brachiaria plantaginea, Echinocloa crusgalli, Pennisetum setosum, Hyparrhenia rufa* and *Rhynchelytrum roseum* are known to harbour *Magnaporthe* species, however, their role in the epidemiology of wheat blast is not very clear (Prabhu *et al.*, 1992). Infection of *Magnaporthe* on triticale was reported by Mehta and Baier (1998) and black oats by Mehta *et al.* (2006).

Prabhu *et al.* (1992) reported that all of the *M. grisea* isolates from rice, wheat and grass weeds were pathogenic to wheat and barley cultivars but none of the 10 isolates from wheat and 7 from grass could infect rice. Host specificity among *M. grisea* isolates has been reported (Mehta and Baier, 1998).

7.3.2.9. Management:

A strict quarantine needs to be followed by not allowing any wheat seed material from the disease prone areas. Crop rotations by skipping wheat for one season will also reduce the blast incidence. Apart from genetic resistance and fungicide application at heading on crop in tropical and sub-tropical areas has helped in blast management. There is a need to destroy the auxiliary grass hosts also.

i) **Avoidance:** Wheat seed from blast infected areas should not be used for sowing. Whenever there is prolonged precipitation a watch has to be kept for the blast. Sprinkler system of irrigation can predispose wheat to blast, hence be avoided.

ii) **Host Resistance:** Complete to date genetic bases of resistance to blast is not very clear due to tricky and variable pathogen. Host resistance to blast is obscure, however, Brazilian wheat cultivars BR18, IPR85, CD113 have shown moderate level of host resistance. Derivative accessions of CIMMYT line Milan have also been shown to possess a high level of resistance (Kohli *et al.*, 2011). Cruz *et al.* (2012) found that among the wheat cultivars tested, a continuum in severity to head blast was observed; cultivars Everest and Karl 92 were highly susceptible with more than 90% disease severity, however, Postrock, JackPot, Overley, Jagalene, Jagger, and Santa Fe showed less than 3% infection. Limited information is available on the identification of quantitative trait loci (QTL) and linked markers associated with blast resistance. Recently a partially blast resistant rice variety Pongsu Seribu 2 was mapped with QTLs linked to resistance genes. QTL linked to the blast resistance was identified and cloned (Fatah *et al.*, 2014).

iii) **Chemical Methods:** Systematic and convincing work on chemical control of wheat blast has not been done. However, it is well known in rice blast to use tricyclazole 75WP at 2g/ kg or carbendazim 50WP 1g/ kg seed for controlling seed inoculum. Initial infections can also be controlled by using need based sprays of carbendazim 50 WP at 1g/ L or tricyclazole 75WP 0.6g/ L or Propiconazole 25EC/ carpropamid 30SC 1ml/ L or Isoprothiolane 40EC 2g/ L as and when infection is noticed at spike initiation or at flowering depending on the situation. Spray at flowering has been effective in controlling the disease.

Fungicides inhibiting melanin production in appresoria have been found effective for the control of blast as sprays or even as seed treatment. Application of silicon containing basic ground granulated blast furnace slag has been found to reduce blast (Agrios, 2005). Among chemicals strobilurins have been found effective. Araki *et al.* (2005), however, observed resistance to strobilurins in rye grass isolate of *Magnaporthe*. A need to use fungicides with different modes of action or alternative use of chemicals would be useful. According to Rocha *et al.* (2014) control of wheat blast can be effective for leaves but not for ears. They found epoxiconazole + pyraclostrobin and tebuconazole + trifloxystrobin to reduce the disease considerably. Cruz *et al.* (2011) stated two fungicide active ingredients as tebuconozole and metconazole, but he opined that fungicides are not effective under warm, rainy weather at heading stage.

7.3.2.10. Future Lines of Research

- Use of improved wheat varieties that carry genetic resistance to *M. oryzae*.
- Global monitoring of disease appearances, movement, and evolution, in coordination with local governments and research agencies, as well as predictive models.
- Advanced studies on potentially effective, safe, and affordable chemical control measures.
- Genetic and epidemiological research to strengthen knowledge of the fungus and its interactions with wheat and other host plants.

7.4. CONCLUSIONS

With the development of agricultural technology as well as biotic and abiotic factors, the crop losses and pest problems are changing continuously. There were outbreaks of peanut bud necrosis disease and peanut yellow spot tospovirus on groundnut; and Karnal bunt, *Tilletia (Neovossia) indica* and blast disease, *Magnaporthe oryzae* on wheat. The emerging and future pest problems have to be tackled with IPM approach as a part of a sustainable crop production technology.

8

Emerging Disease Problems in Fruit Crops

8.1. INTRODUCTION

In developing countries, the increasing intensity of non-traditional crops that are actually being planted has also led to increases in introduction of pathogens in fruits.

Bacterial blight of pomegranate has assumed epidemic proportions (40-85% severity) in northern Karnataka (Bijapur, Bellary, and Bagalkot) extending its spread to border districts of Sholapur, Sangali as well as in other pomegranate areas of Maharashtra seriously threatening its cultivation resulting in enormous economic losses to the growers. In recent years, pomegranate wilt complex caused by the association of fungi (*Ceratocystis fimbriata, Fusarium oxysporum, Rhizoctonia solani*), root-knot nematodes (*Meloidogyne incognita*) and shoot hole borers (*Xyleborus fornicates* and *X. perforans*) is an important disease which results in complete wilting of plant.

Guava is a crop where the Fusarium wilt disease is very serious and it can be said that this is the only disease of guava which is threatening guava cultivation in India. In general, losses due to wilt in guava in Uttar Pradesh vary from 5-60% causing annual loss amounting to Rs. one million.

Papaya ringspot virus is a major limiting factor for commercial papaya production particularly in Hawaii, areas of Thailand, Taiwan, India, Mexico, Brazil, Bangladesh, the Philippines, and the southern region of China. Hawaiian papaya production due to ringspot virus has been severely affected and the production dropped by 94%.

8.2. POMEGRANATE, *Punica granatum*

8.2.1. Bacterial Blight, *Xanthomonas axanopodis* pv. *punicae*

Bacterial blight has been of wide occurrence in India, resulting in economic losses in all major pomegranate growing areas of India. A disease of less economic importance posed its outbreak severity in the year 2001-02 at Bijapur and Bellary districts of northern Karnataka. Bacterial blight of pomegranate has assumed epidemic proportions (40-85% severity) in northern Karnataka (Bijapur, Bellary, and Bagalkot) seriously threatening its cultivation. Afterwards the spread of bacterial blight epidemic was also extended to border districts of Sholapur, Sangli as well as in other pomegranate growing areas of Maharashtra (Yenjarappa *et al.*, 2006). It ruined the pomegranate orchards on a large scale and resulted in enormous losses to the growers valued at crores of rupees.

Disease build up is rapid during rainy season. High disease severity is observed from July to October. Temperatures between 25 to 35° C coupled with humidity above 50%, rains and wind favour rapid disease development. The disease affects all plant parts, but is most destructive on fruits.

Presently, the disease occurs widely and outbreaks have been recorded in all major pomegranate-growing states including Maharashtra, Karnataka, Andhra Pradesh, Tamil Nadu, Himachal Pradesh and Rajasthan. Outside India it is reported from Pakistan and South Africa. Losses caused by bacterial blight were recorded in Hanumangarh district of Rajasthan in 2009.

8.2.1.1. Symptoms:

On leaf the disease appears in the form of regular to irregular dark brown to brown black necrotic spots, surrounded by a yellow halo or water soaked margin (Plate 8.1). The infected leaves turn yellow and drop on the ground. The pathogen also infects stem and branches causing girdling and cracking symptoms.

On fruits disease starts with water soaked lesions on the skin surface, which turn dark brown to black. Spots enlarge and merge with each other to cover large area; however, infection is restricted to the rind. Small L or Y shaped cracks appear on the spots and in severe cases entire fruit splits open along the lesions (Plate 8.1). The spots may be covered with thin shining white encrustation consisting of bacteria.

8.2.1.2. Causes:

The bacterial pathogen, *Xanthomonas axanopodis* pv. *punicae* is the causal organism responsible for this disease infection. Initially brown to black spots with an oily appearance appears on leaves and fruit, therefore, it is also known as 'Oily Spot Disease'. This disease remains prevalent in mild to moderate form throughout year at 9° to 43° C temperatures and also at lower humidity but become more severe under humid conditions (> 80% RH) and moderate temperatures (25° to 35°C) during the rainy season (Jadhav and Sharma, 2009).

Entry is through wounds and natural openings. The first water-soaked lesions develop within 2–3 days and appear as dark red spots. Disease build up is rapid from July to September. Severity increases during June and July and reaches a maximum in September and October and then declines. Bacterial cells are capable of surviving in soil for >120 days and also survive in fallen leaves during the off-season. High temperatures and low humidity or both favour disease development. Optimal temperature for growth of bacterium is 30°C; thermal death point is about 52°C.

Continuous/intermittent rainfall for a longer period, congenial maximum (29.4 to 35.6°C) and minimum temperature (19.5 to 27.3°C) and relative humidity (63 to 87 %) were found favourable for the development and spread of the disease.

8.2.1.3. Impact:

The initial infection started firstly on leaves with the oily appearance of minute, irregular spots with few trees later it turned to dark brown or black spots. Then leaves turned brown to black and dropped off. Secondly the fruit infection was the most damaging stage of bacterial blight. Usually it attacks in the early fruiting stage or when fruits are half grown. Appearance of oily spots on skin of fruits is the major symptom. Affected fruit skin became hard that loses the elasticity or softness that causes 'L' or 'Y' shaped cracks in fruits. Finally, the bacterial pathogen also enters inside through cracks leading to rotting of the entire fruit. The rotted fruits turn black to brown in colour. Hence they became unfit for consumption and marketing.

Bacterial blight is capable to damage the entire orchard within 15 - 30 days. During the field survey, the fruits were often seen in the form of dark brown to black fruit mummies scattered on the floor of severely affected orchards and a few of them hanging over the trees. In this case, trees appeared weak that would be leading on the way of dying in a few days.

8.2.1.4. Survival and Spread:

The pathogen survives in fallen leaves and fruits during off-season for more than 8 months. It can also survive for several years on stem cankers on plants in orchard or in dormant buds. It spreads through planting material, cutting secateurs and wind splashed rains. Temperatures around 31.8-35.6°C with enough rainfall during July-October were found to favour disease development.

Primary source of inoculum is infected cuttings, fruit and twigs. The secondary spread of bacteria is mainly through rain and spray splashes, irrigation water, pruning tools, man and insect vectors.

8.2.1.5. Management

i) Cultural Methods

- Provision of wide row spacing.
- Selection of disease-free seedlings for fresh planting.
- Pruning affected branches, fruits regularly and burning.
- Bahar should be done in Hasta or Ambe bahar.
- Give minimum four months rest after harvesting the fruits.
- Pruning in September-October, hasta bahar cultivation and rest period are highly beneficial to pomegranate plants and suppress the pathogen.
- Adopt orchard sanitation, proper plant nutrition and irrigation, cultural practices and chemical spray schedules to check diseases and insect pests.

ii) Chemical Methods

- Spray before pruning with 1% Bordeaux mixture.
- Spray ethrel for defoliation,
- Paste or smear with 0.5g Streptomycin Sulphate + 2.5g Copper oxychloride + 200g red oxide per litre of water.
- Spray 0.5g Streptomycin Sulphate or streptocycline or Bacterinashak + 2.5g Copper oxychloride per litre of water.
- Spray Bordeaux mixture (0.5% except 1% just after pruning and rest period), alternated with spray of streptocycline (5g/ 10 L) or 2-bromo, 2-nitro propane-1, 3-diol (5g/ 10 L) mixed with copper based formulations

like copper oxychloride or copper hydroxide (2.0-2.5 g/ L) and spreader sticker (0.5 ml/ L) at 10-15 days interval depending on weather conditions. Depending on fungal problems present in the orchard, copper based formulations may be replaced with appropriate fungicides.

- Next day spray with 1 g ZnSo4 +1 g MgSo4 +1 g Boron +1g CaSo4 per litre of water.
- Spray Chlorothalonil 75 WP (2.5 g) or Copper oxychloride 50 WP (3 g) per lit.
- Thiophanate methyl / Carbendazim at 1g/lit were most effective.
- Among the contact fungicides, Chlorothalonil at 2ml/l followed by Mancozeb at 2g/l were more efficacious.

8.2.2. Disease Complex, *Fusarium* spp., *Rhizoctonia bataticola*

It is also known as decline of pomegranate, which refers to dying of pomegranate trees. It was first noticed in Karnataka during 1988 in Kanamadi and Kaladagi areas on a soft seeded Ganesh variety (Somasekhara, 2006). The wilting plants were commonly seen during the field survey. The various types of wilt symptoms were observed like yellowing of leaves in a single branch, vertical stem cracking and blue stain patches on the bark. The disease appears firstly as yellowing and dropping of leaves on one or more branches of the tree. Then it takes a few days to two months for complete drying or wilting of the affected tree.

8.2.2.1. Causal Agents:

Fusarium solani causing root rot and *Rhizoctonia bataticola* causing collar or bark rot are main pathogens responsible for dying of the branches from apex towards the main stem. Hence it is also called as 'die-back of pomegranate'. The wilt pathogens enter stem through holes caused by insect (shot hole borer) leading to stem canker of tree. Similarly, root wounds caused by insect and root knot nematodes help *Fusarium oxysporum* pathogen to enter into the roots resulting in decaying of roots. These pathogens are soil-borne and survive in the infected plant parts up to 190 days and in the soil for a long time.

Though pathogens are primary causal agents for wilt but heavy soils with poor drainage and close plant spacing is other contributing factors for rapid disease built up (Jadhav and Sharma, 2009). In addition, the infected seedlings, budding knives, secateurs, root contact implements are major sources for the spread of the disease (Jamadar *et al.*, 2009).

8.2.2.2. Impact:

The wilt disease was prevalent in 72.34% of orchards. The wilt severity in surveyed orchards ranged from 5% to 50%. All sampled orchards in the high concentration zone of pomegranate (Malegaon, Satana and Deola tehsil) revealed maximum infestation of wilting or dying.

Severely affected respondents (15%) stated that cultivation of this fruit crop became totally uneconomical. The productivity per unit of land significantly declined since more than 50% trees died due to wilt infection. Again there was threat of infection to remaining trees in the orchard.

8.2.2.3. Symptoms

- Affected plants show yellowing of leaves in some twigs or branches, followed by drooping and drying of leaves (Plate 8.2). The entire tree dies in few months or a year.
- When affected tree is cut open length-wise or cross-section, dark greyish-brown colouration of wood is seen.
- Primary source of inoculum is soil and secondary source of inoculum are conidia and moisture. Disease is more in heavy soil and increases with soil moisture.

8.2.2.4. Management

i) Cultural Methods

- Do not allow water to stagnate, improve drainage facility.
- Do not irrigate for 2-3 days after drenching with fungicides.

ii) Chemical Methods

- At initial stage drench soil with 2 ml Propiconazole + 4 ml Chloropyriphos per litre of water at 8-10 lit of solution per tree.
- If any plant had already wilted, drench with above mentioned chemicals, then remove entire plant along with roots and burn it.
- Drench soil with Formaldehyde at 25 ml/lit. of water.

8.3. GUAVA, *Psidium guajava*

8.3.1. Fusarium Wilt, *Fusarium oxysporum* f. sp. *psidii*

Guava (*Psidium guajava*) is an important fruit crop of subtropical countries. It is grown almost in all the states of India. It is a hardy crop and is cultivated successfully even in neglected soils and is attacked by a large number of pathogens, mainly fungi. Wilt is the most destructive disease for guava plant in India and losses due to this disease are substantial. *Psidium guajava* wilt is known to occur in India, Latin America, Malaysia, Pakistan, South Africa, South Asia and Taiwan. Wilt of guava from India was first reported in 1935 from Allahabad. The disease is soil-borne and is difficult to control. Wilt is predominantly caused by the species of *Fusarium*, of which *F. oxysporum* is generally the main cause. The other species of *Fusarium* i.e., *Fusarium solani* also dominates in isolations. Since, the disease results in the complete mortality of the affected plants, the loss is total. Although, severe loss is there in the annual crops also, huge monitory losses occur especially in perennial fruit trees as it is a loss of labour of several years. Guava is a crop where this disease is very serious and it can be said that this is the only disease of guava which is threatening guava cultivation in India. It causes monitory as well as nutritional loss.

Das Gupta and Rai (1947) recorded the disease in the severe form in the orchards of Lucknow, UP for the first time from India. Dey (1948) reported it from Allahabad, Kanpur and Lucknow. After the first few reports from U.P., the disease spread was also observed in the Gangetic alluvium of Baruipur area in the district of 24 Parganas and in the laterite zone of Jhargram and Midnapur of West Bengal (Chattopadhyay and Sengupta, 1955). Chattopadhyay and Bhattacharjya

(1968a, b) reported the disease from Kashakul and Bankura areas. The disease was also been reported from West Bengal, Haryana, Punjab, Rajasthan, Delhi, Jharkhand, Andhra Pradesh, Madhya Pradesh, Odisha, and Tamil Nadu.

Wilted guava plants have also been reported from USA (Florida), Taiwan, Cuba, South Africa, Brazil, Pakistan, Bangla Desh and Australia (Canberra).

8.3.1.1. Economic Importance:

Singh and Lal (1953) estimated 5-15% loss amounting to almost one million rupees due to guava wilt every year in 12 districts of U.P. In West Bengal, the disease reduced the yield by 80% i.e., from 11.35 t ha^{-1} in healthy plantations to about 1.81-2.27 t ha^{-1} in affected orchards (Chattopadhyay and Sengupta, 1955). Chattopadhyay and Bhattacharjya (1968a, b) failed to rejuvenate the affected trees. The new seedlings, grafted or planted in the affected areas showed stunted growth, flowered rarely and succumbed to wilt within a very short time. The land value of 7,000 acres of land in Andhra Pradesh was reduced to half by the presence of this disease (Jhooty *et al.*, 1984). About 150 and 300 acres of guava orchards in Punjab and Haryana, respectively were uprooted during 1978-81 (Jhooty *et al.*, 1984). In general, losses due to wilt in guava around Lucknow area vary from 5-60% (Misra and Shukla, 2002).

8.3.1.2. Symptoms:

The affected plants show yellow colouration with slight leaf curling at the terminal branches, becoming reddish at the later stage and subsequently shedding of leaves take place. Twigs become bare and fail to bring forth new leaves or flowers and eventually dry up. Fruits of all the affected branches remain under developed, become hard, black and stony. The entire plant becomes defoliated and eventually dies. It requires almost sixteen days for complete wilting. Some trees affected linger on even up to 252 days and then die (Misra and Pandey, 2000b). Misra and Pandey (2000b) also studied variations in the symptoms during different time of the year. They noticed yellowing of the leaves with inter-venial chlorosis during the month of August, which drop even with the slight shaking of the plants. During September, general drooping of the leaves takes place. During October complete wilting of plants are seen with almost dried leaves and small dried black fruits hanging on the branch. Few plants also show partial wilting, which is a very common symptom of wilt in guava. It is also recorded that some plants show wilting of variable degree (leaf yellowing, drooping of leaves, drying of terminal branches or partial wilting) during different months but later escape/resist wilting. These plants start recovering from December onwards. It was recorded that out of total wilted plants, around 17% plants, which initially show some symptoms of wilting, ultimately escape/resist wilting (Neeraj *et al.*, 2005; Gupta *et al.*, 2010).

The finer roots show black streaks, which become prominent on removing the bark (Das Gupta and Rai, 1947). The roots also show rotting at the basal region and the bark is easily detachable from the cortex. The cortical regions of the stem and root show distinct discolouration and damage. Light brown discolouration is noticed in vascular tissues (Chattopadhyay and Bhattacharjya, 1968a, b). Wilted plants later show bark splitting. The pathogen attacks young as well as old fruit bearing trees but older trees are more prone to the disease (Misra and Shukla, 2002). New seedlings and grafts also show disease symptoms (Singh and Lal, 1953;

Edward, 1960a). Chakraborty and Singh (1989) identified mainly two types of symptoms i.e., slow wilt (where plant takes several months or even a year or two to wilt after the appearance of initial symptoms) and sudden wilt (where plant takes 15 days to one month to wilt after the appearance of initial symptoms).

8.3.1.3. Causal Organisms:

Several plant pathologists have reported that *Fusarium oxysporum* f. sp. *psidii* and *F. solani* are the important causal organisms of guava wilt (Das Gupta and Rai, 1947; Dey, 1948; Prasad *et al.*, 1952; Edward and Srivastava, 1957). Forma speciales of *F. oxysporum* causing wilt disease in *Psidium guajava* is designated as *F. oxysporum* f. sp. *psidii* (Prasad *et al.*, 1952; Gupta *et al.* 2010). Edward (1960b, c) reported that *F. oxysporum* f. sp. *psidii* penetrate either directly through the root piliferous layer of the guava seedlings or through openings caused by secondary roots. Hyphae were found in the xylem vessels of the roots of the inoculated plants. He also observed that *F. oxysporum* f. sp. *psidii* existed in a variety of forms, which differ in cultural and morphological characters.

8.3.1.4. Management

i) **Chemical Methods:** Jain (1956) found chemotherapeutic action of 0.1% water-soluble 8-Quinolinol sulphate against the wilt pathogen (*F. oxysporum* f. sp. *psidii*). Injection of this chemical in apparently healthy guava plants in a diseased area provided protection against wilt at least for one year and when injected into slightly wilted plants, it was beneficial for their partial recovery. Suhag (1976) reported control of wilt by severe pruning and then drenching with 0.2% Benlate or Bavistin four times in a year and spraying twice with Metasystox and Zinc sulphate. But due to soil borne nature of the disease, pruning does not seem to control wilt. Misra and Pandey (1999b) reported that different fungicides viz. bavistin, topsin M, indofil M-45, thiram, blitox check the various wilt pathogens in laboratory effectively but these pathogens increases their aggressiveness with profuse spore mass production in the soil, once the effect of these fungicides diminishes in soil. Bhargava *et al.* (2003) also found control with thiophanate methyl in laboratory. In Taiwan, carbendazim, captaful and thiabendazole proved effective against wilt pathogen under laboratory experiments but failed *in vivo* (Leu *et al.*, 1979). In South Africa tebuconazole, propiconazole, prochoraz, triforine and carbendazim + flusilazole were effective under *in vitro* evaluation (Joubert and Frean, 1993). Antibiotic actidion (Dwivedi, 1990) and heavy metals Hg, Cd and Cu (Dwivedi, 1991) were found effective for control of wilt. Nematodes are reported to aggravate the wilt incidence in guava. Disinfection of soil with DBCP at 52.8 mL/10 m^2 or Metham sodium at 252.5 mL/10 m^2 was achieved to control nematodes (Rodriguez and Landa, 1977).

Besides fungicides some soil amendment chemicals/cakes/fertilizers were also evaluated for control of wilt. Mathur *et al.* (1964) and Mathur and Jain (1960) found wilt control by soil treatment with 1.82 kg lime or gypsum/tree, although the control mechanism was not well understood. At Central Institute of Sub-tropical Horticulture, Lucknow, wilt was controlled by application of 6 kg neem cake + 2 kg gypsum per plant (Misra and Pandey, 1994). Oil cakes like neem cake, mahua cake, kusum cake supplemented with urea at 10 and 1 kg, respectively also check the disease (Das Gupta and Ghoshal, 1977).

Suhag and Khera (1986) advocated that spread of wilt could be checked by judicious amendments of N and Zn.

ii) **Cultural Practices:** Mathur (1956) advocated that wilt could be controlled by proper sanitation in the orchard. Wilted trees should be uprooted, burnt and trench should be dug around the tree trunk. Edward (1960a) suggested that while transplanting, roots of plants should not be severely damaged. Maintenance of proper tree vigour by timely and adequately manuring, inter-culture and irrigation enable them to withstand infection. The pits may be treated with formalin and kept covered for about 3 days and then transplanting should be done after two weeks. Symptoms of the disease do not appear under green manuring and the disease development is less when organic sources of nitrogen are used (Chattopadhyay and Bhattacharjya, 1968b). Soil solarization with 0.03 mm transparent polyethylene sheet during May-June (Dwivedi, 1993) has been suggested for the control of wilt pathogens. Prasad *et al.* (2003) and Misra *et al.* (2004) reported intercropping with turmeric or Marigold to check the wilting of guava. These cultural practices are useful and should be adopted to escape wilt.

iii) **Varietal Resistance:** None of the guava varieties in India is reported free from wilt incidence and hence these cannot be recommended directly for cultivation in wilt infested areas. No information is available in literature regarding breeding varieties for wilt resistance except the information on relative resistance on the natural incidence, which is provided under heading varietal reaction. To combat the disease, option of resistant rootstock seems to be of great use and some work was done on this aspect and is summarized here. Since interspecies and intergeneric graft compatibility is possible, Edward (1961) suggested guava species *Psidium cattleianum* var. *lucidium* and *Syzigium cuminii* (Jamun), which seldom get attacked with wilt, may be an effective way for the control of wilt disease. Edward and Gaurishanker (1964) in their further studies found *Psidium cattleianum* (*Psidium molle*), *P. quianense*, Chinese guava (*P. friedrichsthalianum*) and Philippine guava were compatible and suggested them for use as rootstocks. A local variety Pei-pa In Taiwan was reported resistant and *Psidium friedrichsthlianum* has been recommended as possible rootstocks (Leu and Kao, 1979). Misra *et al.* (2003) identified F1 population of *Psidium molle* x *Psidium guajava* free from wilt, when grown in wilt sick plot and artificially inoculated repeatedly with *Gliocladium roseum*, *Fusarium solani* and *Fusarium oxysporum*. As graft compatibility is very successful, this resistant rootstock is very useful for the control of wilt.

iv) **Botanicals:** The indiscriminate and inappropriate uses of synthetic fungicides lead to the development of resistance in pathogens, environmental pollution and food contamination by toxic residues. An alternative approach for disease management is therefore, the use of botanical fungicides. Plants are the reservoirs of biodegradable secondary metabolites that are reported to inhibit various phytopathogenic fungi. Botanical fungicides developed from plant products are eco-friendly and found to control many plant diseases (Oros and Ujvary, 1999; Mamatha and Rai, 2004). Patel and Vala (2004) studied that extract of garlic produced maximum inhibition of *F. solani* under *in vitro* conditions Twenty three plant extracts were tested *in vitro* for their biocontrol efficacy to inhibit the growth of *F. oxysporum* f. sp. *psidii* and *F.*

solani isolates. Highest percentage of inhibition for all the five isolates of *F. oxysporum* f. sp. *psidii* was achieved by extracts from *Achyranthes roses*, *Curcuma longa*, *Calotropis gigantea* and *Tagetes erecta*. It was also found that F24 and F30 isolates were comparatively more aggressive among all the isolates of *F. oxysporum* f. sp. *psidii*. Leaf extract of *Calotropis gigantea* and *Cannabis sativa* showed comparatively maximum inhibitory effect to *Fusarium solani* (Gupta *et al.*, 2007).

v) **Biological Methods:** Dwivedi (1992) from Varanasi advocated *Trichoderma* sp. and *Streptomyces chibaensis* for the control of wilt. Seed oil of *Foeniculum vulgare* was also reported to control wilt (Dwivedi, 1993). In Pakistan, combined use of Topsin M sprays and the antagonists *Trichoderma harzianum* and *Arachniotus* sp. added in soil amended with wheat straw controlled decline of guava (Ansar *et al.*, 1994). Logani *et al.* (2002) reported application of wiltnema (seven plant extracts - *Allium cepa*, *A. sativum*, *Ocimum sanctum*, *Azadirachata indica*, *Datura stramonium*, *Cannabis sativa* and *Nicotiana tabacum*) for prevention of wilt and better growth of guava plants. Srivastava *et al.* (2001) found that by use of AMF symbiont at the rate of 5 kg tree^{-1} is beneficial for the control of wilt.

Bioagents like *Aspergillus niger*, *Trichoderma* sp., *Penicillium citrinum* and some bio-dynamic antagonists have shown their effectiveness towards the control of wilt pathogens of guava (Singh *et al.*, 2003; Misra, 2006; Gupta *et al.*, 2009a). When these fungi were tested for the control of wilt pathogen in laboratory conditions, these were found quite effective (Misra *et al.*, 2004). When relative growth of the three bioagents was studied, it was found that *Aspergillus niger* was fastest growing and most effective (Misra and Prasad, 2003). These can be grown easily on any substrate like maize/bajra seeds etc. and can also be multiplied on cheap substrates like *Sacchrum* sp. (grass) and dry and green leaves of *Psidium guajava* (Shukla *et al.*, 2003). It was also found that at village level these bioagents can be multiplied in earthen pots (Misra and Prasad, 2004). Among these, *Aspergillus niger* was found very fast growing, easy to propagate and most effective in controlling the wilt disease in field. Besides this quality, it is also growth enhancer and the plants treated with *Aspergillus niger* developed faster with more height, more thickness and more numbers of leaves (Misra *et al.*, 2000). Dwivedi and Shukla (2002) reported that out of several bioagents tested, *Trichoderma harzianum*, *T. viride* and *Gliocladium virens* are best for the control of wilt. Singh *et al.* (2003) reported bioagent *Aspergillus niger* was most effective in controlling the wilt disease followed by *Trichoderma viride*.

Isolates of bioagents, comprising *Aspergillus niger*, *Trichoderma* sp. (*T. virens*, *T. harzianum* and *T. viride*) and *Penicillium citrinum*, their culture filtrates and volatile compounds were evaluated against isolates of *F. oxysporum* f sp. *psidii* and *F. solani* (Gupta and Misra, 2009; Misra and Gupta, 2009). All isolates of bioagents significantly checked the growth of *F. oxysporum* f. sp. *psidii* and *F. solani*. Per cent inhibition was maximum by direct use of *Trichoderma* sp. in dual culture *A. niger* expressed moderate efficacy against *F. oxysporum* f. sp. *psidii* and *F. solani*. However, *P. citrinum* isolate showed lesser effect. They also observed that in overall, *T. virens* and *T. viride* were superior in inhibiting the growth of both *Fusarium* sp. and *Trichoderma*. *Aspergillus* isolates were effective in 'complete suppression of guava wilt incidence under field conditions.

vi) **Integrated Eco-Friendly Approach:** Considering the complexity of the problem, integrated eco-friendly approach for the control of guava wilt was suggested by Misra *et al.* (2003b, 2004) and Misra (2005) utilizing bioagent *Aspergillus niger* strain AN17, using resistant rootstock (*P. molle x P. guajava*), intercropping of guava with turmeric or Marigold as well as following proper cultural practices to minimize losses due to this disease.

Management of *Fusarium* wilt is not possible when we go for a single control measure. So, an integrated approach based on biocontrol agents, plant products and cultural methods are being adopted for the successful management of *Fusarium* wilt.

8.4. PAPAYA, *Carica papaya*

8.4.1. Papaya Ringspot Virus (PRSV)

PRSV-P (papaya strain) is one of the most destructive diseases of papaya and occurs in nearly every region where papaya is grown. It has been reported to be a major limiting factor for commercial papaya production particularly in Hawaii, areas of Thailand, Taiwan, India, Mexico, Brazil, Bangladesh, the Philippines, and the southern region of China. In addition, PRSV-W (watermelon strain) also causes serious losses worldwide in several important cucurbit crops of tropical, subtropical and temperate regions.

Papaya ringspot virus (PRSV) is a pathogenic plant virus in the genus *Potyvirus* and the virus family *Potyviridae.*

8.4.1.1. History:

Hawaiian papaya production has been severely affected twice by PRSV. The aggressive strain of the disease was introduced around 1950. Within 12 years, the amount of land under papaya production dropped by 94%. Production was then moved from Oahu to the Puna region of Hawaii Island (the "Big Island") under strict quarantine. The virus emerged in commercial farms in 1992 and by 1995 production in Puna was impossible. Hawaiian papaya production was halved by the end of the decade. Transgenic papaya varieties that are resistant to PRSV entered production in 1998 and resuscitated the industry (Tripathi *et al.*, 2008).

8.4.1.2. Types and Distribution:

There are two major types (PRSV-P and PRSV-W) of this virus that are serologically indistinguishable and are so closely genetically related that they are now considered the same virus species. The type PRSV-P infects papaya and several members of the melon family (Cucurbitaceae). The PRSV-W does not infect papaya, but do infect cucurbits such as watermelon, cucumber, and squash and were originally known as *Watermelon mosaic virus 1* (Webb, 1965; Purcifull and Hiebert, 1979).

Both pathotypes are distributed worldwide. PRSV-P, for example, is known to be present in the Middle East, Africa, South and Central America. It has also been found in China, France, Germany, India, Italy, Mexico, Taiwan, and the United States. PRSV-W isolates have been found in the United States, the Caribbean, Mexico, Italy, Germany, Australia, France, India, the Middle East, and South America.

8.4.1.3. Symptoms

i) **PRSV-P:** *Papaya ringspot virus* infects papaya and cucurbits systemically. In papaya, leaves develop prominent mosaic and chlorosis on the leaf lamina, and water soaked oily streaks on the petioles and upper part of the trunk. Severe symptoms often include a distortion of young leaves which also result in the development of a shoestring appearance that resembles mite damage (Plate 8.3). Trees that are infected at a young stage remain stunted and will not produce an economical crop. Fruit from infected trees may have bumps similar to that observed on fruit of plants with boron deficiency and often have 'ringspots', which is the basis for the disease's common name (Plate 8.3). A severe PRSV isolate from Taiwan is also known to induce systemic necrosis and wilting along with mosaic and chlorosis. Cucurbit symptoms tend to be similar to papaya symptoms including blisters, mosaic, yellowing, and leaf distortions (Dahal *et al.*, 1997). *Chenopium amaranticolor* and *C. quinoa* (*Chenopodiaceae*) are serving as local lesion hosts.

ii) **PRSV-W:** In cucurbits, leaves show intense mosaic with a narrowing of the leaves. Severe cases can result in a shoestring effect similar to what is observed in papaya. Plants that become infected at a young age do not develop a crop. Older plants that become infected do produce fruit that often show distinct changes in colour and are deformed. For example, a yellow crook neck squash would have many green blotches and bumps (Plate 8.4). PRSV-W is considered to be one of the limiting factors in the growing of cucurbits in Florida (Webb *et al.* 2003).

8.4.1.4. Virus Structure:

Virions of PRSV are filamentous and flexuous rods measuring 760-800 x 12 nm with a monopartite single-stranded positive sense RNA as its genome. Virus particles typically contain 94.5% protein and 5.5% nucleic acid by weight, and have no outer membrane (non-enveloped). The thermal inactivation point (TIP) of PRSV is 54-60°C and the longevity *in vitro* (LIV) is about 0.3 days.

8.4.1.5. Transmisssion:

Aphids are the predominant means by which PRSV is transmitted and the virus does not multiply in the vector. Many species of aphids (*Aphis gossypii, A. craccivora, A. citricola, Myzus persicae, Acrythosiphon pisum,* and *Rhopalosiphum maidis*) can transmit PRSV, particularly the Peach Aphid and Melon Aphid (Heu *et al.*, 2002). The disease cycle can start with aphids feeding on infected papaya for as little as 15 seconds and subsequently feeding on a healthy papaya. There is no incubation period. The virus does not persist in the vector so transmission to another plant has to occur rather rapidly.

PRSV is also easily transmitted via mechanical inoculation.

Disease transmission can also occur by planting infected seedlings in fields where the virus is not present. Using "clean", virus-free seedlings for planting is of very high importance.

8.4.1.6. Management:

There are four main methods of control for PRSV: quarantine and geographic displacement, rouging and netting, cross-protection, and genetic modification of the host plant. Because PRSV is a non-persistent virus and is consequently transmitted to healthy plants by aphids within a very short time period, insecticidal control is difficult and impractical. Once symptoms have been observed, it is already too late to spray for aphids – the disease has most likely already been transmitted to nearby healthy plants. In order to implement successful vector control, frequent preventative sprays are required, which is rarely done.

i) **Quarantine and Geographic Displacement:** Prevention through quarantine and geographic displacement of cropland is common and has occurred in Hawaii, the Philippines and Brazil (Gonsalves, 1998). When fields become infected, such as the case of Oahu papaya growers, attempts to relocate growing areas to virus-free fields are made. This is usually just a temporary avoidance of the disease, which eventually spreads to the new fields.

ii) **Rouging:** Rouging, or the removal and destruction of infected plants, is a way to control the spread of PRSV. This method was employed unsuccessfully when PRSV began to invade the Puna region of Hawaii. It is difficult to suppress the spread of PRSV through rouging because it is spread very quickly and effectively by aphids.

iii) **Netting:** Netting can also be used to prevent insect vectors from spreading the virus. Production under netting is prohibitively expensive for subsistence and small-scale producers, but was used effectively in Taiwan because geographical displacement was not possible on such a small island (Fuller, 2005).

iv) **Cross Protection:** Cross protection is similar in practice, although not in mode of action, to viral vaccinations in humans. A mild strain of PRSV is introduced into the host plant, which then develops resistance to virulent strains of the virus. Development of cross-protection in papaya was researched in Hawaii starting in 1979. A delay in the onset of symptoms was achieved, as well as a reduction in the severity of symptoms. However, inoculation of the mild strain also caused pathogenesis on the papaya plants (Gonsalves, 1998).

v) **Transgenics:** Pathogen-derived resistance (PDR) is the technique of inserting a gene fragment from the pathogen into the transgenic crop, which then affords the crop plant resistance against the pathogen. This method was employed in the development of transgenic papaya. PRSV's coat protein gene was inserted into highly embryogenic plant tissue by using the newly invented gene gun, which led to the development of the transgenic line 55-1, which was proven to be effective in controlling Hawaiian strains of PRSV.

There are two transgenic varieties of papaya, both of which were introduced for production in May 1998. 'Rainbow' is an F1 hybrid, which is a cross between the yellow fleshed Kapoho and the red-fleshed 'SunUp' (Plate 8.5). Rainbow is produced on 76% of Hawaiian papaya acreage, while SunUp is hardly grown commercially. SunUp is thought to be more resistant to exotic strains of PRSV, while Rainbow has shown susceptibility to such exotic strains of the virus. Transgenic varieties have been shown to be extremely effective against Hawaiian strains of PRSV, showing complete resistance to the virus compared to 100% infection of susceptible strains in some trials (Ferreira *et al.*, 2002).

a) **Sunrise/Sunup:** The Sunrise variety is known locally as 'strawberry papaya'. When ripe the fruits have yellow skin and intensely red-orange pulp. The Sunup variety is resistant to PRSV and is the genetically altered version of Sunrise.

b) **Rainbow:** A hybrid made by crossing the yellow pulped Kapoho Solo with the red pulped Sunup. Rainbow is genetically altered to be resistant to PRSV. It is the most popular variety amongst consumers and is also the most widely cultivated on the main island Hawaii and on the islands Oahu and Kauai. The fruits strongly resemble the Kapoho papayas with golden yellow pulp.

c) **Kamiya/Laie Gold:** Kamiya is rounder and larger compared to the aforementioned varieties. It is cultivated exclusively on Oahu for local consumption and, just like Sunrise, has orange coloured pulp. The Laie gold variety is resistant to PRSV and is the genetically altered version of Kamiya.

8.5. CONCLUSIONS

There has been an outbreak of serious disease problems in the recent years in fruit crops such as pomegranate and guava through dissemination of diseases with infected planting materials. There were outbreaks of bacterial blight, *Xanthomonas axanopodis* pv. *punicae*, *Fusarium* spp. and *Rhizoctonia bataticola* disease complex on pomegranate; Fusarium wilt, *Fusarium oxysporum* f. sp. *psidii* on guava; and ringspot virus on papaya. Hence, there is an urgent need for rapid and accurate detection, diagnosis, and development of management recommendations for these new pests and diseases.

9

Emerging Disease Problems in Vegetable Crops

9.1. INTRODUCTION

In developing countries, the increasing intensity of non-traditional crops that are actually being planted has also led to increases in introduction of pathogens in vegetables.

Potato late blight caused by *Phytophthora infestans* is considered as a re-emerging plant disease which is actually a global threat to food security. The 13_A2 lineage was responsible for severe late blight outbreaks on potato and tomato in South India. In 2009, there was an outbreak of late blight in the U.S. due to Us22 strain of the pathogen which actually moved into tomato transplants.

Over recent years, several viral diseases, including tomato yellow leaf curl virus (TYLCV; genus Begomovirus, family Geminiviridae) and tomato spotted wilt virus (TSWV; genus Tospovirus, family Bunyaviridae) have emerged in tomato crops and are presently impacting fresh-market tomato production in diverse geographic areas worldwide.

Iris yellow spot virus (IYSV; genus Tospovirus, family Bunyaviridae) is an immediate and serious threat to sustainable and productive onion cropping systems around the world, and the recent detection of this disease in numerous onion-producing countries demonstrates that the disease is spreading rapidly in a range of environments.

Watermelon bud necrosis virus (WBNV) has emerged as a serious threat to watermelon cultivation in India recently, with an estimated yield loss of 60 to 100 per cent in Karnataka and spreading to Aurangabad and Paithan regions of Maharashtra.

9.2. POTATO, *Solanum tuberosum*

9.2.1. Late Blight, *Phytophthora infestans*

Phytophthora infestans, the pathogen that caused late blight disease of potato, was responsible for the potato famine, which occurred in 1843, and this epidemic in Ireland led to the death of more than 2 million people and immigration of another several million people. At that point of time, potatoes were the main food crop of the peasant farmers, and they had no other food source.

Late blight has emerged as the most destructive disease of potato and tomato in South India since 2008 (Plate 9.1). All 157 isolates (63 from potato and 94 from tomato) collected from major potato and tomato production areas of South India were found to be equally aggressive, confirming that the same clonal population is infecting both hosts. This study demonstrates that the 13_A2 lineage was responsible for severe late blight outbreaks on potato and tomato in South India and has replaced the prior population represented by the US-1 and other genotypes. Revised management strategies will be required to combat this destructive 13_A2 clonal lineage and monitoring of the population across other potato- and tomato-growing regions of India is warranted (Chowdappa *et al.*, 2014).

Among the most wanted plant pathogens in terms of the armed and dangerous, potato late blight is at the top of the list. Late blight is considered a re-emerging plant disease, and it is a constraint to food security because the disease has increased greatly from its first introductions into Ireland and the United States, and is actually a global threat now to potato production all over the world. The pathogen infects tomato and potato, and also the tubers and the fruit, and this pathogen has re-emerged because it has a very effective dispersal mechanism, it can move in crop material, has a polycyclic nature (meaning repeated spore production over the season), there are fungicide-resistant strains, the pathogen can move into wild hosts, and the genome of the pathogen is very plastic and it is able to overcome host resistance genes. In 2009, there was an outbreak of late blight in the U.S. due to Us22 strain of the pathogen which actually moved into tomato transplants.

9.2.1.1. Symptoms:

Late blight affects the foliage and tubers of potato (Plate 9.2). The characteristic foliar symptoms on any susceptible plant are irregular to circular lesions. Initially, in wet weather, the lesions may appear water-soaked; in dry weather, the centres tend to dry out. The dark centre is often surrounded by yellow, chlorotic tissue or brown, collapsed tissue. These spots often occur at the tips or edges of leaves. If moist weather persists, the leaf spots enlarge rapidly to blight larger areas of the foliage. Dark brown discolouration of leaf stalks and stems is also commonly encountered. The symptoms frequently develop first on lower leaves but can develop at any level in the canopy. The symptoms are usually most dramatic inside the canopy. During humid weather, a white fuzzy or mildew-like growth appears on the leaves (upper or under sides) and on stems. This indicates that the fungus is producing spores.

Tubers become infected when spores of the fungus wash down through the soil and come into contact with the fleshy organs. Infected potato tubers are discoloured by a reddish-brown, granular rot. The rot can be superficial or penetrate more deeply into the tubers (2.5 cm or more). Tubers must be cut open to observe this symptom (Plate 9.2). The rot continues to develop after the tubers are harvested. Further decay of the tubers is often hastened by secondary fungi and bacteria. The soft rot bacteria often turn tubers into a foul smelling liquid.

9.2.1.2. Epidemiology:

The infected seed tubers and refuge piles in the temperate regions and cold stored infected tubers in the sub-tropical regions serve as the primary source of inoculum.

The fungal growth and spores produced could only survive in living tissues. Infected potato tubers used for seed, infected tubers left in cull piles, and unharvested infected tubers left in the ground form the sources of inoculum, or spores, for the disease from one growing season to another.

The development of the disease is favoured by daytime temperatures of 21-27°C and night time temperatures of 10-16°C. The presence of free moisture (such as dew, rain or fog) on the foliage also favours the disease. Free moisture is a prerequisite for spore germination and infection and thus is essential for disease development. As the foliage becomes further blighted, more spores are produced and spread of the disease to adjacent plants becomes more likely. As the season progresses, potato tubers can become infected as spores from the blighted foliage are washed down through the soil and contact developing tubers.

9.2.1.3. Management:

The effective management of late blight of potato can be achieved by reduction of both foliar and tuber infection by adopting three pronged attack i.e. by employing chemicals, cultural practices and host resistance in an integrated manner.

i) **Cultural Methods:** The most effective strategy for managing late blight is to avoid sources of inoculum (spores). Do not save potato seed from year to year. Use healthy certified potato seed. Many states that produce potato seed have seed certification programmes to ensure that the seed meets certain standards for disease levels. Using certified seed also reduces the carryover of a number of potato viruses. Certified seed, however, is not a guarantec that late blight or other diseases will not be present. Examine your seed carefully before you plant, and plant only sound, blemish-free tubers. Examine tubers for late blight symptoms by cutting them open with a knife. Immediately sterilize or thoroughly wash the knife if late blight symptoms are found.

Use disease-free seed tubers for planting. Ridges should be made up high enough to cover all daughter tubers and reduce chances of their infection upon exposure.

Destroy any rejected seed tubers that you do not plant (bury them deeply in an area away from the field, burn them, or discard them in a plastic bag in the trash). Make sure any tubers put into compost piles are completely decomposed. Pull up and destroy potato plants that come up from tubers left in the field from the previous season.

If late blight becomes severe, remove diseased plants by digging them up. Destroy these plants immediately by one of the following: burying them deeply in an area away from the field, burning them, or by bagging them in a plastic bag and discarding the bag. These steps will help avoid production of a larger number of spores. Harvest all potato tubers in the field. If late blight occurs when the tubers have already 'sized up', harvest the crop as soon as possible to avoid post-harvest tuber rot. Again, destroy diseased foliage and stems.

It is best to avoid wetting the leaves when you irrigate; soak the ground around the plants instead of spraying the plants with water. If this is not possible, water your garden mid-day so that the foliage dries off quickly. Eliminate weeds around the field so as to maximize air circulation around the plants. In addition, earthing up the soil around the plants provide a more substantial barrier against tuber infection from spores washing down through the soil.

The following cultural practices may be adopted to reduce the disease incidence:

- Use of disease-free seed.
- Removing/rouging of ground keepers.
- Adjustment in planting time.
- High ridges and proper earthing up to avoid exposure of seed tubers.
- Water management – light or no irrigation when the blight is likely to appear.
- Killing of haulms at 75% crop damage.

ii) **Host Resistance:** There are several potato varieties that show resistance to late blight. These varieties will slow down, but do not prevent, the development of the disease. Elba is currently the most resistant potato variety available. Potato varieties with moderate levels of resistance include: Kennebec, Sebago, Allegany, and Rosa.

In North Indian Plains, use disease resistant cvs. like Kufri Sindhuri, Kufri Jyoti, Kufri Badshah, Kufri Chipsona 1 and 2, Kufri Jawahar and Kufri Sutlej. In hills, use disease resistant cvs. like Kufri Jyoti, Kufri Giriraj in Himachal Pradesh hills; Kufri Kanchan and Kufri Jyoti in Darjeeling hills; Kufri Swarna and Kufri Thenamalai in Nilgiri hills.

a) **Transgenic Potatoes:** Transgenic potatoes have been developed that have RB gene which imparts high resistance to late blight. What is very impressive about this transgenic potato is that generally you have to spray potatoes ten to fifteen times. With one spray you can reduce disease in RB potato. It's not completely immune, but it is highly resistant, and transgenic potatoes can significantly reduce pesticide use.

iii) **Chemical Methods:** Potatoes are susceptible to late blight at any time during the growing season. If the growing season is wet, late blight occurrence is very likely. Fungicides may be necessary to protect potato plants from infection. Continuous fungicide coverage is necessary to protect plants from infection. Spray 1% Bordeaux mixture to control the disease.

When disease first appears, begin treatment with chlorothalonil, maneb or copper oxychloride. Additional pesticides that may be used later in rotation include Heritage or products containing the active ingredients *Bacillus subtilis* or basic cupric sulphate.

The following strategies can be adopted for chemical control:

- Application of prophylactic spray of contact fungicide and subsequently need based fungicide sprays.
- Application of 1 or 2 sprays of systemic fungicides mixtures (metalaxyl, ridomil, etc.) in areas where blight comes every year in epiphytotic form. First and last spray should be of contact fungicides (mancozeb, captafol, Copper oxychloride, Bordeaux mixture) to avoid development of resistant strains of the fungus.

iv) **Biological Methods:** Biological control agents like *Trichoderma viride, Pencillium virdicatum, P. aurantiogrieum, Acremonium strictum, Myrothecium varrucaria* and *Chaetomium brasiliense* were found antagonistic to *Phytophthora infestans* (Arora *et* *al.*, 1992; Roy *et al.*, 1991). *T. viride, P. virdicatum* and *P. aurantiogrieum* inhibited germination of sporangia through zoospores, whereas *A. strictum* caused lysis of zoospores and germ tubes and no further development of the pathogen took place (Arora, 2000) (Table 9.1).

Table 9.1. Biological control of late blight of potato in greenhouse.

Treatment	Greenhouse disease severity (%)		Field disease severity (%)
	Kufri Lalima	Kufri Ashoka	
Control (Unsprayed)	61.6 (51.7)	88.8 (70.4)	90.0
Trichoderma viride (Prophylactic)	3.6 (7.0)	14.0 (22.0)	70.8
Trichoderma viride (Curative)	73.3 (58.9)	71.8 (57.9)	—
Acremonium strictum (Prophylactic)	1.5 (7.0)	6.5 (14.8)	—
Acremonium strictum (Curative)	69.0 (56.2)	77.0 (61.3)	87.5
C.D. (P = 0.05)	**(9.0)**	**(8.4)**	**12.4**

9.3. TOMATO, *Solanum lycopersicum*

Tomato is by far the most important vegetable crop, and many viral species that infect tomato crops have been described. Begomovirus species and other whitefly-transmitted viruses are invading into new areas, and are rapidly spreading over large geographic areas. Over recent years, several viral diseases, including Tomato yellow leaf curl virus (TYLCV; genus Begomovirus), Pepino mosaic virus (PepMV; genus Potexvirus), and Tomato torrado virus (ToTV) have emerged in greenhouse tomato crops and are presently impacting fresh-market tomato production in diverse geographic areas worldwide. The emergence of new viral species and genotypes is frequently reported, probably driven by changes in climate, cultural practices, host plants, and virus vectors, in addition to changes in the viral genome.

9.3.1. Tomato Yellow Leaf Curl Virus (TYLCV)

The worldwide emergence of whiteflies, especially of the *Bemisia tabaci* B biotype, has contributed significantly to the rapid and successful emergence of the tomato yellow leaf curl disease (TYLCD), caused by a complex of viral species and constituting a serious threat to tomato production in many tropical and subtropical regions. The causal agents are members of the genus Begomovirus within the family Geminiviridae, plant viruses with a circular ssDNA genome. Begomovirus species are transmitted by the whitefly *B. tabaci* in a persistent and circulative manner, and their genome generally consists of two DNA molecules. However, nearly all Begomovirus spp. that cause TYLCD have only one genomic component. Many different species have been described within this group, most of which have been given the name TYLCV, or derivatives thereof.

9.3.1.1. Symptoms:

TYLCD symptoms comprise of upward and inward curling of leaflet margins, reduction of leaf surface, interveinal yellowing of leaflets, severe stunting, and flower abortion, resulting in bushy growth of infected seedlings (Moriones and Navas-Castillo, 2000) (Plate 9.3). The virus reduces fruit set considerably, especially when infection takes place before the flowering stage. There are no noticeable symptoms on fruits derived from infected plants (Thongrit *et al.*, 1986). A disease with symptoms similar to TYLCD was already reported in tomato crops in Israel in the late 1930s but the virus was first identified in 1961, and tomato production in the Middle East has been severely affected from the 1970s onward (Cohen and Antignus, 1994; Czosnek and Laterrot, 1997; Cohen and Lapidot, 2007). However, due to a worldwide spread of the B biotype of the vector *B. tabaci*, which has a wider host range than other biotypes and is highly invasive, new TYLCV-like viruses emerged from weed and endemic plant reservoirs in the late 1980s. In addition, with the rapid emergence of whiteflies into new geographic regions, Begomovirus infections have been reported in new tomato production areas over the last decade. The disease was described in Japan, Mexico, and the United States in the late 1990s (Polston *et al.*, 1994; Polston and Anderson, 1997; Kato *et al.*, 1998; Ascencio-Ibáñez *et al.*, 1999).

TYLCV hosts that can serve as virus reservoirs might have a large effect on viral emergence in new tomato production regions. Although TYLCV can produce severe symptoms in tomato, it is also able to establish symptomless infections in both wild and cultivated species. For instance, in many tomato growing areas, bell pepper is grown in close proximity. Because pepper was known as a non-host of TYLCV, whitefly management was not practiced rigorously in pepper plots (Polston and Lapidot, 2007). After former, conflicting reports regarding the susceptibility of peppers to TYLCV (Mansour and Al-Musa, 1992; Reina *et al.*, 1999), it was recently demonstrated that some, albeit not all, cultivars of pepper are, in fact, symptomless hosts of TYLCV (Morilla *et al.*, 2005; Polston *et al.*, 2006). Moreover, despite the lack of TYLCV-induced disease symptoms, whiteflies were able to acquire TYLCV from infected pepper plants and transmit the virus to tomato (Polston *et al.*, 2006). Hence, symptomless infected pepper plants can serve as a virus reservoir for the acquisition and transmission of TYLCV. Other asymptomatic hosts of TYLCV may act as reservoirs in a similar manner.

9.3.1.2. Transmission:

TYLCV is transmitted by the whitefly, *Bemisia tabaci* (Cohen and Harpaz, 1964), which is commonly found in tropical and subtropical countries. *B. tabaci* has a very wide host range of at least six plant families (Cohen and Antignus, 1994). Remarkably, in 2007, a TYLCV outbreak outside the climatic zone in which *B. tabaci* thrives was reported. TYLCV infection was reported in 19 neighbouring greenhouse tomato crops in The Netherlands but the virus has been successfully eradicated and no new outbreaks have been reported since then (Botermans *et al.*, 2009). The source of infection could not be identified but nucleotide sequence analyses revealed a high similarity between isolates from the different crops (>99% identity to the Spanish TYLCV-Alm strain) and, therefore, infection was thought to be initiated from the same source, whereas local spread most likely occurred through *B. tabaci* vectoring (Botermans *et al.*. 2009). Begomovirus spp. emergence in tomato crops has been extensively discussed in several recent reviews (Moriones and Navas-Castillo, 2000; Varma and Malathi, 2003; Seal *et al.*, 2006; Rojas and Gilbertson, 2008).

9.3.1.3. Management:

Control of whitefly-transmitted viruses depends mainly on the availability of efficient insecticides, insect-proof greenhouses, and resistant cultivars (Polston and Lapidot, 2007).

i) **Cultural Control:** In all regions where TYLCV is endemic, seedlings need to be grown and kept in a glasshouse or a screen house in order to prevent early infection by whitefly feeding. In Thailand, farmers grow tomato plants in isolated paddy fields after rice harvesting. In Egypt, the spread of TYLCV in open fields was successfully controlled by rouging infected plants and removing all overwintered tomato crops before the emergence of whitefly populations (Mazyad *et al.*, 1986). Similar measures were undertaken in Cyprus over 3 consecutive years resulting in a decrease in the incidence of the disease from 50 to 5% (Ioannou, 1987). In the Dominican Republic, a mandatory 3 months host-free growing period which included bean, cucurbits, aubergine, melon, okra, pepper and tomato, significantly reduced the inoculum and lowered to a minimum the incidence of TYLCV (Gilbertson *et al.*, 2007). When the incidence of *B. tabaci* is low, early planting of tomato effectively reduced TYLCV infection in Egypt, Israel and Cyprus. Interplanting with rows of TYLCV non-host trap plants such as squash and cucumber have been used to divert whiteflies from tomato in the Jordan Valley, delaying TYLCV infection by 2 months (Al-Musa, 1982). Sticky yellow plastic polyethylene sheets have been used, with limited success, to decrease the pressure of the whitefly population (Cohen *et al.*, 1974). When economically feasible, growing tomatoes in greenhouses covered with insect-proof 50-mesh screens and double-doors constitutes the optimal solution for crop protection against the TYLCV whitefly vector (Cohen and Berlinger, 1986; Ioannou, 1987). UV-absorbing plastic sheets or UV-absorbing 50-mesh screens, used as tunnels or as screen house covers, greatly reduced the infestation of tomato plants by *B. tabaci* and the incidence of TYLCV (Antignus *et al.*, 1998).

ii) **Chemical Control:** Applying systemic insecticides as soil drenches or regular spraying during the seedling stage can reduce the population of the *B. tabaci* vector and the incidence of TYLCV. Many insecticides such as organophosphates, carbamates and pyrethroids effectively reduce the whitefly population, whether sprayed one insecticide at a time, alternately, or together with oil and emulsifiers (Makkouk and Laterrot, 1983). However, these insecticides provided only partial TYLCV control, even when sprayed daily (Cohen and Antignus, 1994). Continued extensive use of insecticides is not only detrimental to the environment but it promotes the development of resistant whitefly populations (Cahill *et al.*, 1995). More potent insecticides with novel modes of action such as imidacloprid were introduced in the 1990s (Elbert *et al.*, 1990). Over the past few years imidacloprid has assumed major importance for control of *B. tabaci* in the field and in greenhouses. To be efficient, the insecticide has to induce the death of all whiteflies within 35-40 min before it succeeds in infecting the plant, thereby preventing inoculation of TYLCV, which is not always the case in the field (Rubinstein *et al.*, 1999). As for other commonly used insecticides, whitefly populations resistant to imidacloprid have been emerging in many places (Castle *et al.*, 2010).

iii) Host Resistance

a) **Genetic Engineering:** Future control strategies for plant virus diseases may include transgenic resistance. In particular, the introduction of an inverted repeat transgene, that is derived from viral sequences and, thus, is able to trigger gene silencing of the viral gene expression, into the plant genome is a promising strategy. Tomato cultivars with resistance to TYLCV, ToTV, and ToMarV are commercially available. Genes controlling resistance to TYLCV and other tomato-infecting Begomovirus spp. (Ty genes) were introgressed from several wild *Solanum* species. The intermediate TYLCV resistance gene Ty1, introgressed from *S. chilense* LA1969 (Michelson *et al.* 1994; Zamir *et al.* 1994), is the most commonly used source of TYLCV resistance, while information on ToTV resistance has not been published. A recent F1 diallele study (Vidavski *et al.*, 2008) demonstrated that several of these Ty resistance genes may interact with each other and, in some cases, result in hybrid plants displaying higher TYLCV resistance compared with their parental lines. This suggests that pyramiding resistance genes originating from different resistance sources can be effective in obtaining cultivars that are highly resistant to Begomovirus spp. (Vidavski *et al.*, 2008).

b) **Pathogen-Derived Resistance:** Virus-resistant transgenics have been developed in many crops by introducing either viral capsid protein or replicase gene encoding sequences. This concept has been called pathogen-derived resistance (PDR). PDR has been the most common approach used to obtain resistance to TYLCV, including viral sequences that generate antisense RNA as well as the expression of full-length and truncated viral genes. The *CP* gene was one of the first TYLCV genes evaluated for the ability to generate pathogen-derived resistance. Tomato plants expressing the TYLCV CP showed a delay in symptoms, a recovery from infection, and resistance upon repeated inoculations (Kunik *et al.*, 1994). Expression of antisense RNA of the TYLCSV *Rep* gene provided a good level of resistance in *Nicotiana benthamiana* (Bendahmane and Gronenborn, 1997). The expression of sense and antisense constructs of truncated TYLCSV *Rep* in *N. benthamiana* also showed resistance (Noris *et al.*, 1996). When a truncated TYLCV-Mld Rep gene was expressed in tomato, the transgenic plants were resistant to TYLCV-Mld, but were susceptible to TYLCV-IL (Antignus *et al.*, 2004). A 2/5 TYLCV *Rep* gene construct conferred high levels of resistance and often immunity in both transformed *N. tabacum* and *S. lycopersicum* (Yang *et al.*, 2004).

c) **Virus Gene Silencing:** Transgenic plants have been developed which exploited the mechanism of silencing via dsRNA (Smith *et al.*, 2000). Using a transformation cassette consisting of the 3'-end of the *Rep* (sense and antisense orientation) as the arms of the hairpin, and a castor bean catalase as the intron, the treated tomato plants showed immunity to TYLCV upon whitefly-mediated inoculation (Fuentes *et al.*, 2006). A similar approach was used with the 5'-end of the TYLCV CP gene and a maize ubiquitin intron, inducing resistance to TYLCV in tomato (Zrachya *et al.*, 2007). The intron-hairpin approach was used to express multiple TYLCV sequences in tobacco and tomato plants (Abhary *et al.*, 2006). The treated plants were symptomless and TYLCV DNA was not detected. In general, silencing was specific to the TYLCV strain; silencing using TYLCV sequences conferred resistance to TYLCV but not to other TYLCV species or strains, such as TYLCSV (Noris *et al.*, 2004).

d) **GroEL:** This novel approach is based on the observation that GroEL produced by whitefly endosymbiotic bacteria binds to the CP of TYLCV in the insect hemolymph, thereby protecting the virion from destruction and ensuring transmission (Morin *et al.*, 1999). TYLCV-GroEL binding has been exploited to generate TYLCV resistant tomato plants. Tomatoes transformed with GroEL under a phloem-specific promoter showed milder to no symptoms in plants in the R_0 through R_2 generation. GroEL/TYLCV complexes were readily detected in resistant plants. It was hypothesized that GroEL/TYLCV complexes formed in transformed plants and that these complexes were interfering in virus movement (Akad *et al.*, 2007).

iv) **Integrated Disease Management:** Optimal strategies to reduce the impact of emerging viruses on current and future tomato production systems will most probably consist of an integrated disease management approach, in which the most efficient control measures are combined depending on the biology of the virus and its vector (Jones, 2004). Consequently, good insight into the biology and epidemiology of these viruses is critical for the development of suitable control strategies.

9.3.2. Tomato Spotted Wilt Virus (TSWV)

TSWV is an emerging threat to tomato and chilli cultivation in eastern India. TSWV incidences on tomato and chilli have been recorded in recent years in Bhubaneswar region. Currently this virus started spreading in the region and its incidence has been recorded in tomato fields near Rourkela in 2013. During the 1980's TSWV caused significant losses on tomatoes in the South-eastern United States, Eastern Europe and South America.

Tospoviruses (Bunyaviridae) are "emerging" viruses not only in the sense of their increasing economic importance as pathogens on a worldwide basis, but also in that our understanding of the complexity of this new viral taxon only began to emerge during the last decade. Tomato spotted wilt virus (TSWV) has been assigned as the type species for the *Tospoviruses*. There are now at least twelve distinct viruses (species) in the *Tospovirus* genus. The genus has been classified in the Bunyaviridae family of viruses. *Tospoviruses* are the only viruses in that family that infect plants.

9.3.2.1. Symptoms:

On tomato, foliage becomes spotted and blackened particularly on the inner leaflets toward the top of the plant. The growing point may also become necrotic and the whole plant may wilt. Fruit symptoms are variable. Small necrotic spots to large circular blotches may develop (Plate 9.4).

9.3.2.2. Host Range:

TSWV has a host range spanning several hundred species in both monocotyledons and dicotyledons. Important hosts of TSWV include:

i) **Field Crops:** Tobacco, peanut.

ii) **Vegetable Crops:** Tomato, bell pepper, potato, eggplant, lettuce, celery, bean, cowpea, spinach, cucumber.

iii) **Ornamental Crops:** Chrysanthemum, China aster, zinnia, dahlia, gladiolus, gerbera, cosmos, geranium, aster, coleus, marigold, petunia.

9.3.2.3. Transmission:

TSWV is spread by small insects called thrips (Thysanoptera: Thripidae) (Plate 9.5). It replicates in their thrips vectors, thus the insects not only spread the virus, but serve as a virus host. Thrips cannot transmit TSWV unless they acquire the virus during their immature stages (Plate 9.6).

When larvae feed on infected plants, ingested virus crosses the midgut barrier and enters the salivary glands. Transmission then occurs when the virus moves into the plant with the saliva during feeding. A midgut barrier in adults prevents virus ingested during this stage from moving to the salivary gland. While the insects remain infective for life, there is no evidence of transovarial passage from one generation of thrips to the next. Seed transmission of *Tospoviruses* is not known to occur, but some of these viruses are commonly spread in infected propagation material when crops are vegetatively propagated.

9.3.2.4. Management:

Cultural practices and varietal selection have proven effective in minimizing losses due to TSWV. A TSWV-infected plant cannot be cured. Management options focus on preventing infection and minimising spread.

- Source control
  - Avoiding contaminated or infected seedlings
  - Weed management within and surrounding crops
  - Rouging and destroying infected plants
  - Removing or cultivating in crops after harvest
  - Screening greenhouses with thrips-grade mesh
- TSWV resistant varieties (bell peppers)
- Monitoring crops for WFT and TSWV
- Biologically based integrated pest management
- Foliar insecticide sprays to control WFT

i) **Source Control:** The most important management option is to reduce the populations of WFT and TSWV. Avoid bringing in seedlings contaminated with WFT or already infected with TSWV. Sourcing seedlings from nurseries that are screened with thrips-grade mesh and actively monitoring for WFT and TSWV will greatly reduce chances of introducing them in the field.

Given that both WFT and TSWV have very large host ranges, management of weeds within and surrounding crops is essential.

If TSWV is noticed in crops, removing the plants showing symptoms will reduce the within-crop infection (Plate 9.7). Early detection and removal can greatly reduce the disease incidence and increase the fruit yield (Plate 9.8).

Crops grown in greenhouses can be protected from WFT using thrips-grade mesh covering all openings and double doors to avoid entering of thrips (Plate 9.9).

ii) **TSWV-Tolerant Varieties:** Some varieties of bell peppers and tomatoes are resistant to TSWV although strains of TSWV that break the resistance can develop in areas of high TSWV pressure. Therefore, when resistant varieties are used, it is still important to reduce the virus pressure through weed management and other sanitation measures.

iii) **Monitoring WFT and TSWV:** Sticky traps, visual monitoring and vacuum sampling are all used to monitor WFT. Blue traps may be preferable for monitoring WFT because they are less attractive to other non-thrips species (Plate 9.10). Tapping flowers over a white tray is recommended when susceptible crops are flowering.

iv) **Biological Controls:** Predatory mites such as *Transieus* (=*Typhlodromips*) *montdorensis* and *Amblyseius cucumeris* (Plate 9.11) are proving to be effective in greenhouses that use only 'selective' chemistry and have adequate ventilation.

Another promising area is the use of entomopathogenic (insect infecting) fungi. Two groups have developed strains of *Beauvaria* and *Metarrhizium* fungi that are effective against thrips and some other sap-sucking insects.

v) **Chemical Methods:** Insecticide applications targeting WFT can reduce WFT populations and reduce within-crop transmission of TSWV.

vi) **Integrated Pest Management:** For TSWV and WFT management, an IPM approach concentrates on reducing the source of both WFT and TSWV. Where it is possible, use of TSWV tolerant varieties is recommended. Monitor crops for both WFT and TSWV levels and when TSWV is observed rouge out and destroy the infected plants. If growing in a greenhouse, use thrips-grade mesh on all openings and use double doors.

If WFT numbers are high enough to be causing physical damage at vulnerable stages or TSWV is present, then insecticide sprays may be needed. Use recommended rates, adhere to the resistance management strategy and apply sprays to maximise spray coverage.

Momol *et al.* (2004) developed a successful IPM strategy to control tomato spotted wilt virus by combing UV-reflective mulch, induction of SAR with acibenzolar-*S*-methyl, and applications of 'soft' insecticides (to preserve the natural enemies of thrips).

Designing IPM Packages for TSWV Management (Plate 9.12)

- Seed treatment with *Pseudomonas fluorescens* at 10g/kg and *Trichoderma viride* at 4g/kg of seeds.
- Soil application of neem cake at 250 kg/ha.
- Soil application of *P. fluorescens* at 2.5 kg/ha.
- Selection of healthy seedlings for transplanting.
- 'Rouging' virus infected plants up to 45 days after transplanting.
- Installation of yellow sticky traps.
- Spraying neem formulations / neem seed kernel extract (need-based).

9.4. ONION, *Allim cepa*

9.4.1. Iris Yellow Spot Virus (Iris Yellow Spot Disease)

Iris yellow spot disease represents an immediate and serious threat to sustainable and productive onion cropping systems around the world, and the recent detection of this disease in numerous onion-producing countries demonstrates that the disease is spreading rapidly in a range of environments.

Iris yellow spot virus (IYSV) is a tospovirus which is closely related to two other serious viruses: *Tomato spotted wilt virus* (TSWV) and *Impatiens necrotic spot virus* (INSV). Tospoviruses belong to a genus of enveloped viruses within the family Bunyaviridae. They are the only group of plant-infecting viruses in this family.

9.4.1.1. Economic Impact:

Tospoviruses are agriculturally important because they cause severe economic damage to various vegetable and ornamental crops and are transmitted by thrips in a circulative and propagative manner.

The economic impact of IYSV can be important in onion crops. The loss of an entire crop has been reported in Brazil, Israel, some states in the USA (e.g., Oregon, Idaho and Texas), Spain (Córdoba-Sellés *et al.*, 2005) and the Netherlands. The incidence of IYSV in onion often reaches up to 60% in Israel, and over 90% in Slovenia when an onion crop was infected by the virus but there was no record of yield loss. In Spain, the impact on onion production was considered potentially devastating. In the Netherlands 50-90% of iris plants became infected with IYSV. The projected economic impact of IYSV in the western USA could reach 60-90 million dollars (for 10-15% yield loss), in addition to environmental and economic costs due to additional pesticide sprays for thrips control estimated at 7.5-12.5 million dollars (for three to five sprays on 48,500 hectares/year) (Gent *et al.*, 2006).

9.4.1.2. Distribution:

The distribution of IYSV in different states of India is presented in Table 9.2.

Table 9.2. Distribution of IYSV in different states of India

Country/State	Distribution	References
India	Restricted distribution	CABI/EPPO, 2014; Kumar and Rawal, 1999; Pawan and Poonam, 2013; EPPO, 2014
-Gujarat	Present	EPPO, 2014; CABI/EPPO, 2014
-Karnataka	Present	EPPO, 2014; CABI/EPPO, 2014
-Madhya Pradesh	Present	EPPO, 2014; CABI/EPPO, 2014
-Maharashtra	Present	CABI/EPPO, 2014; Ravi *et al.*, 2006; EPPO, 2014; Gawande *et al.*, 2014
-Tamil Nadu	Present	EPPO, 2014; CABI/EPPO, 2014
-Uttar Pradesh	Present	EPPO, 2014; CABI/EPPO, 2014

9.4.1.3. Host Range:

IYSV has a relatively restricted host range. Edible *Allium* crops including onion (bulb and seed crops), garlic, chive, shallots, leeks and some cut flower/potted ornamental species including *Alstroemeria*, chrysanthemum, iris and Lisianthus are the most economically important crops affected by IYSV. Wild *Allium* species and ornamental alliums are also potentially at risk. A range of weed species (*Datura stramonium, Nicotiana* spp. and *Amaranthus retroflexus*) can also act as reservoirs.

Six plant species have been successfully mechanically inoculated in experimental host range trials (*Chenopodium amaranticolor, C. quinoa, Datura stramonium, Nicotiana benthamiana, N. rustica* and *Gomphrena globosa*). Ben Moussa *et al.* (2005) reported infection of another three members of the Solanaceae (bell peppers, potatoes and tomatoes) but it is unclear if these are natural hosts or were artificially inoculated. The host range of IYSV is presented in Table 9.3.

Table 9.3. Host range of *Iris yellow spot virus*

Plant name	Family
Allium cepa (onion)	Liliaceae
Allium cepa var. *aggregatum* (shallot)	Liliaceae
Allium fistulosum (Welsh onion)	Liliaceae
Allium porrum (leek)	Liliaceae
Allium schoenoprasum (chives)	Liliaceae
Eustoma	Gentianaceae
Iris (irises)	Iridaceae

9.4.1.4. Symptoms:

Symptoms of IYSV consist of eyespot to diamond-shaped, yellow, light-green or straw-coloured lesions (sometimes necrotic) on the leaves, scape and bulb leaves of onion and other *Allium* host species. In the early stages of infection, lesions appear as oval, concentric rings. Some green islands can be observed within the necrotic lesions. They usually originate around a thrips feeding point. Infected leaves eventually fall over at the point of infection during the latter part of the growing season. Infection at early stages of crop growth results in yield losses. Infection at later stages of development can still cause significant losses due to reduced quality: severely infected fields will senescence prematurely and entire areas will turn brown before they collapse (Plate 9.13). Symptom severity is dependent on host cultivar, timing of infection, overall health of the host at the time of infection, and environmental conditions (Gent *et al.*, 2004). Du Toit (2005) reported that out of 46 onion cultivars tested, all but 3 had a significant yield decrease and reduced bulb size. The incidence of symptomatic plants generally increases after bulb formation (Gent *et al.*, 2006). IYSV does not always kill its host(s); however, the virus reduces plant vigour, disturbs photosynthesis and reduces bulb size. IYSV infection weakens the plants making them more susceptible to other diseases and pests. IYSV-infected onions grown for seed have reduced seed yield and quality (Evans *et al.*, 2009; Pappu *et al.*, 2009).

The symptoms of IYSV on different pant parts are as follows:

Leaves: Abnormal patterns, necrotic areas, wilting, yellowing or death.

Roots: Necrotic streaks or lesions.

Stems: Discolouration of bark, lodging; broken stems.

Vegetative organs: Surface lesions or discolouration.

Whole plant: Early senescence, wilt.

9.4.1.5. Transmission:

IYSV is only transmitted by onion thrips (*Thrips tabaci*). IYSV is transmitted by both larvae and adult thrips, but only the larvae can acquire the virus from infected plants. Virus transmission is persistent and once a thrips has acquired the virus, it can transmit it for the remainder of its lifetime. *Frankliniella fusca* can also transmit IYSV, but at a lower efficiency than *T. tabaci* (Srinivasan *et al.*, 2012).

9.4.1.6. Survival and Spread:

IYSV is likely to overwinter from one season to the next in volunteer onions or weeds found among or around crops. Emerging thrips spread the virus from infected to healthy hosts whilst feeding. The disease has the potential to spread rapidly in fields with large numbers of viruliferous thrips. The distribution of infected plants in the field is associated with feeding activity by the vector. In many cases, the damage is first noticed at the field edges, in areas of stressed plants, or in locations with thin plant stands. The virus is not seed-borne nor does it survive in the soil (Gent *et al.*, 2006; Pappu *et al.*, 2009).

9.4.1.7. Pathway Causes:

The pathway causes are presented in Table 9.4.

Table 9.4. The pathway causes

Cause	Notes	Long Distance	Local	References
Agriculture	*Via* seedlings, infected bulbs, and thrips movement. Virus overwinters in volunteer onions and weeds	Yes	Yes	Gent et *al.*, 2006; Pappu et *al.*, 2009
Nursery trade		Yes	Yes	Gent et *al.*, 2006; Pappu et *al.*, 2009

9.4.1.8. Pathway Vectors:

The pathway vectors are presented in Table 9.5.

Table 9.5. The pathway vectors

Vector	Notes	Long Distance	Local	References
Plants or parts of plants	Infected seedlings	Yes	Yes	Gent et *al.*, 2006; Pappu et *al.*, 2009
Vector/host species	The disease is vectored by the onion thrips	Yes	—	Lewis, 1997
Wind	Wind can spread thrips to new cropping areas	Yes	Yes	—

9.4.1.9. Management

i) **Prevention:** Control measures for tospoviruses need to be deployed within integrated pest management strategies which include phytosanitary, cultural, host-plant resistance, chemical and biological measures. Volunteer onions plants and weeds acting as reservoirs should be destroyed (either by tillage or herbicide) and weeds should be actively managed around cropping systems and within fields. Crop rotation should be implemented to reduce the build-up of thrips populations and transplants free of IYSV and thrips should be used. Studies have shown that thrips are less attracted to green hues than blue hues; green-leaved onion cultivars are less attractive to thrips and therefore less likely to be infected with IYSV (Kirk, 1997).

ii) **Eradication:** Eradication is possible if IYSV is detected in a glasshouse situation. However, an outbreak in field crops may be more difficult to eradicate, especially if there is an abundance of alternative hosts (including weeds) and a large outbreak of *T. tabaci.*

iii) **Cultural Control and Sanitary Measures:** Volunteer onion plants, neighbouring infected weeds and contaminated transplants are the primary inoculum sources and provide an important early-season source of inoculum to initiate outbreaks in neighbouring onion crops. Measures should be put in place to kill volunteers and control weeds and only virus-free transplants should be used. As the vector (*Thrips tabaci*) has a relatively limited host range, rotations of host with non-host crops and spatial isolation of host crops limit spread of the virus within a region. It is recommended that onion bulb and seed crops should be isolated geographically (Gent *et al.*, 2006).

iv) **Movement Control:** Movement of infected onion transplants facilitates the spread of new strains of IYSV and biotypes of *T. tabaci* within and among regions of onion production. Only IYSV-free onion transplants should be used.

v) **Chemical Methods:** The management of thrips is essential. The insecticidal management of *T. tabaci* as an indirect means of controlling iris yellow spot has been an area of study in recent years (Hammon, 2004; Cranshaw, 2006).

vi) **Host-plant Resistance:** The potential value of systemic acquired resistance (SAR) compounds for control of iris yellow spot was demonstrated by Gent *et al.* (2004). The team demonstrated that the application of SAR inducing chemicals (acibenzolar-*S*-methyl) lead to a 34% reduction in the incidence of plants with symptoms of iris yellow spot, compared with non-treated controls in onion crops.

Differences in the susceptibility of onion cultivars to iris yellow spot have been reported (Gent *et al.*, 2004; du Toit and Pelter, 2005). Breeding for resistance to/tolerance of thrips damage and IYSV is a candidate for effective disease management.

vii) **Integrated Pest Management:** An effective integrated management programme has been developed for another tospovirus which causes important losses in tomato crops. Momol *et al.* (2004) developed a successful IPM strategy to control Tomato spotted wilt virus by combing UV-reflective mulch, induction of SAR with acibenzolar-*S*-methyl, and applications of 'soft' insecticides (to preserve the natural enemies of thrips). Similar strategies could be used in the control of IYSV in onion crops (Gent *et al.,* 2006).

9.5. WATERMELON, *Citrullus lanatus* var. *lanatus*

9.5.1. Watermelon Bud Necrosis Virus (WBNV)

The major limiting factor in watermelon cultivation is low productivity due to diseases of diverse etiology. Viral diseases are a serious problem in cultivation of cucurbits in India. A total of 14 viruses have been reported to infect watermelon naturally (Caciagli, 2008). Among them, watermelon bud necrosis virus (WBNV) has emerged as a serious threat to watermelon cultivation in India recently. The occurrence of this disease in India has been reported by many workers. The first report of the disease was by Singh and Krishna Reddy (1995). According to this report the extent of WBNV was 39 to 100 per cent with an estimated yield loss of 60 to 100 per cent in Karnataka. Suresh *et al.* (2013) recently reported high incidence of WBNV in Aurangabad and Paithan regions of Maharashtra.

Watermelon bud necrosis virus is an emerging Tospovirus, (family Bunyaviridae) on cucurbits in India causing up to 100% crop loss in watermelon. The affected plant produces a range of symptoms characterized by mottling and yellowing of leaves, upright and stunting of vines and death of shoots due to necrosis of buds. The disease is wide spread in watermelon growing areas in Karnataka, Andhra Pradesh and Maharashtra with an estimated yield loss of 60-100% (Singh and Krishna Reddy 1995). The virus was initially identified as a Tomato spotted wilt like virus under the genus Tospovirus, family Bunyaviridae based on virion morphology, serology, host range and transmission by thrips. Subsequently, the virus was recognized as Watermelon bud necrosis virus (WBNV) based on the characteristics of nucleocapsid protein gene sequence (Jain *et al.*, 1998).

9.5.1.1. Symptoms:

The symptoms induced by *Tospoviruses* are highly varied and include ring spots, line patterns, wilting, stunting, silvering, mottling, bronzing, chlorosis, necrosis and a range of leaf and stem lesions (Plate 9.14) (German *et al.*, 1992). Yeh *et al.* (1992) reported that the diseased plant shows symptoms such as mosaic, crinkling and yellow spotting on leaves, narrowing of leaf lamina and severe stunting of plant, upright growth of younger branches, tip necrosis and necrotic spots on fruits. Singh and Krishna Reddy (1995) reported the disease first time in India and the symptoms described were crinkling, mottling and yellowing of leaves, necrotic streaks on vines, shortened internodes, upright branches and necrosis and dieback of buds. Bhanupriya (2006) recorded the symptoms as mosaic on leaves, stunted growth and shortened internodes on stem, un-opening of flower buds, bud necrosis, longitudinal brown necrotic streaks on vines, tendrils, petioles and fruit stalks.

9.5.1.2. Causal Agent:

Watermelon bud necrosis virus belongs to genus Tospovirus and family Bunyaviridae (German *et al.*, 1992). The virions are quasi-spherically shaped and enveloped particles of diameter varying from 80-120 nm (Moyer *et al.*, 1999). Further, electron microscopic studies on *Tospovirus* infected watermelon in India revealed that WBNV particles are roughly spherical in shape with a diameter ranging from 80 to 110 nm (Singh and Krishna Reddy, 1995). The nuleocapsid protein gene of a *Tospovirus* infecting watermelon in India was cloned and sequenced. Sequence analysis showed that the gene was most closely related to those of *Watermelon silver mottle tospovirus* (WSMV) from Taiwan and *Peanut bud necrosis tospovirus* (PBNV) from India, the two definitive species of serogroup IV (Jain *et al.*, 1998).

9.5.1.3. Transmisssion

i) **Mechanical Transmission:** Transmission is an important experimental tool to establish the etiology of viral diseases and plays vital role in disease spread. Yeh *et al.* (1992) reported the sap transmission of WBNV using phosphate buffer (pH 7.0) to *Chenopodium quinoa* and *Nicotiana benthamiana* and observed local lesions on *C. quinoa* and systemic symptoms as mottling and wilting on *N. benthamiana.* Singh and Krishna Reddy (1995) extracted *Tospovirus* from leaves of watermelon showing bud necrosis symptom using 0.1 M phosphate buffer, pH 7.0, containing 0.02 M 2 mercaptoethanol. On mechanical inoculation of the extract local lesions were produced in cowpea cv. C-152 and systemic infection was produced on watermelon. Krupashankar (1998) reported 20 to 80 per cent transmission of WBNV among different plant species. The virus produced chlorotic or necrotic lesions on *Vigna unguiculata, Chenopodium amaranticolor* and *Nicotiana rustica.*

ii) **Vector Transmission:** Studies on virus-vector relationship are necessary to find out role of vectors in virus transmission. Pittman (1927) first reported the transmission of *Tospoviruses* by thrips (Thysanoptera; Thripidae). The transmission by infective adults while thc acquisition of virus occurred during the larval stage of the insect was reported first by Linford (1932). Singh and Krishnareddy (1995) reported transmission of WBNV in watermelon by *Thrips flavus* (Thysanoptera: Thripidae). According to this report about 10-15 nymphs of *T. flavus* require 3-4 days acquisition access period and 15-20 days of inoculation access period to transmit the virus successfully. Krupashanker (1998) recorded 30 per cent transmission of WBNV in watermelon plants by *Thrips palmi* (Plate 9.15). According to the report, after the thrips inoculation, 30- 40 days are required for the expression of symptoms. Sreekanth *et al.* (2006) reported that adults of *T. palmi,* could not acquire the virus, while their larvae could acquire. On the other hand, acquired larvae could not transmit the virus, and only freshly emerged adults from these acquired larvae could transmit the virus. According to this report *T. palmi* required two days acquisition access period during larval stage and two days inoculation access period in the adult stage for the virus transmission.

9.5.1.4. Predisposing Factors:

Diseases occur throughout the entire year but the disease incidence is generally higher during dry and hot periods when thrips populations increase rapidly. The virus is transmitted by thrips species, *Thrips flavus* in a persistent (propagative) manner (vector can acquire and transmit the virus after feeding for several minutes to hours: virus replicates inside the vector). Young thrips (larval stage) acquire the virus and adult thrips spread the virus from plant to plant during feeding. The virus is not seed transmitted.

9.5.1.5. Management

i) **Cultural Methods:** Use virus-free seedlings, remove infected plants as early as virus symptoms are observed to prevent/minimize spread of the virus by thrips. Remove crop debris, weeds and other sources of thrips after the cropping season. Ploughing and harrowing may help to reduce vector population in the soil. Control thrips population by using plastic mulch and blue sticky traps.

ii) **Chemical Methods:** The following chemicals have been recommended for the management of the disease:

- Spraying seedlings with Acephate (0.15%) or Monocrotophos (0.1%) prior to transplanting.
- Spraying insecticides like Monocrotophos (0.15%), Acephate (0.15%), at fortnightly intervals after transplanting till flowering stage.
- Chemical spraying followed by neem seed kernel extract (2%) is also effective in rotation with insecticides.
- Spraying imidacloprid (confidor) at 1ml/litre of water to control sucking insects.

iii) **Host Resistance:** The wild relative of watermelon *Citrullus colosynthaes* showed highly resistant response to WBNV under field conditions but showed moderate resistant reaction (40% DI) in glasshouse conditions.

Among the hybrids and lines of watermelon tested, four were moderately resistant (EC-393240, Durgapura Meetha, MHW-6 and Ashey Questo) and three were resistant (EC-393243, Durgapura selection and RHRWH -2) to the pathogen (Pandey and Pandey, 2001).

iv) **Integrated Management:** The treatments involving seed treatment with Imidacloprid (5 g kg^{-1} seed), planting of 3 rows of maize in border, use of silver coloured UV reflective mulch along with alternate chemical sprays of Imidacloprid (0.3 ml L^{-1}) and Thiomethoxam (0.2 g L^{-1}) on crop and/or on alternate hosts were more effective. It is evident from these observations that the later strategies could effectively minimize the vector population and thus minimizing the secondary spread of WBNV on water melon. It is thus inferred that not only the control of vector but also the destruction of weed hosts or spraying insecticides is a useful and effective strategy in the integrated management of watermelon bud necrosis disease (Rajasekharam *et al.*, 2011).

9.6. CONCLUSIONS

In recent years the fungal and viral disease scenario in vegetable crops is changing, and there were outbreaks of late blight *Phytophthora infestans* on potato; yellow leaf curl virus and spotted wilt virus on tomato; Iris yellow spot virus on onion; and bud necrosis virus on watermelon. There is a dire need to redefine management strategies for the management of above pests in vegetable crops.

10

Emerging Disease Problems in Ornamental, Spice and Tuber Crops

10.1. INTRODUCTION

In developing countries, the increasing intensity of non-traditional crops that are actually being planted has also led to increases in introduction of pathogens in flowers, spices and tuber crops.

During the past 20 years, rose downy mildew, *Peronospora sparsa* has become a serious problem, causing significant losses to growers almost every year. Impatiens Necrotic Spot Virus (INSV) is the predominant tospovirus found in ornamentals transmitted by Western flower thrips (*Frankliniella occidentalis*) and more than 600 plant species are known to be susceptible.

Bacterial wilt of ginger caused by *Ralstonia solanacearum* race 4 has been found to be an emerging disease in India (Odisha State), Hawai'i and much of Asia.

Cassava mosaic virus, of the geminiviruses family transmitted by white flies (Anderson and Morales 1993), has caused enormous losses in cassava (manioc, yucca, or tapioca) in sub-Saharan Africa, where it is a staple in the diet of millions. Leaf blight caused by *Phytophthora colocasiae* is the most destructive disease of taro. The collar rot disease of elephant foot yam (*Amorphophallus poeniifolius*) caused by *Sclerotium rolfsii* is a major constraint affecting this crop.

10.2. ORNAMENTAL CROPS

10.2.1. Ornamental Plants

10.2.1.1. Impatiens Necrotic Spot Virus (INSV)

Impatiens necrotic spot virus (INSV) is a plant pathogenic tospovirus of the family Bunyaviridae. Almost all greenhouse crops are susceptible except for poinsettias and roses. The only confirmed vector for INSV is Western flower thrips (*Frankliniella occidentalis*). INSV is the predominant tospovirus found in ornamentals. More than 600 plant species are known to be susceptible. These viruses are spread from plant to plant in greenhouses by the Western flower thrips.

i) **Symptoms:** The original name of INSV was derived from symptoms (necrotic ringspots) on *Impatiens* (Plate 10.1). The virus cause a wide variety of symptoms including etches or ringspots, purple to black lesions ('chicken pox-like' sunken spots) on leaves and stems, flower colour breaking, poor flowering, stunting, death of terminals, and wilt may occur. The stage of growth can also influence symptom development.

Cineraria develops chlorotic ringspots and/or mottling on the upper leaf surface, and purple to black lesions on the veins underneath. On exacum, the virus causes tan to dark brown cankers on stems. Reiger begonias develop necrotic ringspots, mosaic, and necrosis of the leaf veins (Plate 10.2). Symptoms on *Impatiens* and New Guinea impatiens include stunting, leaf distortion and blackened spots or rings on foliage and stems. In some cases, terminals will die and the entire plant may collapse. As with other virus diseases infected plants may remain symptomless for a period of time. Symptoms can be diagnostic in some commonly infected crops. *Gloxinias* infected as young plants develop necrosis of the central leaves, resulting in collapse. Older *Gloxinias* develop necrotic ringspots on the foliage, as well as necrosis along the veins (Plate 10.2.).

The virus has only been noted on bell pepper and tomato among vegetable crops. On pepper the foliage becomes distorted and the fruit develop circular rings that are slightly indented into the skin. On tomato, foliage becomes spotted and blackened particularly on the inner leaflets toward the top of the plant. The growing point may also become necrotic and the whole plant may wilt. Fruit symptoms are variable. Small necrotic spots to large circular blotches may develop.

ii) **Transmission and Spread:** INSV is transmitted by vegetative propagation of infected plant material and by the feeding activity of certain species of thrips. In greenhouses, the most important vector is considered to be the western flower thrips (WFT), *Frankliniella occidentalis*. The virus is associated with thrips in a persistent manner.

Larvae become viruliferous after feeding for about 30 minutes on an infected plant. After a latent period lasting 3 to 18 days, the thrips can infect new plants after feeding for 5 to 10 minutes. During feeding, saliva is injected into the plant cell and the contents withdrawn. When INSV is present in the saliva, it is transmitted to the plant during the feeding process in as little as 15-30 minutes. Only the two larval instars of WFT can acquire the virus. Adult WFT do not acquire the virus from INSV-infected plants because tospoviruses pass right through the gut and do not enter the salivary fluid. Once thrips acquire INSV, virus multiplies within the salivary glands and other tissues of the thrips and is later transmitted to new plants by the feeding of the infected thrips after it has reached the adult stage. Adults however, do not transmit the virus to their young and there is no transmission from adult to egg. Overlapping generations of thrips within a greenhouse may result in continuous virus transmission. In greenhouses the viruses can be perpetuated through successive crops and weeds resulting in losses year-round.

iii) **Host Range:** The host range of INSV include Amaryllis, Aster, Azalea, Begonia, Calendula, Calla lily, Carnation, Cauliflower, China aster, Chrysanthemum, Cineraria, Coleus, Coriander, Cosmos, Cyclamen, Dahlia,

Easter lily, Eggplant, Fuchsia, Geranium, Gerbera, Gladiolus, Gloxinia, Impatiens, Lettuce, Lily, Marigold, Nasturtium, Pea, Periwinkle (*Catharanthus*), Peppers, Petunia, Salvia, Snap bean, Sweet pea, Tomato, Verbena, Zinnia, etc.

iv) **Management:** Elimination of infected plant material and WFTs are the most important management practices. Newly-acquired plant shipments should be inspected for evidence of thrips and symptoms of virus before being introduced into the greenhouse. Since the thrips are very small and stay hidden most of the time, they are difficult to detect. Yellow or blue sticky cards placed at crop height are excellent for monitoring thrips. Blue is effective for thrips but not a good colour for other insects. Symptomatic plants should be discarded or isolated and the supplier should be notified after laboratory confirmation of the disease. Since various weed species can harbour the viruses, thorough weed control is important. If possible, keep vegetatively propagated plants isolated from those produced from seed. Do not grow vegetable transplants in the same greenhouse with susceptible ornamentals. Thrips populations should be monitored with sticky traps. Since only a small number of viruliferous thrips are necessary to start an epidemic. Hence, total eradication of WFT should be the goal.

10.2.2. Rose, *Rosa* spp.

10.2.2.1. Downy Mildew, *Peronospora sparsa*

During the past 20 years, rose downy mildew has become a serious problem, causing significant losses to growers almost every year. The disease has been found in nearly every state across the United States and Canadian provinces. It does not occur each year since conditions that promote it are not always present. Various downy mildew diseases can occur on many plants (including roses) in the nursery or landscape, but not all of these are caused by the same pathogen. The downy mildews on most vegetables and flowers are distinct. They do not cross from one plant to another as a general rule, unless the disease hosts are from the same plant family. The cause of rose downy mildew is *Peronospora sparsa*, which also attacks certain cane berries (blackberry and raspberry) and all types of roses, both wild and cultivated.

Although it may seem that bare-root roses come with downy mildew, they may also be infected from other sources. Roses in landscapes or natural areas around production site, as well as crop material left over from previous seasons, are likely sources of a downy mildew outbreak. The spores can live in roses in a dormant state only to become active when the cool, wet weather of late winter and early spring occurs. It is advisable to check previous year's roses carefully for signs of downy mildew just as often as possible, besides checking the present standing crop. Extensive leaf drop may occur, if checking is not very frequent.

i) **Symptoms:** Rose downy mildew is remarkably variable. In the past 20 years the most common symptom has been angular, tan spots with a very small amount of white crystalline sporulation on leaf undersides (Plate 10.3). The name of the pathogen *Peronospora sparsa* reflects the sparse production of white spores. Sometimes collapsing buds or cankers form on newly emerging rose canes. Different species and cultivars of roses respond differently to downy mildew, making a diagnosis very difficult. Severe leaf drop is common on some cultivars.

Tiny speckles may form that are tan or even bright pink. Such symptoms are difficult to distinguish from early infections with Botrytis, powdery mildew or even rust. Under moist conditions, leaves turn brownish or yellow and drop. A white, downy fungus growth may occur on leaf undersides, but this is usually too hard to see without a microscope. Under drier conditions, leaf spots appear as brown burned areas, mimicking spray injury. Reddish streaks or sunken cankers can also occur on rose canes. Since similar cankers can be caused by other pathogens like *Botrytis* and *Coniothyrium*, it is a good idea to have canker symptoms diagnosed by State Agricultural universities/ ICAR Institutes. Sometimes downy mildew sporulation can be observed directly on discoloured spots on canes if they are first incubated at a high humidity for a few days.

ii) **Predisposing Factors:** The optimal temperatures for infection and colonization of rose leaves were 15 to 20°C and 20 to 25°C, respectively. At optimal temperatures, infection required only 2 hr of leaf wetness, although disease severity increased significantly with an increasing duration of leaf wetness up to 10 hr. Infection of leaves occurred at temperatures as low as 5°C with 8 hr of leaf wetness. The latent period of infection varied from 4 to 7 days (Aegerter *et al.*, 2003).

iii) **Management:** Rose downy mildew can be managed by using cultural and chemical methods.

a) **Cultural Methods:** Symptoms can appear almost overnight and a severe outbreak can occur if preventive measures are not taken. Fans and wind easily move spores from plant to plant but they are best allies in preventing downy mildew outbreaks, since they improve air circulation around plants, dry leaves rapidly reduce relative humidity. Cool to warm, humid and rainy weather is favourable for downy mildew development. Wet leaves and high humidity will trigger sporulation overnight. When the sun comes up, leaves start to dry and spores are released. Most spores spread by fans or wind and infect new leaves before noon. Six hours of constant leaf wetness are enough for spores of many downy mildew fungi to germinate and infect leaves. Even though fans might move spores, you should use them along with venting to reduce humidity and leaf wetness. Watering early in the day or whenever leaves will dry quickly, ensure dry foliage at night. All new plants upon receipt should be checked for symptoms. Early detection is the only way to control downy mildew. Scouting of all plants should be undertaken at least once a week, preferably every two to three days. Balanced fertilization should be followed to protect crops, since hungry plants are more susceptible to downy mildew.

The cultural steps that should be taken when growing garden roses include:

- Destruction of rose debris from previous crops' spores can overwinter in leaves and canes, then the downy mildew can attack new plants.
- Isolation of new shipments away from last year's plants to protect them from infection.
- Proper spacing should be given for plants to allow rapid drying of leaves. If the leaves are touching each other, they are too close. As the canopy closes in, the humidity increases.

- Watering should be done at a time of day that will promote the fastest possible drying.
- Preventive application of fungicides.

b) **Chemical Methods:** Choosing the most effective fungicides in order to prevent or eradicate rose downy mildew. Rotation of chemicals should be followed for prevention of disease.

Over the past 15 years about 15 trials have been performed (and reported) for control of rose downy mildew from Florida to Alabama to California in the greenhouse and in the field. They are summarized in Table 10.1.

Table 10.1. Efficacy of some fungicides in eradication of rose downy mildew on potted roses in a greenhouse

Treatment	Rate/ 100 gal	Result
Water	—	Poor
Fenstop	5 oz	Excellent
Micora	8 oz	Very good to excellent
Segway	3.5-5 oz	Very good to excellent
Stature SC	6.4 oz	Good to excellent
Orvego	14 oz	Very good
Aliette 80WDG	32 oz	Very good
Subdue MAXX	1-2 oz	Very good to excellent

The most consistently effective products for downy mildew on rose have been Subdue MAXX (1 to 2 oz/100 gal) used as a foliar spray, Micora (8 oz/100 gal), Stature SC (12.25 oz/100 gal), Aliette (2.5 oz/100 gal) and Segway (3.5 oz/100 gal). Subdue MAXX and Adorn must be tank-mixed with another product from a different MOA group that is effective on downy mildew (they can be mixed with each other, for example). Rotation of products with the same MOA group number should be avoided. For example, products in MOA group 40 (Stature, Micora and Orvego), although very effective, should not be rotated with one another. Similarly, rotation of a strobilurin (or FenStop) belonging to MOA 11 should be with another MOA group.

Another fact that appeared in reviewing the available trials was that a shorter interval is usually more effective. Thus, using a fungicide once a week is better than using once a month. Using one every 10 days is better than every 14 days. However, for a spray programme to be effective, strict adherence to spray schedule is important. If a 10-day interval is hard to implement, switch to every seven days (curative) or 14 days (preventative). Weekly spraying to prevent downy mildew on sensitive crops is common at times of year when environmental conditions favour disease. Spraying more often may be counterproductive.

10.3. SPICE CROPS

10.3.1. Ginger, *Zingiber officinale*

10.3.1.1. Bacterial Wilt, *Ralstonia solanacearum* Race 4

An important factor in the decline of ginger root acreage and sales since 2006 was bacterial wilt, a plant disease that devastates ginger. Farms with pathogen-infested soils cannot be replanted successfully. Epidemics of bacterial wilt cause major crop losses and discourage growers from farming edible ginger. Disease onset can be rapid and severe, causing great crop loss. *R. solanacearum* race 4 infects ginger in Hawai'i and much of Asia (Nelson 2013).

Bacterial wilt of ginger has also been found to be an emerging disease in Odisha State of India. It is reported to be caused by *Ralstonia solanacearum* race 4 one of the soil- and water-borne bacterium. Ginger has been found to be affected by bacterial wilt disease at DRWA Experimental Farm, Bhubaneswar, during 2008 and also in the farmers' fields in Kandhamal district, Odisha.

i) **Symptoms:** Leaves of infected plants turned yellow followed by browning (Plate 10.4). Progress of the wilting was slow. Wilt progressed upward, affected the younger leaves first, and followed by yellowing and browning of entire shoot. Infested plants were stunted and yellow. Young succulent part of the plant completely rotted (Plate 10.5). Plant dried rapidly and total foliage became yellow brown within a week. Underground rhizomes were also soft and rotten. Water-soaked appearance on infected rhizomes and vascular tissues was noticed. Green ginger leaves roll and curl due to water stress caused by bacteria blocking the water-conducting vascular system of the ginger stems. Diseased plants died before harvest.

ii) **Causal Agent:** The pathogen is a soil- and water-borne bacterium, *Ralstonia solanacearum* race 4. The pathogen is widely distributed in areas where edible ginger has been grown previously. The host range of *R. solanacearum* race 4 is restricted to edible ginger (*Zingiber officinale*).

iii) **Survival and Spread:** *R. solanacearum* spreads by infested soil adhering to hands, boots, tools, vehicle tires, and field equipment; in water from irrigation or rainfall; and within infected ginger rhizomes (Janse, 1996). This bacterium infects ginger roots and rhizomes through openings where lateral roots emerge or through wounds caused by handling, parasitic insects, or root-knot nematodes (Swanson *et al.,* 2005). The pathogen survives in soils within infected plant debris and as free-living bacteria. Ginger crops can be completely lost to the disease in heavily infested soils.

iv) **Management:** The most efficient practices to manage bacterial wilt will integrate some or all of the following practices.

   a) **Site Selection:** The cultivation site should be well drained and without recent history of ginger cultivation (at least 10 years). The best sites have rainfall dispersed evenly throughout growing season, with a drier period near the end, before and during harvest. Excessive rainfall and waterlogging conditions favour bacterial wilt and other soil-borne diseases. Avoid planting downslope from another ginger field if runoff water can carry the pathogen into field. The soil from field should be tested for the presence of pathogen by either PCR or a plant bioassay before planting.

b) **Planting Considerations:** Planting during very wet weather should be avoided, as this promotes dispersal of the pathogen within fields on muddy boots, tools, and vehicles.

c) **Bed Preparation:** Use clean, pathogen-free tractors and equipment. Hose tractor down with clean water (not obtained from streams located below existing ginger fields) before entering the field to be planted with ginger. Preferably spray with 10% bleach solution, particularly tractor wheels and blades. Prepare soil by ploughing and harrowing so that the site and soils drain well after rainfall. Proper soil and bed preparation are essential for ginger production. Prior to planting, soils are typically ploughed to a depth of 45–60 cm. Lime is incorporated to adjust the pH, and furrows are cut with a hand tiller to a depth of 30–45 cm, spaced about 150 cm apart. Fertilizer or organic amendments are placed at the bottom of the furrows and incorporated prior to seeding.

d) **Plant Pathogen-Free Seed:** Growing disease-free ginger starts with planting pathogen-free seed. Bleach can be used to sterilize the surface of ginger seeds: dip them in a 10% bleach solution (1 part commercial bleach to 9 parts water) for 10 minutes. However, this may not eliminate bacterial infections within the rhizomes. A hot water treatment, consisting of exposing seeds to a constant 50°C temperature for 10 minutes, is effective in controlling nematodes but also may not be effective for disease organisms such as *R. solanacearum* that are already present inside the rhizome.

e) **Hilling, Cultivation, and Drainage:** Hill the planting rows to promote aeration of the roots and to allow adequate soil drainage. Hilling is often done by hand, even on commercial farms. The ginger crop, as it grows from the bottom of the furrows, is hilled 3–5 times during the growing season. This results in raised hilled beds that allow proper development of the rhizomes. Hilling of the ginger row, which is done at 6-week intervals, allows the rhizome to grow vertically. Ensure adequate soil drainage and diversion ditches to prevent the runoff of infected water sources into fields that are downslope from an infected field.

f) **Restriction of Site Traffic and Pathogen Dispersal:** Strictly limit non-farm vehicles and visitors into ginger fields, as the pathogen can be in soils on truck tires or boots. Do not bring dirty tools to the farm from other locations. Disinfest boots by steeping into a 10% bleach solution before entering a ginger field. Erect fences to deter pigs and other animals from entering a ginger field.

g) **Composting:** Organic soil amendments (mulch, compost) promote microbial activity that may suppress *R. solanacearum* by competition, antibiosis, or both.

h) **Crop Rotation and Intercropping:** Rotate ginger with crops that are not hosts of the bacterial wilt pathogen, such as sweet potato and taro. Do not intercrop or rotate ginger with Solanaceous crops, including tomatoes, peppers, and eggplant. The pathogen can reproduce in these crops and build up levels of bacterial wilt in the field. Ginger may also be rotated with grain crops such as corn or upland rice. Other possible crops for use in rotation or intercropping include green onion, soybean, sweet corn, cabbage, and sun hemp.

i) **Managing Other Pests:** Control other pests that damage ginger plants, such as the lesser corn stalk borer (*Elasmopalpus lignosellus*) and the banana moth (*Opogona sacchari*), which may be controlled with applications of *Bacillus thurigiensis* (Bt). Injuries are infection courts for R. *solanacearum*, and some insects create wounds that favour the bacterial wilt pathogen.

j) **Biofumigation:** The use of volatile oils to kill or suppress a pathogen is known as biofumigation (Paret 2010). These oils are a natural component of certain green-manure crops such as mint, palmarosa, and lemon grass. When these plants are turned or ploughed into the soil several months before planting, they decompose and release the essential oils. These oils are toxic to the bacterial wilt pathogen and can severely reduce its population in the field soil. However, the oils are very expensive if bought pre-processed, not naturally occurring as a component of green-manure crops, and in this case their use may not be economically feasible.

k) **Timely Harvest:** Harvesting on time minimizes crop exposure to the pathogen.

10.4. TUBER CROPS

10.4.1. Cassava, *Manihot esculenta*

10.4.1.1. Cassava Mosaic Disease (CMD)

Cassava mosaic disease (CMD) is the most serious viral disease of Cassava in India and presently being also observed in Odisha state.

i) **Symptoms:** The first symptom appears on young leaves as chlorotic speck. Gradually, they enlarge and intermix with green tissue to provide a mosaic pattern (Plate 10.6). The pale discolouration may be intensified to yellow colour depending on the varieties. The leaf area is reduced and in extreme cases leaf distortion and shoe string appearance are observed. Intensity of symptoms varies with season, genotypes and virus load present. There may be symptom variation in the same plant. Some plants show symptoms on the younger leaves and recover at the later stages. In serve case of infection, the growth of the plant is affected which ultimately leads to crop loss.

Tuber yield of diseased plants vary with different varieties as well as time of infection. Up to 80% crop losses have been recorded in highly susceptible varieties as against 18% in field tolerant varieties (Plate 10.7). Crop loss is maximum when the disease appears at the time of planting. However, no reduction in yield is recorded when the disease occurred after 5 months of planting.

ii) **Causal Agent:** The disease is caused by single-stranded DNA virus of the genus *Begomovirus* under the family *Geminiviridae*. Two distinct virus species under the genus *Begomovirus*, viz., *Indian cassava mosaic virus* (ICMV) and *Sri Lankan cassava mosaic virus* (SLCMV) are known to cause cassava mosaic in India. The particles typically are bisegmented, approximately 30 × 20 nm.

iii) **Transmission:** A specific biotype of whitefly, *Bemisia tabaci* (cassava biotype) is known to transmit the virus. Insect transmission alone does not cause much spread of the disease— man appears to be the main factor. The virus could be easily transmitted through grafting and dodder.

iv) **Spread:** Primary spread of the disease is through the infected planting material. Secondary spread of the disease is through the insect vector whitefly, *Bemisia tabaci.* The virus occurs in the saliva of the whitefly. During feeding, the insect injects saliva containing the virus into cassava leaves. The virus multiplies and occurs in large numbers in the leaves and stems. Cassava mosaic disease is also spread by planting stem cuttings from plants infected with the disease.

v) **Management:** Cassava mosaic disease can be managed by using cultural, biological, host resistance, and integrated methods.

a) **Cultural Methods:** Use of disease-free planting material and cassava true seed programme can be employed to manage cassava mosaic disease.

Use of Disease-Free Planting Material: Meristem culture combined with thermo and chemotherapy will help virus elimination from the infected plant. However, it is recommended that meristem- raised tissue-cultured plants (Plate 10.8) should be indexed for the absence of virus through molecular diagnostic techniques like PCR or NASH. Wherever possible, the virus-free planting materials may initially be multiplied in the vector-free zones or disease-free areas, and later on the affected areas may be replaced by healthy plants. It should be remembered that the tissue cultured plants are not resistant to disease; hence, they are likely to get infected under field conditions. However, it has been demonstrated that loss due to secondary spread of the disease is minimum compared to primary spread through infected cuttings. Like other viral diseases, it is very difficult to control CMD with any one control measure.

Cassava True Seed Programme: Cassava true seed programme was useful in popularizing cassava in non-traditional areas where the crop was grown mainly for starch purpose. The male sterile line "Ambakkadan" which is a popular cultivar grown in Kottayam and Idukki districts of Kerala had recorded the highest mean number of fruits per plant (168). The weight of 100 seeds ranged from 9 to 12.9 g. Seeds could be stored under ambient conditions up to 6 months without loss of viability. Male sterile lines recorded very good germination (80–90%). Field establishment when the seedlings were transplanted from primary nursery ranged from 75 to 91%. The seedling progenies were free from CMD, compared to the clones. Progenies of CE-647, S-1, and Ambakkadan were good yielders at closer spacing of 70 × 70 cm, and the yields were at par with that of the clones of promising varieties. Ambakkadan seedlings recorded higher tuber yield of 2.2 kg plant^{-1}, and the cooking quality was very good in both the seasons. In the first clonal evaluation of seedling progeny lines, the progeny lines of CE-639, CE-630, CI-453, S2, and Ambakkadan have recorded distinctly higher yield than the control. The clonal selections of Ambakkadan × TCH-2 recorded maximum tuber yield (29.38 t ha^{-1}) followed by Ambakkadan × S-856 (27.80 t ha^{-1}), and the selections were free from CMD.

b) **Biological Methods:** Among four rhizobacteria [*Pseudomonas aeruginosa* (RB9), *Bacillus pumilus* (RB26), *B. subtilis* (EN16), and *Enterobacter cloacae* (EN22)] and 11 different combinations, the incidence of CMV was less in all the treated plants compared to control. About 53 % CMV incidence was noticed in the case of *E. cloacae*- treated plots compared to 90 % in control. There was significant increase in plant height, girth, and tuber yield in all the treated plants compared to control. Among the combination treatments, *B. subtilis* and *E. cloacae* showed significant reduction of CMV (31 %), but it was less than *E. cloacae.*

c) **Host Resistance:** Triploid cassava cvs. Sree Apoorva, and Sree Athulya are tolerant to cassava mosaic disease (CMD). Growing field tolerant varieties—H97, H165, Sree Visakham and Sree Sahya—are also helpful for managing the disease.

GM Cassava: The recent progress in the biotechnology of genetic modification and the continuing importance of cassava in the welfare of impoverished people in Africa make GM orphan crops such as cassava promising tools to significantly reduce poverty and improve food security in the least-developed countries (LDCs) in sub-Saharan Africa.

Transgenic cassava resistant to cassava mosaic disease (CMD) is of particular interest to many African countries due to CMD's prevalence in the continent and the relatively advanced knowledge of the disease in American and European research institutions.

Exploitation of the plant immune system against viruses (Ding and Voinnet, 2007) by expression of hairpin RNA homologous to viral sequences has proven effective to generate virus resistant crops including cassava (Yadav *et al.,* 2011). The modularity of the hairpin RNA technology is suitable to generate plants resistant to multiple viral species. Stretches of viral sequences can be combined into a single hairpin construct (Bucher *et al.,* 2006). Alternatively DNA sequences highly conserved between viral species can be targeted to generate transgenic plants with broad-spectrum resistance. Analysis of the CBSV and UCBSV isolates revealed high DNA sequence conservation in the 3' region of the viral genome, where the gene for the coat protein (CP) sequence is located. The stable resistance to both CBSV and UCBSV can be engineered by using hairpin RNA homologous to the 3'-end of the CBSV-CP sequence in the cassava model cultivar 60444 (previously referred to as TMS 60444). Using an improved cassava transformation protocol, the technology was transferred to a selected Nigerian cassava landrace previously identified to be CMD resistant (Raji *et al.,* 2008). The combination of natural and engineered virus resistance is therefore a promising approach to combat multiple cassava viral diseases in Africa.

d) **Integrated Management:** An integrated approach for the management of the disease has been recommended. The following practices are advocated for effective management of the disease:

- Selection of disease-free meristem derived planting material, followed by clonal multiplication with periodical screening and rouging of freshly infected plants, will be useful to raise a good disease-free crop.
- Use disease-free healthy planting materials by selection.

- Spray 0.05 % dimethoate for control of white fly vector.
- Aqueous leaf and root extracts of *Boerhaavia diffusa* and *Mirabilis jalapa* were found to inhibit the ICMV under glasshouse conditions.
- Disease-free planting material can be multiplied on a large scale at higher altitude where the whitefly population is low or nil.
- Breeding for resistance to CMD.
- Use field-resistant or field-tolerant cultivars like Sree Padmanabha, Sree Vijaya, H-97, H 165, Sree Visakham, Sree Sahya, etc.
- Strict field sanitation and rouging of infected plants and burning.
- Use of true cassava seeds.
- Planting of border rows with crops like maize or sorghum.
- Raising the plants in the nursery at closer spacing prior to transplantation into the main field is a useful step to prevent the primary spread of the disease in the main field.
- Adherence to strict sanitary practices such as timely harvest and prompt disposal of crop residues, and eradication of self-sown plants and weeds which may harbour both the disease and vectors.

10.4.2. Taro, *Colocasia esculenta*

10.4.2.1. Leaf Blight, *Phytophthora colocasiae*

Taro leaf blight caused by *Phytophthora colocasiae* is the most destructive disease of taro (*Colocasia esculenta* var. *esculenta* and *C. esculenta* var. *antiquorum*).

i) **Symptoms:** Affected leaves initially show small dark spots which enlarge rapidly and turn purplish brown with yellowish margins (Plate 10.9). The lesions frequently form concentric zones and exude drops of yellowish liquid. Periphery of the diseased tissues may be covered with whitish fuzz consisting of sporangia. As the disease progresses, the lesions (mostly along the leaf margin) continue to expand and frequently coalesce. Diseased tissues disintegrate forming holes of irregular size and shape on the affected leaves. Occasionally the pathogen may cause water-soaked lesion on the petioles. Infected leaves collapse within 20 days of unfurling, compared to 40 days for healthy leaves. The normal 6-7 leaves/plant is reduced to 3-4 leaves/plant in severe disease incidence.

After harvest, grey-brown to dark-blue lesions occur on undamaged corms. These lesions enlarge rapidly and coalesce. The boundary between the healthy and diseased tissues is distinct and soft. Affected corms are almost completely decayed within 8 days after harvest in wet conditions. Deciduous sporangia with apical papilla are produced on slender sporangiophores which branch irregularly or sympodially with a swelling at the point of branching. Sporangia are ovoid to ellipsoid mostly 45-50 x 23 μm with a length-to width ratio of 1:1.6. Chlamydospores are thick walled, usually 26-30 μm diameter. Oospores averaging 29μm diameter are produced in oogonia with amphigynous antheridia attached.

ii) **Survival and Spread:** The pathogen survives in infected tubers and self-sown *Colocasia* plants. Spores are produced on the leaf spots and are readily spread by rain. Disease spread is primarily by splashing rainwater.

iii) **Predisposing Factors:** Frequent rainfall and high RH increase disease severity. Most favourable weather conditions are 20–22 °C temperature and 100 % RH during night and 25–28 °C and 100 % RH at day with cloudy weather. Disease emergence is favoured by rainy overcast weather with low night temperatures.

iv) **Management:** Taro leaf blight can be managed by using cultural, chemical, biological, host resistance and integrated methods.

a) **Cultural Methods:** Various cultural methods have been recommended for the control of *Colocasia* leaf blight. Removal of infected leaves has been effective during the early stages of disease development in a number of countries (Hunter *et al.,* 2001). According to Jackson *et al.* (1980), regular rouging of diseased leaves in plots affected by a severe blight did not eradicate the pathogen. Disease increased rapidly after rouging ceased and corm yields were greatly decreased. Rouging may delay the start of epiphytotics (Ashok and Mehrotra, 1987). Wide spacing of plants has been reported to reduce disease severity, but this appears to have a negligible effect when conditions favour disease development (Hunter *et al.,* 2001).

Other cultural methods that have been recommended include delayed planting on the same land for a minimum of 3 weeks, avoiding plantings close to older infected ones, and preventing the carryover of corms or suckers which can harbour the pathogen from one crop to another (Jackson, 1999).

Amosa and Wati (1997) reported that disease incidence and severity of leaf blight were lower in *Colocasia* /maize intercropping system than those grown in monoculture. The effect of planting density and relative time of planting on *Colocasia* /rice intercropping system yielded similar results (Agyekum, 2004). Fertilizer treatment may help the plant cope with leaf blight (Tilialo *et al.,* 1996).

Wide spacing minimizes spread of the disease due to rain splashing. Wide spacing together with interplanting and open sides to the *Colocasia* field helps to improve the air flow and thus keeps the humidity down. Farmers should avoid walking through the field when it is wet as this encourages the spread of spores. Where possible the *Colocasia* should be cultivated on higher land. Crop remains should be burnt after harvest. In case of mild infection of young plants (up to 2–3 months), leaf rouging can be effective in controlling or delaying the buildup of the disease. Rouged leaves should be burnt.

Strict hygiene should be observed at planting, and infected leaves should be removed twice a week for 80–90 days after planting, or until a dense canopy is formed.

Crop sanitation and rouging, removing infected plant parts or whole plants from the field, are possible measures against TLB. Used early in the season or during mild infections, they can limit disease severity (Jatala and Bridge, 1993). During moderate to severe infections, however, removing infected leaves becomes counter productive, reducing corm yield even more effectively than leaves lost to the disease. Increased spacing between plants may slow disease spread by lowering relative humidity and wetness in the field and by decreasing spore transfer due to leaf contact. The movement of spores in a field during heavy rains and wind, however, quickly leads to the same level of disease. Total yield is reduced with wider spacing as fewer plants are grown in the same area (Jatala and Bridge, 1993).

The planting of *Coleus blumei* (a plant whose nectar and pollen attract adult parasitoids) in the taro fields favours the development of natural enemies.

b) **Chemical Methods:** The use of fungicides such as copper and copper metalaxyl-based compounds is the most reliable and popular with farmers because of their quick and effective action (Adejumo, 1997). Jackson (1996) reported that blight disease can be controlled by spraying with Copper fungicides. Ashok and Mehrotra (1987) observed in field trials that excellent control of *Colocasia* leaf blight was obtained when plants were treated with chloroneb and captafol, good control with metalaxyl, fair control with copper oxychloride, and poor control with thiophanate methyl and zineb.

Field experiments conducted to study the effect of fungicides in controlling leaf blight caused by *P. colocasiae* in *C. esculenta* revealed that 0.2 % metalaxyl and mancozeb as Ridomil MZ-72 was the most effective treatment, followed by 0.2 % captafol, Bordeaux mixture (1% copper sulfate and lime), and 0.25 % mancozeb (Ashok and Saikia, 1996). A significant increase in yield was recorded for all treatments over the untreated control.

The frequency and time of spray application have been reported to affect the effectiveness of fungicides (Adegbola, 1993). Bergquist (1974) confirmed the effect of fungicide rate, spray interval, timing of spray application, and precipitation in relation to control of leaf blight disease of *C. esculenta.* In an experiment conducted by Bergquist (1974), *C. esculenta* was sprayed with mancozeb at rates of 4.48, 2.24, or 1.12 kg/ha at intervals of 5, 7, 10, or 14 days at drier and wetter sites. Rate of fungicide had no effect in the drier sites, while at wetter sites, the highest rate of 4.48 kg/ha was the most effective. Spraying every 5 days was significantly more effective than spraying every 14 days. Applications of fungicide at 7-day intervals gave substantial disease control.

Tribasic copper sulphate and copper oxychloride have been used successfully at intervals of 7–15 days. Zineb and mancozeb have been shown to give effective control. The best control was obtained with sprays of mancozeb at 2.24–4.48 kg/ha at intervals of 5–7 days. It has been suggested that copper oxychloride sprays and phytosanitary measures could be used together in a practical control programme.

Four sprays of 0.25 % zineb at 15-day interval was found effective in reducing disease intensity (23.24 % as compared to 56.15 % in control) and increasing the tuber yield (20.91 t/ha as compared to 15.49 t/ha in control).

Two sprays of Ridomil MZ at 0.25 % at 20-day interval starting just after the appearance of the disease were found significantly superior to other treatments in terms of per cent disease incidence (12.06 %) and per cent disease index (8.13 %) yielding 20.3 t/ha of corms which was at par with Indofil M-45 and Captaf (Table 10.2).

Table 10.2. Field efficacy of fungicides for the control of leaf blight disease and yield of *Colocasia*

Fungicide	Conc. (%)	Disease Incidence (%)	Disease index (%)	Corm yield (t/ha)
Indofil M-45	0.25	21.88	21.25	19.028
Blitox-50	0.30	20.31	18.12	17.500
Captaf	0.20	18.75	17.75	19.514
Ridomil MZ	0.20	12.06	8.13	20.347
Bordeaux mixture	0.50	23.44	20.00	17.222
Control	—	46.88	32.50	14.790
CD (P = 0.05)	—	**5.505**	**4.487**	**0.9465**

c) **Biological Methods:** Several potential biocontrol agents have been reported on various plants. These include *Aspergillus niger*, *Penicillium* spp., and *Trichoderma viride* (Frais and Garcia, 1981), *Bacillus* spp. (Odigie and Ikotun, 1982), and *Anoplolepis longipes* (McGregor and Moxon, 1985). Effect of soil application, seed treatment, and foliar spray of rhizobacterial cultures that were isolated from *C. esculenta* on *Phytophthora* blight reduced the disease incidence and severity and increased the yield, compared to untreated pathogen-inoculated control plants (Sriram *et al.,* 2003).

Biological control agents may be used judiciously as a complement to chemical application and cultural practices. In such a situation, compatibility with the synthetic fungicide would be desirable, as it is often possible to schedule both in control programmes (Coffey, 1991).

Prophylactic spray of *Pseudomonas* 1–2 % checked the incidence of leaf diseases like blight. *Trichoderma viride*, *T. harzianum*, and *Pseudomonas fluorescens* were found effective against *Colocasia* leaf blight.

Under polyhouse conditions, the antagonistic rhizobacterial cultures S1B3, S11B4, S13B5, and S23BS, when applied as seed tuber treatment, reduced the *Phytophthora* blight disease severity. In these treatments there was no disease incidence compared to control where the disease severity was 2.92 on a 0–5 disease rating scale.

Soil application of rhizobacterial cultures S4BS, S13B5, and S23BS gave complete control of leaf blight disease compared to control where the disease severity was 2.83 on a 0–5 disease rating scale. Foliar application with S1B4 and S11B3 reduced the disease severity to 0–0.33 rating compared to 2.66 in control.

Under field conditions, tuber treatment with S1B3, soil application of S13B5, or foliar application with S1B4 and S11B3 reduced the disease severity and increased the yield compared to untreated pathogen-inoculated control plants. Seed treatment with S1B3 resulted in tuber yield of 255 g/plant compared to 95.42 g in control. Soil application with S13B5 resulted in 232.65 g/ plant, while in foliar application with S1B4 or S11B3, yields were 274 g and 605 g per plant, respectively. These treatments promoted the plant growth also.

The field application of bacteria in combination (seed treatment, soil treatment, and foliar spray) helped in reducing the leaf area damaged due to blight by 41 % during the first peak of the disease spread and by 28 % during the second peak of the disease spread. Rhizobacteria treatment also helped in reducing the storage losses. The storage loss of tubers harvested from rhizobacteria- treated plots ranged from 4.14 to 21.24 % compared to 26.02 % in fungicide- treated plots, resulting in 18–36 % increased yield in the field trials (Sriram and Misra, 2007).

Potent biocontrol agents of vermicompost origin against taro leaf blight were identified as *Bacillus subtilis*, *B. cereus*, *Providencia rettgeri*, and *Trichoderma asperellum*. The bacterial isolates expressed N_2 fixing, P and K solubilizing, and growth promotion properties. Disease suppression varied with the origin of vermicompost.

Minimum concentration of 10% was required for pathogen suppression. The application of vermicompost reduced taro leaf blight incidence by 10–30 % in pot culture. Vermin wash reduced the lesion development by *P. colocasiae* on detached taro leaves in vitro. The production of defense enzymes, chitinase, glucanase, and total phenols, increased with the application of vermicompost in taro plants.

d) **Host Resistance:** Colocasia cvs. 'Jhankri' and 'Muktakeshi' are tolerant to leaf blight disease. Taro cvs. P1, P5, P7, P10, and P20 (tiapula planting sets) were found resistant to taro blight and most popular among farmers in American Samoa.

e) **Integrated Methods:** The following practices are recommended for integrated management of taro leaf blight:

- Use of healthy planting material.
- Growing resistant varieties—Muktakeshi and Jankhri.
- Suitable adjustment in planting dates to escape disease/crucial stage.
- Field sanitation and removal of infected leaves.
- Intercropping with non-host crops.
- Removal of self-sown plants from the vicinity.
- Preventive sprays with mancozeb (protective) and metalaxyl (systemic) as a last resort. Spraying of metalaxyl (0.05 %), folpet (0.2 %), and dithane M-45 (0.25 %) is also effective.
- Early planting (in May) with early maturing variety like Telia.
- Use of biocontrol agents— *Trichoderma viride.*

Biological and Chemical Methods: Tuber treatment with *T. viride* + 0.25 % mancozeb spray after the first appearance of the disease + 0.2 % Ridomil spray 20 days after mancozeb spray was effective in reducing the per cent leaf area damaged and increased the yield.

Host Resistance, Cultural, and Biological Methods: IDM package (selection of healthy tubers + cormel treatment with *Trichoderma viride* at 5 g/kg mixed with cow dung slurry before planting + application of *T. viride* enriched compost before planting + mulching with paddy straw after planting + spray of CTCRI formulation at 60 and 90 DAP) recorded minimum incidence of taro leaf blight (12.9 % compared to 20.3 % in control) and intensity (7.8 % compared to 10.8 % in control) with highest sprouting (99.4 % compared to 94.74 % in control) and cormel yield (13.5 t/ha compared to 9.5 t/ha in control) in Bihar.

In Andhra Pradesh, seed tuber treatment (cv. Telia) with *Trichoderma viride* (5 g/kg cow dung slurry) and soil application of *T. viride* – enriched compost at 1 kg/pit gave maximum sprouting (97.93 %) and yield (16.5 t/ha) and least disease incidence (PDI—17.36) (Table 10.3).

Table 10.3. Effect of cultural methods, bioagents, botanicals, and host resistance for the management of leaf blight disease and cormel yield

Cultivar	Treatment	Sprouting (%)	% Disease Incidence on leaves (PDI)	Corm yield (t/ha)
Telia (susceptible)	Mulching with dry leaves (15 cm thickness)	95.83 (78.27)	24.09 (29.34)	12.7
	Mulching with black polythene	94.98 (77.19)	23.71 (29.08)	13.4
	Seed tuber treatment with *T. viride* (5 g/kg cow dung slurry) and soil appln. of *T. viride* – *enriched* compost (1 kg/pit)	97.93 (81.78)	17.36 (24.47)	16.5
	Foliar spray with CTCRI formulation* at 60 and 90 DAP	94.23 (76.10)	23.46 (28.97)	12.8
	Control untreated	94.98 (77.09)	30.55 (33.51)	11.0
Muktakeshi (resistant)	—	96.55 (77.32)	11.01 (18.60)	14.2
CD (P = 0.05)	—	**2.075**	**5.544**	**2.998**

* CTCRI formulation: Water - 10 L; neem cake - 1 kg; and neem, bhel, custard apple, eucalyptus, and castor leaves, 21 each were taken in an earthen pot and mixed thoroughly, and the mouth of the pot is sealed with polythene sheet and allowed to decompose for 3–4 weeks. Finally dilute the mixture with 100 L of water and filter through two layers of muslin cloth and use for spraying

v) **Positioning of Control Methods:** In terms of the development cycle of the disease, positioning of control methods are as follows:

(a) **Choice of the Plot before Planting:** Choose plots that are isolated from other taro fields, if possible surrounded by high vegetation (forest clearing). When the plots are close, synchronize the taro planting dates if possible, which prevents the young plants from being contaminated by affected plots that represent a significant stock of inoculum.

(b) **At Planting:** Use cultivars that are resistant to the *Phytophthora* (Muktakeshi, Jankhri) obtained from varietal selections or creations carried out by agronomic research. Resistant material can be obtained in particular from the Regional Germplasm Centre of the Pacific Community based in Fiji.

(c) **Throughout the Plant Cycle:** Remove the diseased leaves as soon as the symptoms of wilting appear and burn them. To do so, it is necessary to regularly inspect the plots, in particular 3–4 days after strong rainfall or in the event of morning dew. Apply fungicides on the leaves by spraying as soon as the symptoms appear and following the removal of the affected leaves.

(d) **After the Harvest:** To avoid rotting due to the *Phytophtora* (and to the *Pythium*), the corms will be treated before being put on the market by soaking them in a solution with a 1 % bleach (sodium hypochlorite) content for 2 min. Dry the corms well before packing in a polyethylene bag.

10.4.3. Elephant Foot Yam, *Amorphophallus poeniifolius*

10.4.3.1. Collar Rot, *Sclerotium rolfsii*

The collar rot disease of elephant foot yam (*Amorphophallus poeniifolius*) caused by *Sclerotium rolfsii* is a major constraint affecting this crop. This disease is generally observed in later part of crop growth but it can infect the crop at any stage of crop growth.

i) **Symptoms:** The symptoms include water soaked patch on the collar region of the pseudostem and finally the entire plant falls due to rotting of the collar region. Injury to the collar region during intercultural operations and infestation of roots by *Meloidogyne incognita* (nematode) predisposes the plants to collar rot disease. The disease causes damage by rotting of the collar region of the stem (Plate 10.10) which ultimately topples down and is responsible for heavy reduction in yield and qualitative degradation of the crop. The symptoms include water-soaked patch on the collar region of the pseudo-stem and finally the entire plant falls due to rotting of the collar region. The white mycelial mat and lot of sclerotia could be seen on the collar region of the affected plant.

ii) **Management:** Collar rot on elephant foot yam can be managed by using cultural, chemical, biological and integrated methods:

a) **Cultural Methods:** Disease can be managed by the use of disease-free planting material, removal of infected plants, improving drainage, and application of neem cake in soil. Thick mulching with paddy straw or other organic waste reduces the frequency of weeding and other intercultural operations, thereby avoiding injury to the plants. Mulching with paddy straw/polythene sheet reduced the collar rot disease.

Seed treatment and drenching of soil with vermiwash showed lower collar rot incidence (10–30 %). Soil application combined with spraying of vermiwash 90 and 120 DAP resulted in tuber yield increase of 44.0 and 121.4 % over package of practices (POP) recommendations and control, respectively.

The minimum collar rot incidence, i.e., 47.22 %, was recorded in the crop sown on March 28, being at par with March 15 and February 15 sown crop (Table 10.4). Maximum yield of the crop which was recorded in March 23 sown crop being at par with February 15, March 15, and April 30 sown crop indicated that the crop sown on early dates get more time for its vegetative growth which may have resulted in good yield. The lowest incidence and severity of collar rot in March 28 sown crop may be another reason for enhancing the crop yield (Singh *et al.,* 2006).

Table 10.4. Effect of different dates of sowing on collar rot and corm yield (Singh *et al.*, 2006)

Date of sowing	% collar rot incidence*	Yield (t/ha)
February 15	52.08 (46.19)	12.326
March 15	49.99 (44.99)	12.586
March 28	47.22 (43.39)	13.338
April 30	56.25 (48.60)	11.486
May 16	73.33 (59.21)	9.982
CD (P = 0.05 %)	**8.835**	–

* The figures given in parentheses are angular transformed values

b) **Biological Methods:** Application of *Trichoderma* in soil and treatment of planting material in cow-dung slurry mixed with *Trichoderma* have been found very effective against collar rot. An isolate of *Trichoderma* sp. (Td-1) isolated from the parasitized sclerotia of *S. rolfsii* was highly effective in suppressing the pathogen on elephant foot yam under storage conditions (Mukherjee and Raghu, 1997).

Potent biocontrol agents of vermicompost origin against elephant foot yam collar rot were identified as *Bacillus subtilis, B. cereus, Providencia rettgeri,* and *Trichoderma asperellum.* The bacterial isolates expressed N_2 fixing, P and K solubilizing, and growth promotion properties. Disease suppression varied with the origin of vermicompost. Minimum concentration of 10 % was required for pathogen suppression. Application of vermicompost reduced collar rot incidence by 10–30 %.

c) **Chemical Methods:** Disease can be managed by drenching soil with captan 0.2 %.

d) **Integrated Methods:** Management of *Amorphophallus* collar rot can be achieved by integrating cultural, chemical, and biological methods in different combinations.

Biological and Cultural Methods: Application of neem cake and biological control agents like *Trichoderma viride, T. harzianum,* and *Pseudomonas fluorescens* was found effective against amorphophallus collar rot.

Biological and Chemical Methods: Seed corm treatment followed by soil application/ drenching with *Trichoderma viride* and Carboxin at 600 g/ha is recommended for the management of collar rot of *Amorphophallus.*

Combination of corm and soil treatment with *T. harzianum* lowered disease incidence to 12.9 % followed by corm + soil treatment with captan (0.2 %) recording 14.8 % disease incidence as against 83.3% in control. Both *T. harzianum* and *B. subtilis* multiplied more in soil when applied as soil treatment than used as corm treatment. Antagonists and chemical treatment significantly lowered the rhizosphere population of *S. rolfsii* (Table 10.5) (Gogoi *et al.,* 2002).

Table 10.5. Effect of corm and soil treatment with antagonists and chemicals on the incidence of collar rot in elephant foot yam (Gogoi *et al.*, 2002)

Treatment	% Disease Incidence (PDI)*	Population density ($x10^3$ cfu/g of soil)**		Population density of *S. rolfsii* ** (viable sclerotia/50 g of soil)
		T. harzianum	*B. subtilis*	
Corm treatment				
T. harzianum	38.8 (38.3)[b]	26.3	—	5.2
B. subtilis	49.9 (44.8)	—	31.6	6.0
Captan	33.3 (35.0)	—	—	5.5
Soil treatment				
T. harzianum	31.4 (32.3)	31.1	—	4.9
B. subtilis	40.7 (37.8)	—	34.1	6.1
Soil drenching				
Captan	27.7 (31.2)	—	—	4.4
Corm + soil treatment				
T. harzianum + *T. harzianum*	12.9 (17.6)	46.2	—	3.5
B. subtilis + *B. subtilis*	33.3 (35.1)	—	43.2	5.0
Captan + Captan	14.8 (21.6)	—	—	3.8
T. harzianum + Captan	33.3 (35.1)	23.6	—	5.0
B. subtilis + Captan	24.0 (28.8)	—	31.7	4.8
Captan + *T. harzianum*	25.9 (27.5)	33.9	—	4.0
Captan + *B. subtilis*	31.4 (33.8)	—	36.5	4.8
Control	83.3 (80.9)	16.9	19.8	8.2
CD (P = 0.05)	**7.6**	**1.7**	**2.0**	**0.7**

* Figures within parenthesis are angular transformed values

** Average of three replications pooled over two seasons

Cultural, Chemical, and Biological Methods: The maximum reduction in collar rot incidence (75.99 %) was exhibited by the crop treated with *Trichoderma harzianum* (T.h.) + *Pseudomonas fluorescens* strain (P.s.c.) + FYM with 88.47 % yield increase, which was significantly higher than the check (Table 10.6). The use of antagonists, their various combinations, and chemicals showed significant reduction in the collar rot incidence by lowering the number of infected plants as compared to check (Singh *et al.*, 2006).

Table 10.6. Effect of the application of antagonists, their combinations, and chemicals on collar rot and corm yield (Singh *et al.*, 2006)

Treatment/dose/mode of application	Collar rot		Yield t/ha	% yield increase
	% Disease incidence	% disease control		
Trichoderma harzianum (Th) - 100 g/10 L of water-seed treatment	18.75 (25.45)	63.99 (53.12)	21.354	70.83 (57.31)
Pseudomonas fluorescens (Pf) - 100 g/10 L of water - seed treatment	27.08 (31.33)	48.00 (43.85)	17.708	41.66 (40.20)
Pf (a) + Pf (b)-(1:1) 50 + 50 g/10 L of water—seed treatment	22.91 (28.55)	56.00 (48.45)	20.607	64.85 (53.64)
Th + Pf (c) - (1:1) 50 + 50 g/10 L of water - seed treatment	64.85 (53.64)	68.01 (55.56)	22.743	81.94 (64.85)
Carbendazim - 10 g/10 L of water - seed treatment	37.50 (37.72)	27.99 (31.94)	14.756	18.04 (25.13)
Ridomil MZ - 25 g/10 L of water — seed treatment	29.16 (32.66)	44.00 (41.55)	16.718	33.74 (35.51)
Mancozeb - 30 g/l0 L of water — seed treatment	33.32 (33.99)	36.02 (36.88)	15.451	13.60 (29.06)
Pf (a) + Pf (b) + FYM - 75 + 75 g + 300 g/pit	25.00 (27.11)	51.99 (46.14)	20.052	60.41 (51.01)
Th + FYM - 150 g + 300 g/pit	24.58 (20.70)	52.80 (46.61)	21.927	75.41 (60.27)
Pf (c) + FYM - 150 g + 300 g/pit	25.00 (28.55)	5 l.99 (46.14)	18.836	50.68 (45.39)
Th. + Pf (c) + FYM - 75 + 75 g + 300 g/pit	12.50 (20.70)	75.99 (60.66)	23.559	88.47 (70.15)
Check - treated with plain water - seed treatment	52.08 (48.59)	0.05 (1.28)	12.500	0.05 (1.25)
CD (P = 0.05)	**4.191**	**5.004**	—	**5.11**

* The figures given in parentheses are angular transformed values

After one application, the maximum reduction in collar rot incidence was recorded with the application of Ridomil MZ + neem cake and *Trichoderma harzianum* which were similar (53.85%). In the case of two applications of fungicides, bioagents, and neem cake, it was found that maximum reduction in collar rot incidence (64.01%) was achieved due to the application of Ridomil MZ + neem cake. The use of bioagents, fungicides, and neem cake showed significant reduction in disease incidence by lowering the number of infected plants as compared to control.

The highest yield was recorded in the case of neem cake + Ridomil MZ. The yield increase may be due to the lowering down of the inoculum density and capacity which resulted in the improvement of the plant health, with the application of antagonists.

10.5. CONCLUSIONS

Emergence of crop pests has been affecting crop production from time immemorial. There were outbreaks of downy mildew, *Peronospora sparsa* on rose; and Impatiens necrotic spot virus on ornamental plants. Among spice crops bacterial wilt, *Ralstonia solanacearum* on ginger is emerging as a serious disease; while mosaic virus on cassava; leaf blight, *Phytophthora colocasiae* on taro; and collar rot disease, *Sclerotium rolfsii* on elephant foot yam are emerging as devastating diseases on tuber crops. Hence, there is an urgent need for rapid and accurate detection, diagnosis, and development of management recommendations for these new diseases.

Section IV
NEMATODE PATHOGENS

11

Emerging Nematode Problems in Field Crops

11.1. INTRODUCTION

The root-knot nematode, *Meloidogyne graminicola* prevalent in eastern, north-eastern and southern states of India has emerged as a serious threat for the successful rice production in nurseries, uplands, deep water and irrigated fields due to rice cropping intensification and increasing scarcity of water. In India, the losses due to *M. graminicola* have been estimated to be 16-32% and yield loss due to poorly filled kernels to be 17-30%. The rice white tip nematode, *Aphelenchoides besseyi* is also emerging as a serious pest causing 17-54% yield reduction in rice.

The cereal cyst nematode, *Heterodera avenae* is now recognized as one of the most important diseases limiting wheat yields in south-eastern Australia and, in future years, a similar situation may be recognized in Rajasthan, India.

The pigeon pea cyst nematode, *Heterodera cajani* is considered to be one of the most important economic pests of pigeon pea which recorded yield losses of over 30% in pigeon pea.

The root-knot nematodes, *Meloidogyne arenaria* and *M. javanica* are emerging as important constraints to groundnut production in Gujarat which are responsible for 13-50% and 10-23% of yield losses, respectively.

11.2. RICE, *Oryza sativa*

11.2.1. Root-Knot Nematode, *Meloidogyne graminicola*

The water saving rice production systems, such as direct wet seeding, intermittent irrigation, cultivation on raised beds, and the cultivation of aerobic rice varieties, are being developed and increasingly implemented. However, observations increasingly indicate that the large scale introduction of these techniques is favouring the development of high populations of *M. graminicola*, drastically increasing its economic significance.

Among various pests and diseases of rice and wheat crops in India, the root-knot nematode, *Meloidogyne graminicola* has become a constraint in rice production due to rice cropping intensification and increasing scarcity of water (Soriano and Reversat, 2003; Prasad and Somasekhar, 2009; Somasekhar and Prasad, 2009). *M. graminicola* has been prevalent in eastern, north-eastern and southern states of India.

However, its occurrence has also been reported from northern states. Recently with the introduction of resource conservation technologies (RCTs), the root-knot nematode, *M. graminicola* is now spreading under rice-wheat cropping system (RWCS) in the Indo-Gangatic plains (IGP) conditions with least disturbance in the soil in rice-wheat cropping system (Pankaj *et al.*, 2006).

In upland conditions and shallow intermittently flooded land, *M. graminicola* is considered to be by far the most damaging *Meloidogyne* species on rice. Second-stage juveniles (J2s) invade rice roots in upland conditions just behind the root tip (Rao and Israel, 1973). *M. graminicola* is well adapted to flooded conditions enabling it to continue multiplying in the host tissues even when the roots are deep in water. J2s cannot invade rice in flooded conditions but quickly invade when infested soils are drained (Manser, 1968).

11.2.1.1. Distribution:

M. graminicola, is widely distributed in the countries of S.E. Asia, Burma, Bangladesh, Laos, Thailand, Vietnam, India, China, the Philippines, USA both on upland and deep water rice. The decrease in water supply have induced the large scale adoption of direct wet seeding method coupled with chemical weed control and intermittent irrigation due to economic reasons in the Philippines (Bhuiyan *et al.*, 1995). These operations favour the development of *M. graminicola* and have drastically increased economic significance of the nematode.

The nematode is prevalent in all rice growing areas in most of the states in India (Prasad *et al.*, 1987; Prasad and Somasekhar, 2009). Severe damage occurs in upland and sometimes transplanted rice in north-eastern states, West Bengal, Orissa, Bihar, Eastern Uttar Pradesh, Chhattisgarh, parts of Madhya Pradesh and Karnataka. Infestations are particularly severe where two crops of rice are taken in a year, or where graminaceous weeds are abundant between two rice crops. Pockets of heavy infestation of rice nurseries and transplanted crop have been noticed in north-Indian plain zones including Jammu (J and K), Punjab, Himachal Pradesh, Haryana, Delhi and Uttar Pradesh (Gaur *et al.*, 1996; Pankaj *et al.*, 2006; Singh and Singh, 2009). These infestations are relatively more frequent in fields under rice-wheat cropping system. The frequency and intensity of the root-knot disease was high in eastern India and low in northern India. However, with favourable soil and agro-climatic conditions and continuous practicing of rice-wheat cropping systems led to increase the distribution and disease intensity of this nematode in northern states (Haryana, Punjab and Uttar Pradesh).

In Northern India, there is a shift of cultivation of rice in the wheat growing belt of Punjab, Haryana and Uttar Pradesh. The farmers prefer to grow more of premium variety of rice compared to other varieties because of their high export quality. Moreover the scarcity of water has forced farmers' to grow these varieties under water scarce situation. Most of the farmers' of non-traditional areas in Haryana Khadar belt grow rice in sandy soil with low water holding capacity, thus rice is grown under shallow water conditions, which is more conducive for the multiplication of the *M. graminicola*. Recently, heavy nematode infestation has been observed in Bulandshahr and Gautam Budh Nagar districts of Uttar Pradesh.

Of late, Rice root-knot nematode has become a potential menace and cause serious damage to cereals, especially rice in Karnataka. Rice root-knot nematode appeared in devastating form in parts of major rice growing areas of Shimoga during 2001, which was a first report from Karnataka and subsequently reported from Mandya district of the state (Krishnappa *et al.*, 2001). Severe outbreak of root-knot nematode is also observed in Shimoga, Karnataka (Sehgal *et al.*, 2012). Initially, it was noticed only in aerobic conditions. Since 2011, it has been observed in anaerobic conditions also and appearing in all types of rice cultivation situations. Among the districts, more severe incidence of root-knot nematode was observed with root knot index of 5 in Shimoga and 4 in Chickmagalur, Mandya and Hassan districts. Moderate infection was noticed in districts viz., Davanagere, Dakshina Kannada, Udupi, Uttar Kannada, Mysore, Kodagu and Haveri with root knot index of 3.

11.2.1.2. Economic Importance:

M. graminicola was reported to cause up to 21% yield loss in rain-fed and well drained soils throughout the country (Prasad *et al.*, 1987). On upland rice the nematode was reported to cause 16-32 % loss in grain yield due to incomplete filling of kernels (Biswas and Rao, 1971; Rao and Biswas, 1973). Some of the fields were so poor in crop stand that the farmers had to plough the fields in midseason. The major factor responsible for such high incidence of nematode infestation in these areas was the presence of light-textured soil and non-availability of ample irrigation water. At the highest initial population density (Pi), grain yield was reduced by 97%, but even at an initial population density of 0.1 J_2/g soil a reduction of 31% occurred.

In flooded rice, damage by *M. graminicola* is caused in nurseries before transplanting, where the tolerance limit of seedlings is less than 1 J_2/ml of soil (Plowright and Bridge, 1990). The damage also occurs prior to flooding where rice is sown directly in well-drained soil. In light soil, *M. graminicola* was a major factor for yield failure, whereas no evidence for a major contribution of soil-borne pathogenic fungi was found (Kreye *et al.*, 2009).

In *M. graminicola*-infested upland rice fields, nematicide application resulted in a yield increase of 12%–33% in Thailand (Arayarungsarit, 1987) and 28%–87% in Indonesia (Netscher and Erlan, 1993), whereas under simulated upland conditions, yield losses from *M. graminicola* ranged from 20% to 80% (Plowright and Bridge, 1990; Prot and Matias, 1995). In *M. graminicola*-infested lowland rainfed rice, nematicide application resulted in a yield increase of 16%–20% in Bangladesh (Padgham *et al.*, 2004), and in simulations of intermittently flooded rice, yield losses ranged from 11% to 73% (Soriano *et al.*, 2000).

11.2.1.3. Symptoms:

Infested rice fields show patches of stunted and chlorotic rice plants with less vigour, yellowing and curling of leaves (Plate 11.1). The chlorophyll content of leaves is reduced (Swain and Prasad, 1988). Infection by second stage juveniles in roots causes disruption and hypertrophy of cortical cells due to their migration, movement and secretions. Hypertrophy of phloem and abnormal xylem proliferation causes swelling of stele at the sites of nematode establishment. As a result most characteristic symptom i.e. root galls are formed. The galls incited by *Meloidogyne* spp. on rice are terminal, spiral or horse shoe shaped (Plate 11.2). The crop has poor stand, less tillering and few number of effective tillers. The spikes become smaller in size and less in numbers with poorly filled grains, thus yield is reduced.

In deep water rice the nematode infested seedlings remain stunted, unable to grow above flood water and perish due to continuous submergence. When flooding occurs, submerged plants with serious root galling are unable to elongate rapidly, and do not emerge above the water level (Bridge and Page, 1982). Examination of the dissected infected roots under a microscope at 40x or more shows the presence of small white, pearly somewhat spheroid root-knot nematode females, 0.5 - 1.0 mm in diameter, often in clusters with or without a gelatinous egg matrix embedded in the aerenchymatous cavities of the root cortex. The egg sacs may be seen protruding outside the root galls on very fine roots of rice, wheat and other plants lacking enough empty space. A number of thread-like males 1-2 mm long may be found entangled in the gelatinous egg matrix. Staining the root in 0.1% acid fuchsin shows the presence of red coloured females, males and numerous tiny eggs.

M. graminicola infestation causes reduction in phenols in the shoots and roots. The nematode infection also results in nutritional disorders within plants after 45 days, reducing the levels of chlorophyll and sugars (Swain and Prasad, 1988).

11.2.1.4. Spread:

The root-knot nematodes infecting wheat and rice are carried passively to newer places through following methods (Reversat and Soriano, 2002):

- Transportation of infected rice seedlings with adhering soil.
- Flowing irrigation and runoff water.
- Movement of soil containing eggs and bits of dried infected root with cattle and human feet, farm machinery, weed, wind, water etc.
- Through suckers of banana.

The root-knot nematodes do no spread with seed or chaff. Weeds also play a very important role in the ecology and survival of plant-parasitic nematodes in rice-wheat-legume cropping systems. These weeds serve as reservoir hosts in case one of the crops in the system is a non-host. In addition the nematode survives and able to complete the life cycle on stubs and un-harvested rice plants left in the field.

11.2.1.5. Impact of Wheat on *Meloidogyne graminicola* Populations in the Rice-Wheat System:

The impact of wheat on population densities of *Meloidogyne graminicola* in a rice-wheat rotation revealed widespread infestations of *M. graminicola* following wheat harvest. Root-galling severity ratings on a 1 to 9 scale for all 18 fields tested ranged from 1.5 to 7.0 with a mean of 4.8. Root-galling severity in rice nursery seedbeds ranged from 1.0 to 4.4 with a mean of 3.3. All wheat varieties commonly cultivated in rice-wheat rotation areas of Bangladesh were determined to be excellent hosts to *M. graminicola*. The reproduction on wheat roots was 26.4 to 57.9 times the initial nematode population density. The results indicate that wheat is contributing to high population densities of *M. graminicola* in soil which can negatively impact rice, and possibly wheat, productivity in a rice-wheat rotation (Plate 11.3) (Padgham *et al.*, 2004a).

11.2.1.6. Effect of *Meloidogyne graminicola* on Wheat Crop:

The effects on wheat crop, grown in *M. graminicola* infested field revealed that plant height was reduced in infested field (21.0 cm) as compared to the healthy field (31.6 cm). Fresh shoot weight recorded was 9.8 g/plant in healthy as compared to 5.7 g in the infested one. Root weight also registered 64.3% decrease due to nematode infection. There were 5.0 tillers per healthy plant as compared to 3.2 in the infected one. Average number of galls per plant was recorded to be 27.4. Data on *Phalaris minor*, an obnoxious weed of wheat crop and a very good host of *M. graminicola*, were also recorded from the same fields from which data on wheat crop were recorded. Plant height, shoot weight, root weight and number of tillers per plant in the infected ones recorded a decrease of 32.9%, 64.2%, 48.8% and 40%, respectively compared to those of healthy plant. Average numbers of galls were 49/plant which indicated that it is a very good alternative host of *M. graminicola* (Dabur *et al.*, 2010)

11.2.1.7. Predisposing Factors:

The rice root-knot nematode has become a constraint in Asian rice production. The major factor for such high incidence of nematode infestation in these areas was attributed to the presence of light-textured soil and availability of ample irrigation water and transplantation of infected seedlings. The nematode completes its life cycle in 26-51 days in different periods of the year. A temperature of 22-29°C is suitable for the prevalence of the nematode under Indian conditions.

11.2.1.8. Management

i) **Physical Methods:** The soil solarization in nursery bed may effectively reduce nematode populations. Prior to sowing, the nursery bed is covered with a plastic polyethylene tarp for 4-6 weeks. Soil solarization raises temperature by 4°C as compared to non-solarized soil to reach 40-50°C at 10 cm depth. The nematodes become undetectable in the upper 15 cm soil layer. It can give a reduction of *Meloidogyne* spp. population in rice beds by over 80% and improve seedling growth (Ganguly *et al.*, 1996).

ii) **Cultural Methods:** Selection of root-knot nematode free area for nursery is important. The seedlings which are to be transplanted must be free from nematode galls.

   a) **Crop Rotation:** Certain crops are resistant or poor hosts of *M. graminicola* and could be used in rotation to reduce nematode populations. Crop rotation with non-host crops like jute, mustard and chickpea, castor, cowpea, sweet potatoes, soybeans, sunflower, sesame, onion, cauliflower, turnip, beans, okra and resistant varieties (TKM-6, Patna 6, Dumai, Ch 47, Hamsa) reduces *M. graminicola* infestations. (Rao *et al.*, 1984; Rao, 1985; Rao *et al.*, 1986). Long rotations, more than 12 months, will be needed to reduce *M. graminicola* soil populations to low levels. Introducing a fallow into the rotation will also give control of the nematodes but, to be effective, it needs to be a bare fallow free of weed hosts.

   Rice following two crops of cowpea had increased yield by 34% when treated with carbofuran and by 26% in untreated plants. A single crop of cowpea increased the yield of the following rice crop by 85%. Cowpea cv. Iron Clay was shown to be an effective rotation crop to control some root-knot species because of its high degree of resistance. One or two consecutive crops of cowpea or seasons of fallow before a rice crop lowered nematode populations and improved rice yield by 30-80%. It is recommended that, to ensure higher rice yields, *M. graminicola* populations should be maintained at low density by non-host crop rotations or fallows, ideally for two seasons before planting rice (Soriano and Reversat, 2003).

An alternate rotation of rice with a non-host crop or a season of fallow seems to be the only solution at the moment for managing *M. graminicola* populations. Rice-mustard-rice or a rice-fallow-rice was effective in reducing *M. graminicola* development (Soriano and Reversat, 2003).

Rice-mustard-rice crop sequence, followed by rice-maize-rice and rice-fallow-rice, were reported to be effective in reducing nematode development (Kalita and Phukan, 1990). *Tagetes* sp. and *Crotalaria* spp. (*C. incana* and *C. mucronata*) showed reduction in the population and development of *M. graminicola*. Planting Tagetes or Crotalaria species in nematode infested soil is therefore feasible and can be used in the overall management of *M. graminicola*.

In addition to zero-till rice and wheat, attempts are also being made to diversify the cropping systems by introducing other crops into new rotations that help break disease and insect cycles and provide more income and diversity for farmers (Hobbs *et al.*, 2008).

b) **Fallowing:** One season of fallow (120 days) increased rice yield by 41% and two consecutive fallows (240 days) by only 31%. Two seasons of fallow, with carbofuran treatment before rice, proved to be the most efficient strategy to increase rice yields in the third season (Soriano and Reversat, 2003).

   Two consecutive fallows are necessary to sustain rice yields nearest to the potential (Soriano and Reversat, 2003).

c) **Flooding:** *M. graminicola* will survive normal flooding but damage to the crop can be avoided by raising rice seedlings in flooded soils thus preventing root invasion by the nematodes (Bridge and Page, 1982). Continuous flooding is highly effective in controlling *M. graminicola* in Vietnam (Kinh *et al.*, 1982). The yield losses may be minimized when the rice crop is flooded early and kept flooded until a late stage of development (Garg *et al.*, 1995; Soriano *et al.*, 2000).

   The root-knot problem in paddy can be managed by growing healthy and disinfested nursery and following summer ploughing and puddling of main fields before transplanting. Good puddling before transplanting helps in water retention in soil which reduces aeration and also reduces nematode movement and invasion of fresh roots by infective juveniles of *M. graminicola* and *M. triticoryzae* in rice soil (Garg *et al.*, 1995). The population of *Meloidogyne* spp. decreased drastically when the field was puddled every year over a 10 year period (Gaur, 2003; Pankaj *et al.*, 2006). Widespread infestations of the root-knot nematode (*M. graminicola*) on rice were found when direct-seeded rice was grown instead of puddled rice in Bangladesh (Padgham *et al.*, 2004a).

d) **Organic Amendments:** Soil amendments with decaffeinated tea waste or water hyacinth compost (300 or 600g/4.5 kg soil) reduced root-knot nematode infestation and increased plant growth (Roy, 1976).

Marigold treatment (grown and incorporated into soil before planting rice) also suppressed nematode root galling, and increased rice grain yield by 46%.

Poultry manure can be utilized at least occasionally as a nematicide cum fertilizer to reduce nematode infestation level in rice plants and to improve soil fertility (Amarasinghe, 2012).

Seed treatment with neem-based pesticides can reduce populations and damage caused by *M. graminicola* (Das and Deka, 2002).

iii) **Chemical Methods:** Several studies have supported the effectiveness of carbofuran to control *M. graminicola* in rice. Application of carbofuran increased rice yield by 10% for the first rice crop. During the second season, yield increase was observed when rice was planted after cowpea (85%); after cowpea with carbofuran application (75%); after fallow (41%); and after fallow with carbofuran (48%). In the third season, when all plots were planted to rice, 31 and 49% yield increases were obtained, without and with carbofuran, respectively, from plots that had been left fallow for two consecutive seasons (Soriano and Reversat, 2003).

Carbofuran was found to be most effective as seed treatment for control of *M. graminicola* in upland rice (Fademi, 1984). Similarly, carbofuran at 2.4, 4.8 and 7.2 g a.i./kg seed reduced galls by 57, 79 and 80%, respectively (Segeren van den Oever *et al.*, 1985). Seed soaking at 0.1-0.2% for 12 hr was found to be effective in reducing the root-knot nematode population (Rahman and Das, 1994). Carbofuran, phorate, isazophos, cartap, carbosulfan or quinalphos at 0.5, 1.0 and 2.0 kg/ha given as soil application, significantly reduced galling of *M. graminicola* at 1 kg/ha or above, the greatest reduction in galling (82%) occurred with carbosulfan at 2 kg. Reduction in gall numbers and final nematode population with phosphamidon 85 SL at 0.2% followed by triazophos at 0.2 % for 12 hr seed soaking has been recorded (Deka and Das, 2003). With the non-availability of other nematicides, the possible treatments in the present scenario could be:

- Seed soaking in 500 ppm carbosulfan at 0.1% for 12hr; root-dip in 200 ppm of carbofuran for 12 hr.
- Carbofuran at 1 kg a.i./ha in two equal split doses 15 and 45 days after sowing or transplanting.

iv) **Biological Methods:** The biological control agents, *Trichoderma harzianum* and *T. virens* at 4 and 8 g/kg soil respectively, when applied to soil one week after nematode inoculation were significantly effective in improving plant growth and reducing number of galls and larval penetration per root system. Maximum mortality of (> 96%) of *M. graminicola* juveniles occurred when exposed to culture filtrates of *T. harzianum* (100 and 50% concentration) (Pathak and Kumar, 1996). *T. virens* was more suppressive to nematode population in host tissue than *T. harzianum* (Pathak and Kumar, 2003).

The extent of galling (12.80), gall index (1), yield (3260 kg/ha) and final soil population (209.28) were significantly best in *Trichoderma viride* compared to other bioagents. While other bioagents, *Pseudomonas fluorescens* and *Bacillus subtilis* ranked second and third, respectively in managing *M. graminicola*. The study also revealed that bioagents improved the plant growth and reduced the nematode infestation in aerobic rice. Moreover, *T. viride* can be used as an alternative in biomanagement of root-knot nematode in aerobic rice (Shanmuga Priya, 2015).

Catenaria anguillulae parasitized and killed the eggs and second stage juveniles of *M. graminicola* under natural conditions. The higher parasitism of eggs and J_2 was recorded from those fields in which plants were severely infected with *M. graminicola*. The degree of parasitism of eggs and J_2 by *C. anguillulae* varied with severity of root-knot disease (Singh *et al.*, 2007).

In rice root tissues treated with *P. fluorescens* and challenge inoculated with *M. graminicola*, the nematode population was significantly lower. There was 40.7% reduction in the nematode population 7 days after inoculation in the *P. fluorescens* treated plants. Similarly there was 68.4% and 75.1% reduction in the nematode population 14 and 21 days after inoculation respectively in the treated plants (Anita and Samiyappan, 2012). The present study implies that earlier and higher accumulation of phenols and defence enzymes viz., peroxidase, polyphenoloxidase, phenylalanine ammonia lyase and chitinase in rice root tissue treated with *P. fluorescens* in response to invasion by *M. graminicola* collectively contribute to induced systemic resistance and decrease in nematode infection (Anita and Samiyappan, 2012).

Seedlings raised from seeds treated with, *Pseudomonas fluorescens, Trichoderma viride, P. fluorescens* + *T viride* and *Bacillus subtilis* were superior at early stages. At the time of harvest *B. subtilis* showed superior effect in enhancing the shoot and root characters. *B. subtilis* was found to be equally superior as compared to carbofuran in controlling the nematode population in root and soil. (Renitha Govind and Ushakumari, 2010).

v) **Host Resistance:** *M. graminicola* was reported to reproduce on all the 10 wild *Oryza* species tested and showed different levels of susceptibility. *Oryza australiensis* and *O. brachyantha* showed by far the greatest infestation (5855 and 10,235 J_2s/g root, respectively) compared with *O. officinalis*, which was the lowest at 240/g root (Gergon and Prot, 1993). Soriano *et al.* (1998) observed that one accession of *O. longistaminata* represented by two individuals (WL02-2 and WL02-15) and three accessions of *O. glaberrima* (TOG7235, TOG5674 and TOG5675) were resistant to *M. graminicola*. There are number of cultivars reported resistant to *M. graminicola* (Roy, 1973; Jena and Rao, 1976; Ramakrishnan *et al.*, 1984; Prasad *et al.*, 1986; Rao *et al.*, 1986).

The rice varieties namely At353, Ld356, Bg304, At5, Bg305, Bg450 and Bg379-2 can be used for fields that are prone to *M. graminicola* infestation (Amarasinghe, 2012).

There was increase in phenol, peroxidase, polyphenol oxidase and phenylalanine ammonia lyase content in *M. graminicola* infested resistant rice varieties and it was highest in the variety TKM II compared to susceptible check (Senthilkumar *et al.*, 2010).

Ravindra *et al.* (2013) screened large number of rice varieties and identified root-knot nematode resistance in several cultivars (Table 11.1).

Table 11.1. Field evaluation of rice cultivars against rice root-knot nematode, *Meloidogyne graminicola* (Ravindra et al., 2013)

Reaction	Varieties	Root-knot index
Highly resistant	Nazarbad, Bheemasaale, Nazarbaik, Kichadi samba, Kichadi samba, kattaru, B.P.T, Mouruda, Bagashaparimala akki, Kempu kaalu, Malgudi sanna, Delhiboga bhattha, Sannakki, Padma rekha, Chitiga, Sharavathi kempu, Andra basumati, Orrisa bhattha, Kasubai, Bangara kaddi, KH4, Meese bhattha, Kempu tadi, Kavadhari, Jeerige samba, Tunuru, Dappa valya, Rasakadari, Nagabhatha, Baigan manja, Bud bhattha, Sidda sanna.	1
Resistant	Chinnaponni, Bangara sanna, Jasmine, H.M.T, Siggikaimai, Sannavallya, Kalalajeera, Aravatellu, Karijeddu, Vijayanandeda, Raichur sanna, Bangaru gundu, Yelatakkigidda, Hasudi, Marudi, Aanekombina bhattha, Sambhamasana, Sanna muttige, Karigaja vilai, Bilidaddi budda, Kagga, Kuduvekalaje, Hole saala, Chippiga, Dappa bhattha, Kempudadde gidda, Gandha saale, KA1, Eppattu, Manja kai mai, Sasti, Kari basumati, kallunundigar, Malabar, Bili jeddu, Neragula bhattha, Bangla rice, Madaikar, Togarsi, Maranellu, Champakali, Red jasmine, Doddiga, Neti bhattha, Baredabina saale.	2

a) **Transgenic Nematode Resistant Plants:** Transgenic resistance against a wide range of nematodes have been developed, based on an antifeedant approach that uses plant cystatins. Cystatins are naturally occurring proteinase inhibitors which are involved in plant defences against insects and pests. Furthermore cystatins, are not involved with mammalian digestive enzymes (Atkinson *et al.*, 1993). Cystatins are also found in a wide range of foods such as those found in rice seeds which are eaten daily by thousands of people, other sources include maize and sunflower. Cystatins are not normally produced in the roots where the nematodes attack, but plants can be modified by the use of root specific promoters driving the expression of cystatin in areas of the plants where the nematode infest. This has the added advantage of ensuring the cystatin is not expressed in the edible plant parts or green tissue. Transgenic expression in plant roots reduces nematode growth, development and fecundity.

The anti-nematode defence offers major practical benefits for use in the developing world. It offers the ability to control several nematodes attacking a crop such as rice concurrently or at different locations without any need for awareness of nematode problems at the grower level. This is useful, as rice roots tend not to gall upon nematode infestations. Using a genotype independent biolystic transformation system, twenty-five transformed clones containing an engineered cysteine proteinase inhibitor Oryzacysatin-IDD86, hygromycin resistant (aphIV) and ß-glucuronidase (gusA) genes were recovered from four varieties of rice. Up to 0.2% total soluble protein was detected by western analysis.

vi) Integrated Nematode Management:

Following methods could be used in combination along with the judicious use of chemicals against root-knot nematode in rice-wheat cropping system.

a) Nursery-Bed Treatment

- Selection of root-knot nematode free area for nursery.
- Soil solarization of area for nursery beds, for 3-4 week in April-May (summer months).
- Application of carbofuran at 3.3 g/m^2 of nursery bed, before sowing.
- Soil solarization of nursery beds for 3 – 4 weeks in summer and nursery-bed treatment with carbofuran or phorate at 2 kg a.i./ha was found very effective in reducing root-knot nematode infestation in rice nursery (Gaur, 2003).

b) Main Field

- Proper puddling of main field before transplanting.
- Summer ploughing of main field twice at 14 days interval.
- Avoid direct sowing of rice in the main field.
- Crop rotation with mustard or other non-host crops.
- Late sowing of wheat up to late November till mid-December using suitable variety.
- Green manuring with *Crotolaria* spp. or *Sesbania* spp. if time permits.

Recommendations

- Breeding of root-knot nematode resistant rice varieties using *Oryza glaberimma* should be pursued vigorously.
- A network project for mapping of nematode resistance genes may be formulated for external funding.
- Robust green-house based or *in vitro* screening techniques are needed for germplasm evaluation for nematode resistance.
- The spectrum of *Meloidogyne graminicola* species complex/races needs to be studied in different geographic populations.
- Solarization of nursery beds may be popularized.

11.2.2. White Tip Nematode, *Aphelenchoides besseyi*

A. besseyi is very widely distributed and now occurs in most rice growing areas (Ou, 1985). Its wide distribution has resulted from dissemination through seed.

11.2.2.1. Symptoms:

It is easy to detect the presence of nematode (*A. besseyi*) within the rice seeds. In the field, the initial appearance of symptoms as the leaf tip (up to 5 cm) becomes pale yellow or whitish at tillering stage and subsequently, the leaves get dry (Plate 11.4). These symptoms are found for a short period in the plant. The tip of the flag leaf are often twisted (and reduced in size, as are the grains) which may obstruct the emergence of panicles (Plate 11.4). Infested panicles are shorter and lighter in weight as compared to healthy panicles.

11.2.2.2. Host Range:

The main host plants are strawberries and rice, which are the main crops at risk. *A. besseyi* has also been found on *Boehmeria nivea*, numerous ornamental plants (including chrysanthemums, *Ficus elastica, Hibiscus, Polianthes tuberosa* and *Saintpaulia ionantha*), and grasses (*Panicum, Pennisetum, Setaria, Sporobolus*).

11.2.2.3. Economic Impact:

Nematode attack results in poorly filled or unfilled and chaffy grains. The proportion of the chaffy grains may be as high as 40% in some cultivars, but early-ripening cultivars are relatively less affected. Yield reductions in rice of 17-54% and 0-24% have been recorded for susceptible and resistant cultivars, respectively.

11.2.2.4. Biology:

On rice, infested seed is the primary inoculum source. When the crop is sown, the nematodes become active and move to the growing points of leaves and stems where they feed ectoparasitically. *A. besseyi* may reproduce parthenogenetically. The optimum temperature for development is 21-25°C, the life cycle taking 10 days at 21°C and 8 days at 23°C, and there are several generations in a season. Up to 14 nematodes, mostly pre-adults, have been found on a single seed, where they remain coiled up inside the palea. The grain itself is never invaded. *A. besseyi* can withstand desiccation, retaining viability for 2-3 years on dry grain, but dies in 4 months on grain left in the field; the nematode is not thought to survive long periods in the soil between crops.

11.2.2.5. Spread:

The pest is liable to be carried on rice seed and accompanying chaff, and also on strawberries and other host plants.

11.2.2.6. Management:

Control in rice by seed disinfestation (warm water and chemical treatments) and soil disinfection (chemical treatments) has been successful (Todd and Atkins, 1959a). Rice seeds can be soaked in water at 55-61°C for 10-15 min to destroy the nematode. Various chemical seed and soil treatments are recommended. Other methods which can contribute to a reduction of nematode populations include the use of resistant or tolerant cultivars, early planting when the rice season is preceded by a cooler period, and low seed-bed planting densities.

Nematicides, applied either as foliar sprays or as soil treatments, can reduce the nematode population sufficiently to prevent symptom expression.

i) **Prevention and Control:** Preventing the dispersal of *A. besseyi* in rice requires the elimination of nematodes from seed; hot water or chemical seed treatments are commonly used. Resistant cultivars and cultural methods have been used to reduce infection below damage thresholds, and tolerant cultivars avoid yield loss without nematode control. Stubble burning prevents transmission of *A. besseyi* in straw and chaff but would have to be used in conjunction with other control measures.

ii) **Hot Water Treatment:** There are numerous variations of methods for the hot water treatment of rice seed (Borokova, 1967). The most effective control requires seed to be pre-soaked in cold water for 18-24 hours, then immersed in water at 51-53°C for 15 minutes. Higher temperatures (55-61°C for 10-15 min) are required if seed is not pre-soaked. The temperature and duration of treatment must be closely monitored and, after treatment, the seed must be dried at 30-35°C, or sun dried if stored, but otherwise can be sown directly in the field. For quarantine purposes at the International Rice Research Institute, seed is soaked in cold water for 3 hours followed by hot water at 55°C for 15 minutes.

iii) **Chemical Methods:** Various chemical seed treatments have been used for the control of *A. besseyi* including nicotine sulphate, demeton, malathion and fensulfothion. Lee *et al.* (1972) reported effective control by treating water or by root dipping with diazinon and nemagon. *A. besseyi* control with carbosulfan sprays has been reported (Rao *et al.*, 1986), but pre-harvest chemical treatment alone is only partially effective (Aleksandrova, 1981).

The timing of application is important. Benomyl applied as a seed treatment alone was less effective than when combined with spraying at one and 15 days after exposure to nematodes (Gergon and Prot, 1993).

iv) **Host-Plant Resistance:** Resistance to *A. besseyi* is widespread. In the USA, *A. besseyi* has been controlled principally through the use of resistant cultivars. Resistance to *A. besseyi* has subsequently been reported from Japan (Goto and Fukatsu, 1956), Korea (Park and Lee, 1976), India (Rao *et al.*, 1986), Brazil (Silveira da *et al.*, 1982), USSR (Popova *et al.*, 1980) and Italy (Orsenigo, 1954). Resistance to *A. besseyi* is thought to be genetically controlled and carried by the Japanese cultivar Asi-Hi (Nishizawa, 1953).

Symptomless but susceptible (i.e. tolerant) cultivars (Goto and Fukatsu, 1956) are common. Symptom expression in the field is particularly variable (Atkins and Todd, 1959) and variations between plants of a cultivar also occur (Orsenigo, 1954). In Thailand, all local cultivars are considered tolerant of *A. besseyi* (Buangsuwon *et al.*, 1971).

v) **Cultural Methods:** Irrigating seed beds (Yamada *et al.*, 1953) or direct seeding into water (Cralley, 1956) reduces infection by *A. besseyi*. In these conditions, nematodes emerge and lose vigour before seed germination. High seedling rates in the seed bed (Kobayashi and Sugiyama, 1977) and high numbers of seedlings/hill (Yamada *et al.*, 1953) tend to increase infection by increasing the number of infection loci in the field. Such problems are thought to be responsible for the re-occurrence of *A. besseyi* in Japan (Inagaki, 1985). In the USA (Cralley and Adair, 1949) and Japan (Yoshi and Yamamoto, 1951; Yamada *et al.*, 1953), early planting, presumably in cooler conditions, reduced or eliminated *A. besseyi* infection.

The following practices are recommended for the management of white tip nematode on rice:

- Destruction of infested stubbles/weeds/straw/debris followed by ploughing to keep the fields dry helps in reduction of the nematode population.
- Rice variety Rayada 116, identified resistant to rice stem nematode, can be recommended under nematode-infested conditions.

- Seed treatment of rice with carbosulfan (25 DS) at 3% a.i. weight by weight (w/w) and foliar spray with carbosulfan (25 EC) at 0.1% a.i. at 40 and 120 days after transplanting was found effective.
- Soaking of paddy seeds in carbosulfan (25 EC at 0.1% a.i. for 6 hr) followed with foliar spray with 0.02% carbosulfan 40 days after transplanting reduced *A. besseyi* populations considerably.
- Hot water treatment of white tip contaminated/infected rice seeds for 10 min at 50^0 to 55 ^{0}C before sowing.
- Sun-drying of rice seed on concrete floor on bright sunny days at least for six consecutive days, for four hours/day.
- Use of healthy/certified seeds to check the spread of white tip nematode problem.

11.3. WHEAT, *Triticum* spp. AND BARLEY, *Hordeum vulgare*

11.3.1. Cereal Cyst Nematode, *Heterodera avenae* (Molya Disease)

The Molya disease of wheat and barley is caused by the cereal cyst nematode, *Heterodera avenae*. It is prevalent in isolated pockets in Rajasthan, Haryana, Delhi, Punjab, Himachal Pradesh. Jammu and Kashmir, Uttar Pradesh, Madhya Pradesh and Gujarat. Its incidence is higher in loamy and sandy soils as compared to medium and heavy soils. *H. avenae* is now recognized as one of the most important diseases limiting wheat yields in southeastern Australia and, in future years, a similar situation may be recognized in Rajasthan, India.

Optimum conditions for infection are temperature of 20-22^0 C and soil moisture of 60 per cent field capacity. The losses depends on the infection level of the soil, soil type and soil moisture status.

11.3.1.1. Economic Losses:

H. avenae is responsible for 32.4-66.6% in wheat and 70.2% in barley.

11.3.1.2. Symptoms:

The nematode infected fields exhibit patchy growth with stunted and yellowish plants (Plate 11.5). Infested plants show thin narrow leaves, reduced tillering, fewer leaves and small size of ear heads with reduced number of grains. Under heavy infection, the plant may not develop beyond 15 cm in height. The roots of nematode infected plants appear bushy and bunchy due to the emergence of rootlets at the site of infection, and slight swelling of the root tips may be encountered. The above-ground symptoms are often confused with general deficiency symptoms. However, the presence of cysts is the only confirmatory evidence for nematode infection.

Cereal crops infected with this nematode are more susceptible to infection by fungal diseases such as Rhizoctonia root rot.

11.3.1.3. Host Range:

Besides wheat (*Triticum aestivum*), oats (*Avena sativa*), barley (*Hordeum vulgare*), and rye (*Secale cereale*) are also hosts of *H. avenae*.

11.3.1.4. Life Cycle:

The eggs are oval and the vermiform larvae moult four times. The second instar larvae are mobile and can travel distances of up to thirty centimetres, looking for and invading roots of suitable host species. The second stage juveniles (J_2) infect the growing tips of roots and upon feeding develop specialized syncytial cells for their growth and development. After 3 months, the juveniles achieve lemon-shaped sedentary female which is found to attach with roots. Eggs are laid inside the female body, and after the death of the female, the body cuticle transforms into brown cyst. These cysts are dislodged from the roots in the soil and become the inoculum for the next season. The larvae may remain in anabiosis within this cyst for several years until suitable host plants become available. The activity of the nematode perfectly matched with the growing season of wheat crop grown in the rabi season (October to March/April) in India. During off-season (April to September), eggs and J_2 undergo diapauses induced by the rising of temperature during the end of April and remain dormant inside the cyst in soil. In India, nematode completes one generation during winter (October to March). The duration of its life cycle depends on the temperature and development of white female. Under Indian conditions, December-sown crop may take 8-9 weeks and February-sown crop 5 weeks on wheat (Bajaj and Kanwar, 2005).

11.3.1.5. Biotypes:

There are 11 biotypes of *H. avenae* known world-wide based on international assortment of wheat, oat and barley cultivars. In India, the occurrence of five biotypes from Rajasthan and Haryana (Mathur *et al.*, 1974) and biotype 3 is most common. Swarup *et al.* (1979) reported only 2 biotypes from India. Rajasthan and Haryana populations are one biotype and the Punjab population is different biotype.

11.3.1.6. Spread:

Man, in his migration and cultivation of new land for cereals, was largely responsible for the long-distance transport of cysts--either in soil attached to boots, tools, and equipment, or through his deliberate movement of soil, alone or with plants.

Once established within a field, cysts or larvae may be disseminated by cultivation or the movement of water. In Australia, however, the rapid dissemination of cysts appears to have been due largely to movement by wind.

11.3.1.7. Survival:

The eggs within mature cysts can survive in air-dry soil for at least 13 months (Meagher, 1970), or in a dry atmosphere (75-40% RH) for at least 5.5 years (Meagher, 1974a, b).

11.3.1.8. Management:

The following practices have been recommended for the management of cereal cyst nematode:

- Direct-drilling that did not disturb the soil below seeding depth and reduced the incidence of cereal cyst nematode.
- Deep summer ploughings 2 to 3 times at an interval of 10 to15 days between May and June.

- Early sowing of wheat in first fortnight of November.
- Use of non-host crops like radish, marigold, mustard, chickpea, green gram, carrot, coriander, and resistant varieties continuously for 2 to 3 years.
- Growing wheat cv. C-306 as trap crop early in October, before growing main crop, proved effective
- Use of resistant varieties like C 164, BH-64 and Raj Kiran of barley and Raj MR-I of wheat in nematode-infested areas.
- Use of rhizospheric rhizobacteria, *Azotobactor chroococcum* (Strain HT-54) as seed treatment.
- Application of carbofuran at 1.0 -1.5 kg a.i./ha in soil at sowing time (Kaushal *et al.*, 2001; Kanwar and Bajaj, 2010).

Integrated Management: Early sowing in the month of November along with field application of carbofuran at 2 kg a.i./ha is quite effective in increasing yield and reducing cyst population in soil.

11.4. PIGEON PEA, *Cajanus cajan*

11.4.1. Cyst Nematode, *Heterodera cajani*

In India, *H. cajani* is considered to be one of the four most important cyst nematodes and is an economic pest of pigeon pea. Although accurate estimates of economic crop loss are not available, yield reduction and decline of plants have been reported by many workers such as Sharma *et al.* (1993), who recorded yield losses in pigeon pea of over 30%.

11.4.1.1. Geographical Distribution:

H. cajani is widely distributed in India. *H. cajani* is found mainly in India and Pakistan and has a limited distribution in Egypt. Quarantine measures are recommended against this and other cyst nematodes (Mathur, 1986).

11.4.1.2. Economic Losses:

H. cajani infection suppresses plant growth of pigeon pea by 28% and reduces grain yield by 24%.

11.4.1.3. Symptoms:

The symptoms of nematode injury include stunting, reduced leaf lamina size and yellowing on cotyledonary leaves (Gaur and Singh, 1977). Flowers and pods are reduced in size and number and the root system may also be poorly developed. Sharma *et al.* (1993) cautions that foliage symptoms are generally not apparent even in heavily infested soils, but a reduction in height and vigour of the infected plants can be discerned by careful comparison with healthy plants. The disease is recognized by the presence of pearly white females on roots or brown cysts in the soil (Plate 11.6).

11.4.1.4. Host Range:

Pigeon pea (*Cajanus cajan*), *Lablab purpureus*, *Vigna radiata*, cowpea (*V. unguiculata*) and sesame (*Sesamum indicum*) were the most favoured hosts and showed extensive damage from *H. cajani* attack (Koshy and Swarup, 1973).

11.4.1.5. Biotypes:

Fourteen populations of *H. cajani* from different localities in seven districts of Uttar Pradesh (India) were tested on 9 hosts revealing the presence of 3 races: Race 1 reproduced on all the hosts tested; Race 2 did not reproduce on *Cyamopsis tetragonoloba*; Race 3 did not reproduce on *C. tetragonoloba* or *Crotalaria juncea* (Siddiqui and Mahmood, 1993).

11.4.1.6. Life Cycle:

The development and life cycle of *H. cajani* were studied by Koshy and Swarup (1971c) on pigeon pea plants at an average soil temperature of 29°C (range 27-36°C). Second-stage juveniles (J_2) penetrate the roots within 48 hours of inoculation. Both tap and lateral roots were penetrated, not necessarily near the root tips. Moulting began on the third day beginning from the anterior part and completing on the fourth day. Fourth-stage juveniles with well-developed reflexed ovaries were found on the tenth day. Adult lemon-shaped females were observed on the 12th day. Males were found on the 12th or 13th day. On the 14th day, many eggs were seen in the egg sacs as well as inside the white females. Second-stage juveniles were collected from soil as well as from cysts from which they hatched on the 16th day. A few males, apparently embedded in the egg mass, also emerged from the cysts. The white females turned yellowish from the 20th day onwards; the change from white to bright yellow took 8 days. The egg sac turned opaque and yellowish, occasionally purple. Brown cysts were noticed on the 38th day after inoculation. Although males are thought to be necessary for reproduction, females sometimes reproduced without them.

The life cycle was completed in 16 days at 29°C but during cooler conditions (10-25°C), the life-cycle took 45-80 days for completion (Koshy and Swarup, 1971a). On one crop of pigeon pea, 8-9 generations were recorded (Koshy and Swarup, 1971b).

11.4.1.7. Spread:

Transportation results mainly from flooding, drainage or transfer of infested seeds and plants, from soil washings, and from soil attached to farm machinery, livestock, tools or people (Mathur, 1986; Sharma, 1998).

11.4.1.8. Interaction with Other Micro-organisms:

When *H. cajani* is associated with the fungus *Fusarium udum* (*Gibberella indica*) there is a significant increase in wilting (Hasan, 1984).

11.4.1.9. Management

i) **Chemical Methods:** Cyst nematodes are highly resistant to chemical control because the female, after death, transforms into a tough brown sac protecting the eggs which remain within.

   Fensulfothion, when evaluated as a seed treatment for pigeon pea, reduced nematode populations, with high concentrations and long soaking period being the most effective (Zaki and Bhatti, 1986a).

   Certain chemicals stimulate juvenile emergence and can be used with proper knowledge of the biology of the parasites to control them.

The following non-chemical methods have also provided some control:

- Essential oils of Mentha piperita, *Ocimum sanctum, Cymbopogon martini, C. nardus, C. winterianus, C. flexuosus* and *O. basilicum* (Gokte *et al.*, 1993).
- Root extract of *Xanthium strumarium* (Malik *et al.*, 1987).
- Green manures of *Vigna radiata*, cowpea (*Vigna unguiculata*), *Crotalaria juncea* and *Sesbania bispinosa* (Devi and Gupta, 1995); fenugreek (*Trigonella foenum-graecum*), wild methi (*Senji* species), berseem (*Trifolium alexandrinum*) and drum stick (*Moringo pterygosperma*) (Devi, 1997).
- Powdered leaf extracts of *Terminelia arjuna* on pigeon pea (Singh and Singh 1992); *Argemone mexicana, Cannabis sativa, Datura metel, Nerium indicum* (Mojumdar *et al.*, 1989); and *Solanum xanthocarpum* (Bhatti *et al.*, 1997).
- Oil cakes of neem, mustard and mahua (Rai and Singh, 1995, 1996; Devi and Gupta, 1996).

ii) **Biological Methods:** Plant parasitic nematodes have many natural predators and parasites in the soil which provide opportunities for biological control.

Combinations of biocontrol agents [*Bacillus subtilis, Bradyrhizobium japonicum* and *Glomus fasciculatum* (Siddiqui and Mahmood, 1995a); *Paecilomyces lilacinus, Verticillium chlamydosporium* and *Gigaspora margarita* (Siddiqui and Mahmood, 1995b); and *Trichoderma harzianum, V. chlamydosporium* and *Glomus mosseae* (Siddiqui and Mahmood, 1996)] and vesicular arbuscular mycorrhizal fungi have been used to manage the wilt disease complex of pigeon pea caused by *H. cajani* and *Fusarium udum* (*Gibberella indica*).

Application of *P. fluorescens* at 2.5 kg/ha at sowing significantly decreased *H. cajani* population in roots at 45 DAS by 18.32 % and final nematode population by 29.55 % and thereby gave 35.93 % and 32.46 % higher pod and grain yield, respectively.

iii) **Resistant Varieties:** The use of resistant varieties is the most economical method of controlling plant parasitic nematodes. A greenhouse technique to screen pigeon pea for resistance to *H. cajani* is described by Sharma *et al.* (1991).

Several screening tests have been carried out to identify resistance to *H. cajani* in Pigeon pea and its wild relatives *Cajanus platycarpus* and others (Sharma *et al.*, 1993; Sharma, 1995; Singh and Singh, 1995; Elyas and Sharma, 1997; Siddiqui *et al.*, 1998).

iv) **Cultural Practices:** *H. cajani* has a very restricted host range and is mostly confined to the Fabaceae. Growing the host crop in rotations every 4-5 years with non-hosts can help to maintain populations below threshold levels. The significance of double crop (intercrop and sequential crop), single crop (rainy season crop fallow) and rotations on the densities of *H. cajani* and other nematodes was studied by Sharma *et al.* (1996). Mean population densities of *H. cajani* were about eight times lower in single crop systems than in double crop systems, with pigeon pea as a component intercrop. Plots planted to sorghum, safflower and chickpea in the preceding year contained fewer *H. cajani* eggs and juveniles than did plots previously planted to pigeon pea, cowpea or *Vigna radiata.* Continuous cropping of sorghum in the rainy season and safflower in the post-rainy season markedly reduced the population density of *H. cajani.*

Intercropping sorghum with a tolerant pigeon pea cultivar could be effective in increasing the productivity of traditional production systems in *H. cajani* infested regions (Sharma *et al.*, 1996).

v) **Integrated Management:** An integrated control approach involving summer ploughing, crop rotation, seed treatment, limited applications of chemicals and the use of resistant varieties should be adopted for the control of cyst nematodes (Yadav, 1986).

Seed treatment with neem seed kernel powder at 10 % w/w + *P. fluorescens* at 10 g/kg seed was highly effective which recorded 30.11 % reduction in nematode population and 29.10 % increase in yield over control. Soil application of neem cake at 10 g/m^2 + *P. fluorescens* at 2.5 kg/ha was also highly effective which recorded 31.63% reduction in nematode population and 17.88 % increase in yield over control.

The following practices are recommended for the management of pigeon pea cyst nematode:

- Growing of non-host crops like wheat, maize, rice etc. reduces the population of pigeon pea cyst nematodes.
- Seed dressing with carbosulfan (25 DS) at 3.0% a.i. (w/w) has been found effective.
- Seed treatment with *Pseudomonas fluorescens* and *Trichoderma viride* formulations each at 5g/kg seeds decreased pigeon pea cyst nematode and wilt complex and increased yield significantly.
- Inter-cropping of pigeon pea with sorghum (row ratio 2:1) has been found effective to reduce the population of pigeon pea cyst nematodes.

11.5. GROUNDNUT, *Arachis hypogaea*

11.5.1. Root-knot Nematodes, *Meloidogyne arenaria, M. javanica*

Several species of root-knot nematodes are pathogens of peanut worldwide, the peanut root knot nematode, *Meloidogyne arenaria* is the most common and severe species infecting peanut (Dickson, 1998). In addition, *M. javanica* also has been reported to infect peanut from several parts of the world including India (Patel *et al.*, 1988; Sakhuja and Sethi, 1985).

Meloidogyne arenaria and *M. javanica* are the major constraints to groundnut damage in Gujarat, India (Patel *et al.*, 1996).

11.5.1.1. Economic Importance:

The damage threshold, or the initial nematode population density at which peanut yields begin to decline, has been reported to be in the range of 1 to 10 eggs and second-stage juveniles per 500 cm^3 soil for *M. arenaria* on susceptible peanut cultivars (Abdel-Momen and Starr, 1997).

Meloidogyne arenaria and *M. javanica* are responsible for 13-50% and 10-23% of yield losses, respectively.

11.5.1.2. Geographical Distribution in Gujarat:

The root-knot nematode, *M. arenaria* is widely distributed in the groundnut growing areas in Supedi, Sanala, Kathrota, Upleta, and Patanvav regions in Rajkot district; and Khadia and Dhoraji in Junagadh district of Saurashtra region. *M. javanica* infests groundnut fields in the Khalvada, Telnar, Betawada, Lalpur, Pariana, Muvada, and Narsipur regions of Napadvanj mandal of Kaira district. A pathotype of *M. javanica* which reproduces on groundnut and produces severe galls is present in Kapadvanj area (Patel *et al.*, 1993). It is evident that root-knot nematodes are important constraints to groundnut production in Gujarat.

11.5.1.3. Symptoms:

Areas of root-knot nematode infected peanuts are usually round to oblong in shape. The plants are stunted, less green, and will wilt more readily in the heat of the day than plants in less infested areas of the field. Rows of infected plants may never, or not so quickly, meet as those of healthy plants (Plate 11.7).

Root-knot nematodes cause galling on roots, pegs, and pods. Symptoms of peanut root-knot nematodes include galls (knots) on both roots and pods (Plate 11.8). The presence of galls on the pods appears as single or multiple wart-like growths that may or may not be discoloured. As root-knot nematode infection progresses, secondary root and pod rotting causes further damage and eventual death of the plants.

Root-knot nematode affected fields show stunting, reduced number of branches and internodes, burning of leaf margins, reduced leaf size, and yellowing of foliage. The damage thresholds of *Meloidogyne* spp are 0.1-1.0 juveniles cm^{-3} soil at the time of sowing.

11.5.1.4. Interaction with Other Micro-organisms:

The presence of *M. arenaria* along with *Fusarium solani* advances wilting of groundnut plants (Patel *et al.*, 1985), and reduces the number of Rhizobium nodules.

11.5.1.5. Management

i) **Cultural Methods**

a) **Crop Rotation:** Where practical, crop rotation is an excellent practice for managing plant-parasitic nematodes. Rotation of a peanut crop with non-hosts, or less susceptible hosts, to root-knot nematodes is recommended. Grasses such as Bahia grass, Bermuda grass, millet, and sorghum are among the most effective crops in reducing soil populations of peanut root-knot nematodes and should be grown for at least one year, preferably two years, before planting fields to peanut. Though some root-knot nematode reproduction could be expected on field corn, this crop is considerably less susceptible than peanut and is generally effective in the reduction of root-knot nematode soil populations. It is a suitable crop to grow in a year following a root-knot nematode-infected peanut crop. Cotton is not a host for peanut root-knot nematodes and is a very good rotational crop with peanut. Favourably, peanut is not a host for the southern root-knot nematode that infects cotton. Thus a two-year monoculture of peanut and cotton is helpful in maintaining the root-knot nematode pests of both crops at manageable population levels.

A two-year Bahia grass rotation is sufficient to manage plant-parasitic nematodes in a future peanut crop provided weeds in the Bahia grass planting are controlled early and regularly in the first year Bahia grass and this continued throughout the cycle of the rotation.

Two year rotations with cotton, Bahia grass, or velvet bean (Rodríguez-Kábana *et al.*, 1991a, b) are also effective.

b) **Cover Crops:** The use of winter cover crops is helpful to provide competition against volunteer peanuts and spring weeds and also the cover crop planting process helps destroy those plants growing in the fall. Cover crops, however, must be a poor or non-host of the problem nematode. Winter cereals are most suitable for managing root-knot nematode in this regard.

c) **Organic Amendments:** Organic cakes of castor, mustard and neem (1 ton ha^{-1} and above) significantly reduce the root-knot disease and increase pod and haulm yields. Neem cake is more effective than other cakes. Soil application of neem cake at 1000 kg/ha and above gave effective control of root-knot nematodes.

ii) **Chemical Methods:** Management of root-knot nematodes on peanut has traditionally relied primarily on treatment of infested fields with nematicides such as granular aldicarb (Temik) or the fumigant 1, 3-dichloropropene (Telone). A split application of Temik 15G with a band or in-furrow application at planting followed by a band application at peg initiation is acceptable usage.

Seed treatment with 3-6% aldicarb sulfone or 6% carbofuran reduces *M. arenaria* damage (Patel *et al.*, 1986). Groundnut seeds soaked for 12 hr in phenamiphos, monocrotophos, phosalane or oxamyl (each at 125, 250 and 500 ppm) inhibits the penetration of *M. arenaria* larvae in roots and increases root biomass. Phenamiphos at 250 and 500 ppm is highly effective.

Application of carbofuran at 2 kg ha^{-1} in *M. arenaria* or *M. javanica* infested soils reduces the root-knot disease by 16-96% and increases the pod yields by 15-33% (Patel, 1983).

iii) **Biological Methods:** The biocontrol agent, NemOut, is a formulation of spores of the fungus, *Paecilomyces lilacinus*, and this product has shown a moderate level of nematode suppression in University of Florida trials. It should only be used in conjunction with other recommended nematode management practices including good crop rotation and weed control. For best activity of the product, NemOut should be used both an in-furrow and as an at-pegging treatment.

iv) **Host Resistance:** The source of resistance against root-knot nematodes on groundnut in Gujarat has been identified (Table 11.2) which can be used in breeding programmes.

Table 11.2. Groundnut lines identified as resistant to *Meloidogyne arenaria* or *M. javanica* pathotype 2 in Gujarat, India

Root-knot nematode species	Highly resistant	Resistant
Meloidogyne arenaria	None	A3, Abuarbaa, Ah 25, Ah 3328, Ah 4515, Ah 6902, Ah 6719, Ah 7188, Ah 7299, Ak 10-2, 55-437, C 83, C 149, EC 24118, Kigung, Khargaon 3, NC Ac 50.
M. javanica	ICG 5341, ICG 6330	Apexy, B 1, C 162, C166, Dangi, EC 85994, ICGs 411, 852, 859, 1268, 2248, 2496, 3053, 3104, 6323, 6826, 10047, JH 223, KG 61-22, PI 268594, PI 270787, S 7-2-1, S 7-24-3, U 4-7-3, No 923, No 523.

Groundnut cv. Tifguard, is available in Florida, USA, that contains resistance to the peanut root-knot nematode and also has good resistance to tomato spotted wilt and other diseases prevalent to Florida growing conditions. The variety has a good yield potential and governed by single dominant gene which provides excellent resistance to the peanut root-knot nematode.

The cultivar COAN with resistance to *M. arenaria* and *M. javanica* was released by the Texas Agricultural Experiment Station (Simpson and Starr, 2001). In 2002, the cultivar NemaTam, which has greater yield potential than COAN with the same level of resistance to *M. arenaria* and *M. javanica*, was released by the Texas Agricultural Experiment Station (Plate 11.9). The availability of these nematode-resistant cultivars gives growers an additional option for management of *M. arenaria* and *M. javanica*, and will reduce the growers' reliance on nematicides.

Neither COAN nor NemaTAM have the yield potential of the best susceptible cultivars in the absence of root-knot disease. Therefore, these resistant cultivars are recommended only for fields known to be infested with *M arenaria* or *M. javanica.*

An additional benefit of the resistance of COAN and NemaTAM is that reproduction of the root-knot nematode is inhibited relative to reproduction on a susceptible cultivar (Plate 11.10) (Koenning and Barker, 1992). Therefore, these resistant cultivars not only protect the peanut crop from yield loss due to nematode parasitism, but they also suppress the nematode's population density at crop harvest (Plate 11.11). Any crop planted after the resistant peanut will be subjected to less disease pressure than it would experience following a susceptible peanut crop.

None were highly resistant to *M. arenaria* while ICG 5341 and ICG 6330 were highly resistant to *M. javanica.*

v) Integrated Management

Use of organic amendments, neem or castor cakes at 1 ton/ha seven days prior to sowing has been found to reduce root-knot nematode damage in groundnut. Their combinations with seed treatment, with carbosulfan (25 DS) at 3% a.i. (w/w) further improved efficacy in reducing the nematode population and enhancing the yield significantly.

The following practices have been recommended for the integrated management of root-knot nematodes in groundnut:

- Farmers of South Saurashtra region of Gujarat have been advised to sow groundnut with castor as an intercrop (row ratio 2:1) along with soil application of carbofuran at 1.0 kg a.i./ha to reduce the population of root-knot nematodes infecting groundnut.
- Late maturing varieties have more potential for damage than short-season spanish market types cultivation of varieties

11.6. CONCLUSIONS

New and re-emerging plant pests have raised fear of their potential impact on livelihoods, food security and global markets. There were outbreaks of the root-knot nematode, *Meloidogyne graminicola* and white tip nematode, *Aphelenchoides besseyi* on rice; the root-knot nematode, *Meloidogyne graminicola* on wheat; the cereal cyst nematode, *Heterodera avenae* on wheat, barley and sesame; the pigeon pea cyst nematode, *Heterodera cajani* on pigeonpea; and root-knot nematodes, *Meloidogyne arenaria* and *M. javanica* on groundnut. There is an urgent need to redefine management strategies for the management of above nematodes in field crops to meet future food requirements.

12

Emerging Nematode Problems in Fruit Crops

12.1. INTRODUCTION

The nematode problems have taken their toll and with years passing some of them gaining momentum to decimate the entire cropping systems. The root-knot nematode, *Meloidogyne indica* is devastating the acid lime in Gujarat. *Meloidogyne incognita* is severely infesting mulberry plants which adversely affect the silk industry in Karnataka.

There has been an outbreak of serious root-knot nematode problems in the recent years associated with pomegranate (*Meloidogyne incognita* Race 2) and guava (*Meloidogyne enterolobii*) disseminated through infected planting material and capable of killing the tress outright within three months after the onset of symptoms. Besides causing direct damage, the root-knot nematode species are responsible for causing wilt disease complexes in association with soil-borne fungal plant pathogens. The root-knot nematode complex has similarly created serious situation with *Ceratocystis fimbriata* on pomegranate in Maharashtra, Karnataka and North Gujarat states.

12.2. GUAVA, *Psidium guajava*

12.2.1. Root-Knot Nematode, *Meloidogyne enterolobii* (= *M. mayaguensis*)

The worst nematode problem affecting guava is the root-knot nematode, which is recognized as a limiting factor in commercial guava production in Central and South America. Guava production has declined during the past quarter century in Cuba, Puerto Rico, Mexico, Venezuela and Brazil due to increasing pressure of *Meloidogyne* spp. (El Borai and Duncan, 2005). In Brazil, *M. enterolobii* was originally detected in guava orchards in 2001 in Pernambuco and Bahia states (Carneiro *et al.*, 2001). Since then, this nematode has been a matter of grave concern in the country because it has been spreading rapidly with infected seedlings, making the cultivation of guava unviable in the heavily infested areas (Carneiro *et al.*, 2007; Siqueira *et al.*, 2009). Unfortunately, the Ministry of Agriculture (MAPA) has not yet established legislative measures to prevent the spread of this pest in Brazil. *M. enterolobii* was reported from regions in China from guava (Xu *et al.*, 2004). The presence of this nematode has increasingly been detected in different geographical regions from a wide range of hosts, including crops carrying genes for resistance to the main *Meloidogyne* spp. (Blok *et al.*, 2002; Brito *et al.*, 2007; Carneiro *et al.*, 2006a).

In and around Ayakudi village, Palani taluk, Dindigul District, and Sirumugai village, near Mettupalyam, Coimbatore district, Tamil Nadu, young guava trees are exhibiting symptoms of yellowing, shedding of leaves, drying of branches, reduced vigour, less productivity etc. On examination of affected trees, root-knot nematode was found to be the causal agent for this menace. Further, the nematode was identified as *M. enterolobii,* a species of root-knot nematode so far not encountered in India. The unseen hidden enemies of root knot nematode associated with guava are capable of killing the tress outright within three months after the onset of symptoms. It is considered that the nematode might have spread from other countries through saplings.

12.2.1.1. Geographical Distribution:

Meloidogyne enterolobii has been reported from several countries in North, Central and South America, Africa and Asia (CABI, 2000). Its present distribution in warmer climates suggests that this species will not survive outside greenhouses in northern countries of Europe, but it might be able to establish itself in the Mediterranean region (Castagnone-Sereno, 2012).

In the San Francisco (Pernambuco state of Brazil) region, the area cultivated with guava has decreased from 6,000 to 1,669 ha in seven years (2000-2006), a reduction of more than 70% in guava production (Plantec/Codevasf), a reduction of 10% per year was estimated (Carneiro *et al.*, 2007). Pereira *et al.* (2009) verified direct damages of R$ 1127 in guava production in the Brazilian states of RJ, CE, RN, PE and BA. In addition to these damages, 3,703 jobs were lost due to the decline and death of the orchards in these regions. As these regions are also an important seedling producers, this nematode has spread to other Brazilian states (Carneiro *et al.*, 2007; Siqueira *et al.*, 2009). However, this species was detected in the state of Rio de Janeiro in native areas of Atlantic forest (Lima *et al.*, 2005), suggesting that the nematode has not been introduced in Brazil from other countries, as suggested when it was reported for the first time in the country (Carneiro *et al.*, 2001). In Paraná state it was also detected in guava without introduction from other states (Carneiro *et al.*, 2006b).

12.2.1.2. Symptoms:

Meloidogyne enterolobii affects growth, yield, life span and tolerance to environmental stresses of infected plants. Severely infected trees decline rapidly, culminating in the death of the plants. Moderate infestations are associated with general chlorosis, nutrient deficiency symptoms and reduced flowering and fruiting (Plate 12.1). Roots of infected trees show multiple galls and root necrosis, causing a drastic decrease in feeder roots (Cetintas *et al.*, 2007). The root-knot nematode infects the whole root system, from small rhizoids to more lignified roots (Plate 12.1). The decline and death of guava trees parasitized by *M. enterolobii* is a complex disease involving other micro-organisms like *Fusarium solani* (Gomes *et al.*, 2010), *Verticillium dahliae, Pythium aphanidermatum,* and *Trichoderma* sp. (Avelar-Mejía *et al.*, 2001). Overall, damage due to *M. enterolobii* may consist of reduced quantity and quality of yield. This syndrome is making cultivation of guava in the infested areas unfeasible, causing serious economic problems to growers and the economy of the region (Carneiro *et al.*, 2007).

12.2.1.3. Host Range:

M. enterolobii has an exceptionally wide host range comparable to that of *M. incognita* and includes many vegetables, guava, acerola, ornamentals, and weeds (Humphreys *et al.*, 2012). It has been found on many different host plants including economically important crops such as: *Capsicum annuum* (pepper), *Citrullus lanatus* (watermelon), *Coffea arabica* (coffee), *Glycine max* (soybean), *Ipomoea batatas* (sweet potato), *Lycopersicon esculentum* (tomato), *Nicotiana tabacum* (tobacco), *Phaseolus vulgaris* (bean), *Psidium guajava* (guava), *Solanum melongena* (aubergine), ornamental plants (e.g. *Ajuga, Brugmansia, Clerodendron, Tibouchina*) and wild plants (e.g. *Bidens pilosa*). Experiments carried out in the Netherlands have also shown that *Cactus, Ficus, Syngonium, Rosa* and *Vitis* can also be host plants of *M. enterolobii.*

Only a few crops have been reported as non-hosts or very poor hosts for *M. enterolobii*, including *Brassica oleracea* (cabbage), *Allium sativum* (garlic), *Citrus paradisi* (grapefruit), *Zea mays* sub sp. *mays* (maize), *Arachis hypogaea* (peanut), *Citrus aurantium* (sour orange) and *Allium fistulosum* (welsh onion) (Castagnone-Sereno, 2012).

12.2.1.4. Spread:

As is the case for other plant-parasitic nematodes *M. enterolobii*'s own movement is limited at most to a few tens of centimetres in the soil. The main routes for nematode dissemination are by infested planting material and soil, such as traded host plants or cuttings with roots, traded soil bearing products such as potatoes, soil attached to equipment and machinery and irrigation water (Castagnone-Sereno, 2012).

12.2.1.5. Management

I) Prevention

- Use resistant varieties and rootstocks such as Horana White, Horana Giant, Pubudu and Kanthi (all are moderately resistant).
- Use nematode-free planting material.
- Use sterilized potting media (e.g. sterilised by steam or roasting).
- Use clean implements and tools.
- Do not use previously infested fields for planting. Do proper land preparation. Deep ploughing expose nematodes to sunlight.
- Use recommended fertilizers to improve crop vigour.
- Use trap crops like marigold.
- Green manuring with repellent crop before planting sun hemp and mustard.
- Add burnt paddy husk to planting hole.

ii) Monitoring

- Regularly observe nursery plant roots for the characteristic symptom of root galls.
- Sample the side roots of mature plants to observe for root galls.
- Observe above-ground symptoms like leaf yellowing, withering and scorching of leaf edges with purple patches followed by examination of root systems for galls.

iii) Direct Control

- Avoid surface irrigation from infested fields to healthy fields.
- Add compost to the soil to reduce the effect of nematodes (after infestation).
- Remove weeds and infested crop residues to protect nearby health tree.
- Intercrop with Crucifers like mustard, cabbage and incorporate into soil (before flowering) to suppress nematodes.
- Add poultry manure, cow dung, goat dung, coconut and rubber pomace, neem cake and green manure (Adathoda, Tithonia, Gliricidia, neem and Kalawel) in planting pit as a soil amendment.
- Remove infested plants from the field

iv) Intercropping with Non-Host Cover Crops: An effective alternative control of this nematode is the use of non-host cover crops. Among 38 cover crops assessed, 26 were resistant: peanut (*Arachis hypogaea*) 'Cavalo vermelho', white oat (*Avena sativa*) 'IAPAR 126', black oat (*Avena strigosa*) 'IAPAR 61', rapeseed (*Brassica napus*) 'CAN 420', 'CAN 401', foxtail millet (*Setaria italica*), millet (*Eleusine coracana*), rattlepod (*Crotalaria anguroides, C. apiclolice, C. grantiana, C. juncea, C. okraeluka*), cowpea (*Vigna unguiculata*), hyacinth bean (*Dolichos lablab*), castor bean (*Ricinus communnis*) 'IAC 80', grey velvet bean (*Mucuna cinerea*), green velvet bean (*Mucuna* sp.), black mucuna (*Mucuna aterrima*), forage radish (*Raphanus sativus* var. *oleiferus*) 'AL 1006', 'Seletina Nova', 'N4', 'Jesuíta', white tephrosia (*Tephrosia candida*), timbó (*Ateleia glazioveana*), triticale (*Triticum aestivum* × *Secale cereale*). Ninc species/genotypes of plants were immune: peanut 'IAC Oirã', 'IAC Poitã', 'IAC Tatuí', Italian ryegrass (*Lollium multiflorum*), rye (*Secale cereale*) 'IPR 89', butterfly pea (*Clitoria ternatea*), mungo bean (*Vigna radiata*), cooper (*Glycine wightii*) (Table 12.1). Only three species of plants were susceptible: hairy vetch (*Vicia villosa*) 'Ostssat', ricebean (*Vigna unbellata*) and jack bean (*Canavalia ensiformis*).

Table 12.1. Summer and winter cover crops immune (RF = 0.00) to *Meloidogyne enterolobii*

Crop	Cultivars
Peanut, *Arachis hypogaea*	IAC Poitã IAC Tatuí IAC Oirã
Butterfly pea, *Clitoria ternatea*	—
Cooper, *Glycine wightii*	—
Italian ryegrass, *Lollium multiflorum**	—
Forage radish, *Raphanus sativus* var. *oleiferus**	'N 4'
Rye, *Secale cereale**	IPR 89
Mung bean, *Vigna radiata*	—

* Winter plants; those not marked are summer plants.

The summer and winter cover crops which were resistant to *M. enterolobii* are presented in Table 12.2.

Table. 12.2. Summer and winter cover crops resistant (RF = < 1.00) to *Meloidogyne enterolobii*

Crop	Cultivars	Reproduction factor
Peanut, *Arachis hypogaea*	Cavalo Vermelho	0.03
Timbó, *Ateleia glazioveana*	—	0.02
White oat, *Avena sativa*	IAPAR126	0.02
Black oat, *Avena strigosa**	IAPAR 61	0.04
Rapeseed, *Brassica napus**	CAN 420	0.36
	CAN 401	0.36
Rattlepod, *Crotalaria anagyroides*	—	0.02
Rattlepod, *Crotalaria apioclice*	—	0.15
Rattlepod, *Crotalaria grantiana*	—	0.03
Sun hemp, *Crotalaria juncea*	—	0.50
Rattlepod, *Crotalaria ochroleuca*	—	0.02
Field bean, *Dolichus lablab*	—	0.03
Finger millet, *Eleusine coracana*	—	0.05
Black velvet bean, *Mucuna aterrima*	—	0.27
Green velvet bean, *Mucuna* sp.	—	0.06
Grey velvet bean, *Mucuna cinerea*	—	0.01
Forage radish, *Raphanus sativus* var. *oleiferus**	AL 1006	0.80
	Seletina Nova	0.04
	Jesuíta	0.01
	IPR 116	0.01
Castor, *Ricinus communis*	IAC 80	0.13
Foxtail millet, *Setaria italica**	—	0.03
White tephrosia, *Tephrosia candida*	—	0.20
Triticale, *Triticum aestivum* × *Secale cereale**	IPR 111	0.07
Cowpea, *Vigna unguiculatta*	Australian	0.04
	—	0.60

* Winter plants; those not marked are summer plants.

Some crop systems alternating non-host plants in summer and winter or only planting in the summer may work to control *M. enterolobii*, taking into account different climatic regions where this nematode causes damage (Table 12.3).

Table 12.3. Reaction of 'Paluma' guava to *Meloidogyne* spp.

Populations	RF[1]	Status[2]
Meloidogyne incognita race 1	0.03	R
Meloidogyne incognita race 2	0.05	R
Meloidogyne arenaria race 2	0.0	I
Meloidogyne javanica	0.0	I
Meloidogyne enterolobii	182.4	HS

[1]RF = Final population/Initial population.

[2]Status: I = Immune (RF=0), R = Resistant (RF1.0), HS = Highly susceptible.

Six genotypes of corn ('NB-7361', 'SHS-5080', 'GNX-1020', 'GNX-3010', 'BRS-1031' and 'BM-1115') were highly resistant (FR >1.0) to *M. enterolobii* and can be used in crop rotation systems to reduce or eliminate nematode populations (Dias *et al.*, 2010). Field experiments are necessary to validate these greenhouse results.

Fourteen fruit trees are non-hosts to *M. enterolobii* (atemoya, avocado, cashew, citrus rootstocks, coconut and mango) or poor hosts (assai palm, common mulberry, jaboticaba, grape rootstocks, passion fruit, sapodilla and soursop); only four fruit trees are good hosts (fig, banana, 'Chardonnay' and 'Solferino' grapes and melon). More studies should be done in infested fields to confirm the poor host status of these fruit trees tested in greenhouse conditions.

Growing non host crops or black fallow are the most efficient methods for reducing *M. enterolobii* populations. Unfortunately, the list of non-host plants is very small, including cabbage, garlic, grapefruit, maize, peanut, sour orange and welsh onion.

v) **Chemical Methods:** Application of carbofuran at 15 g a.i./ m^2 in the rhizosphere region before flowering, pulverize the soil in the basin area to mix the chemical in soil followed by irrigation was found effective for the management of guava root-knot nematode.

vi) **Biological Methods:** Application of *Paecilomyces lilacinus/Psudomonas fluorescens*) at 20 g/m^2 (about 9 m^2 around the tree trunk) in the rhizosphere region at the time of flowering gave effective control of *M. enterolobii.*

vii) **Host Resistance:** The susceptibility of 46 and 112 wild *P. guajava* to *M. enterolobii* was observed in Brasília and Petrolina, respectively. The same was detected by other authors for *M. enterolobii* (Almeida *et al.*, 2009; Carneiro *et al.*, 2007; Maranhão *et al.*, 2001; Scherer, 2009). In disagreement with this work, Cassasa *et al.* (1997) evaluated 7 accessions of *P. guajava* in relation to *M. enterolobii*. The resistance was observed only in *P. guajava* accession S3.

Milan (2007, 2010) also reported resistance in guava accessions. In the first work, *P. guajava* "B-12" was classified as resistant to *M. enterolobii* (FR=0.88) having the potential to control the nematode, so that it could be used as a rootstock for commercial guava clones. A total of more than 2000 F1 seedlings were produced in the hybridization programme and the F1 seedlings are in the nursery stages for further evaluation. The results also showed that *P. longipes* and *P. arayan* showed potential resistant characters to the root-knot nematode, but their potential as rootstocks for guava cultivation is still questionable. In the second work, this author observed that three accessions of *P. guajava* (K-10, A-06, and J-16) were resistant to *M. enterolobii*, with IG more than 2. Unfortunately, the RFs were not measured. These accessions are graft-compatible with local commercial clones and thus could be used as a rootstock. No more information about the reaction of 2000 F1 seedlings to *M. enterolobii* was available.

P. friedrichsthalianium (Costa Rican wild guava) was considered resistant to *M enterolobii* (=*M. mayaguensis*), *M. incognita* race 1, *M. arenaria, M. javanica*) in different countries: Venezuela (Casassa *et al.*, 1997, 1998; Molero *et al.*, 2010), Colombia (Villota and Agudelo, 1997) and Brazil (Carneiro *et al.*, 2007).

The performance of guava grafted onto tolerant *P. guajava* rootstock and onto resistant *P. friedrichsthalianium* was evaluated. In the nursery, 100% successful grafts were obtained with guava and Brazilian wild guava (*P. guineense*) rootstocks, while success with *P. friedrichsthalianium* was 90% ('Arrayán', 'Criollo'). Initial vegetative growth was significantly higher in the guava rootstock treatment. In the field this treatment also showed more vigorous growth. 'Arrayán' rootstock clearly showed a strong dwarfing effect while the 'Brazilian' and 'Criollo' wild guava rootstocks exhibited an intermediate effect on vegetative vigour. Number and weight of fruits was higher when guava was used as a rootstock and no differences were observed with wild guava rootstocks (Bogantes-Arias and Newcomer, 2010). In order to evaluate the susceptibility of the rootstocks to nematodes of adult plants, an additional control treatment was included in the field, which consisted of the application of nematicides to the guava rootstock treatment. The results indicated that the root-knot nematode was able to colonize the roots of the guava rootstocks with and without nematicides, as well as colonizing the roots of Brazilian wild guava. No nematodes were detected in the roots of the 'Criollo' and 'Arrayán' rootstocks, confirming the resistance of *P. friedrichsthalianium* (Bogantes-Arias and Newcomer, 2010).

Only Costa Rican wild guava *P. friedrichsthalianium, P. cattleyanum* (yellow wild guava), *A. sellowiana* and *P. rufum* showed resistance or immunity to *M. enterolobii. P. acutangulum* and *P. guineense* (Brazilian guava) were susceptible. However, Lee *et al.* (1998) classified the only *P. guineense* accession tested as immune. The same results were observed by Costa *et al.* (2012). Almeida *et al.* (2009) classified *P. friedrichsthalianum* (probably an incorrect identification) as susceptible to this pathogen (FR=13.03). Additionally, these authors verified that three non-identified *Psidium* spp. accessions and were resistant to *M. enterolobii*, with FRs less than 1. *P. friedrichsthalianum* is not originally from Brazil, and this Brazilian rootstock is probably another *Psidium* species. Ten plants of *P. friedrichsthalianum* and nine of *P. cattleyanum* were compatible with 'Paluma' guava when grafted by budding technique in greenhouse conditions. These grafted plants were maintained in field conditions. One year later, five Costa Rican plants were still alive and producing buds, flowers and fruits, while the plants grafted on *P. cattleyaum* presented an incompatibility reaction (enlargement in grafting region); consequently all nine plants died during the first year.

a) **Resistant Rootstocks:** Carneiro *et al.* (2012) reported that only one accession of *P. friedrichstalianium* (Costa Rican wild guava) was resistant and compatible as rootstock with *P. guajava* 'Paluma', in field conditions.

Abdul Karim *et al.* (2002) found that the Red strawberry guava (*P. littorale* var. *longipes*) which showed apparent tolerance.

Cuadra and Quincosa (1982) found that Costa Rican guava (*P. friedrichsthalianum*) was highly resistant to *Meloidogyne* spp. in Cuba.

Milan (2007) found that only *P. guajava* acc. B-12 has the potential to control the root knot nematode and graft-compatible with commercial clone of guava because they are from the same species. *P. guajava* acc. B-12 had the potential to control the nematode and could be used as a rootstock for guava commercial clones.

vii) **Integrated Management:** The guava growers are advised to avoid nematode affected saplings, removal and destruction of nematode affected trees, application of farm yard manure at 100 kg, neem cake at 250g and nematode egg parasitic fungus *Paecilomyces lilacinus* at 25g and carbofuran at 60g/tree at early stage of nematode infestation to overcome the nematode problem and to enhance fruit yield.

Application of 1 kg neem cake enriched with bioagents + carbofuran at 7g a.i./ m^2 just before flowering was found effective against *M. enterolobii.*

12.2.1.6. Prospects and Future Challenges

- There is an urgent need to strengthen and enforce domestic quarantine mechanism to suspend spread of plant parasitic nematodes with vegetative propagules, especially through rooted plants – along with soil, from nurseries/ sick plots/ hot-spot areas to disease free niches.
- Develop nematode screening methods with more *Psidium* spp. with correct rootstock botanical classification.
- Improve nematode screening methodology, using the Reproduction Factor (RF) as an obligatory parameter.
- Improving grafting techniques in *Psidium* spp.
- Develop interspecific hybridization with *Psidium* spp. to obtain nematode-resistant rootstocks.
- Study in vitro production of resistant plants by somatic embryogenesis.
- Study the mechanisms and molecular characterization of resistance in *Psidium* spp.
- Develop molecular markers related to resistant genes in *Psidium* spp.

12.2.1.7. Future Lines of Work

- Enforce domestic quarantine
- Production of nematode-free planting material (use of solarization, organic amendments enriched with bioagents).
- Use of resistant rootstocks (*P. guajava* Accession B-12, *P. friedrichsthalianum*).
- Intercropping with non-host cover crops (peanut, corn, cowpea, sun hemp, finger millet, foxtail millet, castor, rye, mung bean, rapeseed, field bean, velvet bean).
- INM using enriched neem/Mustard cake with biocontrol agents (*Pseudomonas fluorescens, Paecilomyces lilacinus, Pochonia chlamydosporia, Trichoderma harzianum*).

12.2.2. Guava Decline Disease Complex, *Meloidogyne enterolobii, Fusarium solani*

12.2.2.1. Interaction Studies:

M. enterolobii and *F. solani* have a synergistic effect on disease complex development. Whereas *M. enterolobii* alone was able to induce galling and a limited amount of root rot, *F. solani* did not cause root rot when in isolation. When combined, nematode and fungus led to drastic root rot and typical guava decline symptoms, which suggests that guava decline is a disease complex between *M. enterolobii* and *F. solani* (Gomes *et al.*, 2010, 2012).

The root-knot nematode parasitized trees develop rotten roots as the disease progresses, suggesting the possibility that a soil-borne pathogen could be involved. From the 448 root samples received, fifty per cent were positive for *Fusarium solani*, which confirm the nationwide role of *F. solani* in guava decline. Guava decline, caused by the interaction between the phytonematode *M. enterolobii* and the fungus *F. solani*, has caused direct and indirect losses to the whole productive chain of guava (Gomes *et al.*, 2010, 2012).

Declining trees show chlorosis, scorching of margin, wilting and shedding of leaves (Anon, 2011). Gomes *et al.* (2008) have shown that these symptoms are associated with foliar deficiency of nitrogen, potassium and phosphorus, a near-deficiency of calcium and magnesium, and an imbalanced accumulation of manganese, chlorine and sodium. Declining trees develop progressive rotting of the entire root system. Usually, declining trees do not recover, and death occurs within weeks to a few months.

The onset or worsening of shoot symptoms is often associated with (i) an increase in the number of root galls in the entire root system, (ii) following a high production season or (iii) following drastic pruning, which is used by some growers to synchronize production.

No root rot and virtually no effect on all variables were observed in the seedlings inoculated with the fungus isolates, with or without physical injury. *M. enterolobii* is also not highly aggressive to guava when inoculated alone (Gomes *et al.*, 2011; Almeida *et al.*, 2011). Major root rot and a negative effect on all variables were observed in the seedlings inoculated with *M. enterolobii* and all four *F. solani* isolates. The penetration and initial development of the fungus were reinforced by the morpho-physiological alterations caused by *M. enterolobii* infection in guava plants, which increased the susceptibility to *F. solani*. In this complex disease, parasitism by *M. enterolobii* predisposes guava plants that are immune to *F. solani* to extensive root deterioration caused by this fungus, which leads to loss in production and to death of plants in a short period of time. This characterizes guava decline as a complex disease caused by the synergistic effect of these organisms, in which parasitism by the nematode predisposes the plants to root decay caused by the fungus.

Significant ($P < 0.05$) root rot and damage occurred when *M. enterolobii* and *F. solani* were combined (Plate 12.2). The synergistic effect of *M. enterolobii* and *F. solani* in causing guava decline is clear, and this combination reflects the complex nature of the disease.

The effect of *Meloidogyne enterolobii* and *Fusarium solani* was studied in greenhouse on development of the root and aerial part of guava plants with root system divided into two plastic container (I and II), in greenhouse, inoculated with 500 eggs and second-stage juveniles of *Meloidogyne enterolobii* and/or 5 g of wheat grains colonized by *Fusarium solani* isolate UENF/CF 163. The evaluation took place 120 days later, and the results are presented in Table 12.4 and Plate 12.3.

Table 12.4. Effect of *Meloidogyne enterolobii* and *Fusarium solani* on development of the root and aerial part of guava plants with root system divided into two plastic container (I and II), in greenhouse

Treatments applied to the root system		Fresh shoot mass (g)	Root system volume (cm^3)		Fresh mass of root system (g)	
I	II		I	II	I	II
Non-inoculated	Non-inoculated	93.6a	25.8a	24.4a	27.8a	28.0a
Non-inoculated	*M. enterolobii* (M.e.)	98.9a	20.8a	20.4a	23.2a	26.9a
Non-inoculated	*F. solani* (F.s.)	95.5a	23.8a	24.0a	22.8a	22.5a
Non-inoculated	M. e. + F. s.	74.7b	20.4a	8.3b	21.2a	10.4b
M. e.	F. s.	92.8a	18.6a	23.6a	25.6a	28.4a
CV%	–	**23.4**	**23.8**	**18.4**	**21.2**	**16.2**

Values followed by the same letter in the column are not different according to T test at 5%.

12.2.2.2. Management

i) **Host Resistance:** Misra *et al.* (2003) identified F1 population of *Psidium molle* × *Psidium guajava* free from wilt, when grown in wilt sick plot and artificially inoculated repeatedly with *Gliocladium roseum*, *Fusarium solani* and *Fusarium oxysporum*. As graft compatibility is very successful, this resistant rootstock is very useful for the control of wilt.

ii) **Biological Control:** Singh *et al.* (2003) reported bioagent, *Aspergillus niger* was most effective in controlling the wilt disease followed by *Trichoderma viride*. *Aspergillus niger* was found very fast growing, easy to propagate and most effective in controlling the wilt disease in field. Besides this quality, it is also growth enhancer and the plants treated with *A. niger* developed faster with more height, more thickness and more numbers of leaves (Misra *et al.*, 2000).

iii) **Chemical Methods:** Propiconazole was significantly superior as compared to other chemicals. Propiconazole (0.2%) treated plants were recovered 105th day after treatment, followed by difenconazole (0.2%) treated plants which were recovered at 120th day. These two chemicals were effective than remaining chemicals (Sonyal, 2010).

iv) **Integrated Management:** Integrated eco-friendly approach for the control of guava wilt was suggested by Misra *et al.* (2004) and Misra (2005) utilizing bioagent *Aspergillus niger* strain AN17, using resistant rootstock (*P. molle* × *P. guajava*), intercropping of guava with turmeric or marigold as well as proper cultural practices and can be integrated to minimize losses due to the disease complex.

Integration of *T. harzianum* + *P. flourescence* each at 1 × 10^6 gave effective management of the disease complex (Sonyal, 2010).

12.3. POMEGRANATE, *Punica granatum*

12.3.1. Root-Knot Nematode, *Meloidogyne incognita*

The root-knot nematode, *Meloidogyne incognita* Race 2 is a serious pest of pomegranate (in Maharashtra, Karnataka, Gujarat and Tamil Nadu) which primarily cause galls on root system that eventually are seriously hampered in their main functions of uptake and transport of water and nutrients, reduced plant growth and productivity of the pomegranate (Khan *et al.*, 2005; Nasira *et al.*, 2011).

The area under pomegranate was increasing steadily in Coimbatore, Erode, Dindigul and Tirunelveli districts due to its higher monitory returns and import value. Growers of late in Sivagiri and Modakuruchi areas, Erode district, Tamil Nadu are encountering problem of yellowing of leaves, stunting and less productivity of fruits. Such trees were found to be severely infested with root knot nematodes on assaying of soil and root samples. As the presence and association of root knot nematode in pomegranate has not been documented so far in Tamil Nadu, it is presumed that the nematode could have spread from other States through saplings.

12.3.1.1. Symptoms:

Severely infested plants by root-knot nematode exhibited yellowing of foliage resulting in stunted plant growth (Plate 12.4). These plants produced less number of fruits or no fruits which might be due to the nematode induced nutritional deficiency. In severe cases the galls were predominantly found on entire root system (Plate 12.5). The galls were white in colour turned to light brown and hardy when they became old. The intensity of root knot nematode damage increased with increase in age of the plant. In general more than five year old plants were severely affected by root-knot nematode. Egg masses were observed inside as well as outside of the galls. More numbers of females were found in a single compound gall.

12.3.1.2. Management

i) **Intercropping:** Intercropping with sun hemp, marigold, mustard, fenugreek, onion, and garlic during crop rest period was found effective against root-knot nematode on pomegranate.

ii) **Botanicals:** Application of marigold flowers gave the highest reduction percentage in root galling with value of 98.8 %, followed by colocynth fruits extract which averaged 96.4 % without significant differences between them. The lowest egg-mass index (EI) was obtained from plants receiving marigold extract (0.0) with reduction percentage value of 100 % (Table 12.5) (Nour El-Deen *et al.*, 2016).

Table 12.5. Impact of certain plant extracts on development of *M. incognita* infecting Taify pomegranate under greenhouse conditions.

Treatment	No. of galls*	Root gall index (RGI)	No. of Egg-masses	Egg-mass index (EI)
Citrullus colocynthis	3.0d	1.3d	2.3d	1.7b
Tagetes erecta	1.0d	0.7e	0.0e	0.0c
Rosmarinus officinalis	9.3c	2.0c	5.3c	2.0b
Thymus vulgaris	17.3b	3.0b	9.0b	2.0b
Withania somnifera	23.0b	3.0b	11.0b	2.7a
Control (Nematode lone)	82.3a	4.0a	24.0a	3.0a

* Means in each column followed by the same letter(s) did not differ at $P<0.05$ according to Duncan multiple- range test.

iii) **Biological Methods:** *Pseudomonas fluorescens* at 20 kg/ha was found to be most effective in reducing *M. incognita* population (31.28%) and number of root galls/5 g roots (29.28%) and increasing the fruit yield of pomegranate (18.99%) with 2.37 B:C ratio; followed by *Trichoderma viride* at 20 kg/ha (reduction in *M. incognita* population - 28.3%, number of root galls/5 g roots -23.9%, increasing the fruit yield -16.78%, and B:C ratio - 2.33) (Table 12.6) (Pawar *et al.*, 2013).

Table 12.6. Effect of bioinoculants on the management of root-knot nematode on pomegranate

Treatment	% Decline in nematode population	%Reduction in root galls	Yield (tons/ha)	B: C ratio
Pseudomonas fluorescens at 20 kg/ha	31.2	29.2	18.4	2.37
Trichoderma viride at 20 kg/ha	28.3	23.9	18.0	2.33
Carbofuran at 3 kg a.i./ha	29.8	25.1	17.8	2.27
Control (Untreated)	—	—	15.4	2.27
CD (P = 0.05)	**2.99**	**2.30**	**1.17**	—

The symbiotic bacterium of entomopathogenic nematode *Xenorhabdus* treatment significantly inhibited formation of galls on roots with reduction percentage of 97.9 %, followed by those treated with *Photorhabdus* with value of 96 %. Regarding egg-masses numbers, *M. incognita* was barely reproduced on plants treated with *Xenorhabdus* filtrate with reduction percentage value of 95.8 %, followed by *Photorhabdus* with value of 91.7 % (Table 12.7) (Nour El-Deen *et al.*, 2016).

Table 12.7. Effect of symbiotic bacteria of entomopathogenic nematodes on development of *Meloidogyne incognita* infecting Taify pomegranate under greenhouse conditions

Treatment	No. of galls*	Root gall index (RGI)	No. of Egg-masses	Egg-mass index (EI)
Steinernema spp.	6.0bc	2.0b	3.0c	1.7b
Heterorhabditis spp.	10.0b	2.3b	6.0b	2.0b
Xenorhabdus sp.	1,7cd	1.0c	1.0c	1.0c
Photorhabdus sp.	3.3c	2.0b	2.0c	1.0c
Control (Nematode alone)	82.3a	4.0a	24.0a	3.0a
Steinernema spp.	6.0bc	2.0b	3.0c	1.7b

* Means in each column followed by the same letter(s) did not differ at $P<0.05$ according to Duncan multiple- rage test.

iv) **Chemical Methods:** Application of carbofuran at 2 kg a.i./ha in nursery, and 4 kg a.i./ha in the main field at transplanting and at flowering gave effective management of *M. incognita* on pomegranate.

v) **Integrated Management:** Application of farmyard manure at 20 tons/ha, neem cake at 250 g/tree followed by bioagent, *Paecilomyces lilacinus* at 25 g/plant was found effective to manage root-knot nematode problem in pomegranate and to increase crop productivity.

Two applications of FYM enriched with *P. lilacinus* at 15-25 kg around each plant at 3 months interval along with application of liquid formulation of *P. lilacinus* through drip; showed recovery from symptoms, and putting forth new flushes. Saprophytic microbial population in soil has multiplied many fold, reduced egg masses and females in roots and juveniles of root-knot nematodes were not seen in soil.

The farmers of North Gujarat region cultivating pomegranate are advised to apply talc based formulation of *Pochonia chlamydosporia* (10^8 spores/g) or *Paecilomyces lilacinus* (10^8 spores/ g) at 100g/ plant along with mustard cake at 1.5, 2.0 and 2.25 kg/ plant in first, second and third year, respectively; two times i.e., first during November and second during April for the effective management of root- knot nematodes.

Application of bioagents (*T. viride, P. lilacinus, Ps. fluorescens*) at 10-20 g/m^2 in nursery, and 5 kg/plant of FYM/compst enriched with bioagents in the main field at transplanting and at flowering was found effective against *M. incognita.*

12.3.1.3. Prospects and Future Challenges

- Intercropping with sun hemp, marigold, mustard, fenugreek, onion, and garlic during crop rest period.
- Use of biocontrol agents such as *Pseudomonas fluorescens, Paecilomyces lilacinus, Pochonia chlamydosporia, Trichoderma harzianum, T. viride* at 20 kg/ha.
- Use of entomopathogenic nematodes (*Steinernema* spp., *Heterorhabditis* spp.) and their symbiotic bacteria (*Xenorhabdus, Photorhabdus*).
- INM using enriched neem cake/ Mustard cake/FYM/Compost/ Vermicompost with biocontrol agents (*Pseudomonas fluorescens, Paecilomyces lilacinus, Pochonia chlamydosporia, Trichoderma harzianum, T. viride*).
- Enforcement of domestic quarantine.

12.3.2. Wilt Disease Complex (Fungi, Nematodes and Insect pests)

In recent years, pomegranate wilt complex is becoming an important disease which results in complete wilting of plant. The wilting severity is increasing at faster rate. It is caused by the association of fungi (*Ceratocystis fimbriata, Fusarium oxysporum, Rhizoctonia solani*), root-knot nematodes (*Meloidogyne incognita*) and shot hole borers (*Xyleborus fornicates* and *X. perforans*). It was first noticed in two areas of Bijapur district of Karnataka, India, in 1990 (Somasekhara and Wali, 1999). Around 1993, rapid spread of this disease was observed in the entire Bijapur district. The severity of wilt complex observed was 49.2 per cent in Maharashtra, 61.11 per cent in Karnataka and 8.69 per cent in Andhra Pradesh. In Maharashtra, among 92,185 ha area under pomegranate, 23,858 ha (25.88%) are reported to be wilt affected (Jadhav and Sharma, 2009). Despite many factors conducive for the high severity of disease, seedlings selection for planting, soil-borne nature and also association with shot hole borer and plant parasitic nematodes might be the reason for the current rampant spread of the disease in south Indian states.

12.3.2.1. Distribution:

The disease has been reported from Karnataka, Andhra Pradesh, Tamil Nadu, North Gujarat, and Maharashtra states (Somasekhara and Wali, 1999).

12.3.2.2. Symptoms:

When the roots and lower portion of the stems were split open the following can be observed:

- Shows dark bluish-black/grey/brown discolouration of the wood if the pathogen is *Ceratocytis fimbriata.*
- Pin holes are observed in the bark and wood when shot hole borers are associated with wilt.
- Root galls are observed on fine roots in early stage and other roots in advance stage if it is the root-knot nematode infestation.

i) **Wilt Inducing Fungi:** Wilt symptoms initiate as yellowing of leaves of one or more branches, followed by drooping of foliage and complete wilting of plant within 2-3 months period (Plate 12.6). Plants may reveal drooping of leaves of the entire plant all of a sudden (within 1-2 days) resulting in complete wilting. *C. fimbriata* infection causes discolouration of vascular bundles and adjoining tissues (Plate 12.7).

ii) **Shot Hole Borer:** Shot hole borer infest the collar region of the plant by making innumerable pin or shot holes (Plate 12.8) causing discontinuity in the conducting vessels affecting conduction of water to the upper portion of the plant. Consequently, the drying of the twigs became prominent. The drying progressed from the tip towards base and in severe cases entire plant dried.

iii) **Root-Knot Nematode:** In general, *M. incognita* infected plants were exhibited yellowing of foliage resulting in stunted plant growth. These plants produced less number of fruits or no fruits which might be due to the nematode induced nutritional deficiency. In severe cases the galls were predominantly found in entire root system (Plate 12.9). The intensity of root knot nematode damage increased with increase in age of the plant. In general, more than five year old plants were severely affected by root-knot nematode. Egg masses were observed inside as well as outside of the galls. More numbers of females were found in a single compound gall.

12.3.2.3. Predisposing Factors:

The wilt diseases are generally aggravated due to biotic stresses particularly drought as well as excessive rain, boron deficiency in soil resulting in increased severity of *C. fimbriata.* Wounds, natural or due to insect/nematode or human activity like pruning and intercultural operations predispose the plants to severe infections, as the pathogen is more devastating in overindulged orchards rather than in orchards with little human activity. Stress due to flowering and fruit bearing trees also results in sudden death of the entire plant. All commercial cultivars are susceptible to wilt and it can attack plants of all ages.

12.3.2.4. Interaction of Fungal Pathogen with Plant Parasitic Nematodes:

Disease complexes involving nematode and fungi have gained momentum in the recent years. There was no conspicuous gall production in nematode alone. In combination with fungi, the affected roots showed brownish discolouration in fungus alone and fungus + nematode treatments. In uninoculated treatment there was no brownish discolouration.

The two pathogens in combination (*M. incognita, C. fimbriata*) adversely affected plant growth parameters like shoot and root length, and fresh and dry shoot and root weight. Among the two pathogens inoculated individually, maximum reduction in plant growth parameters was noticed in fungus alone (*C. fimbriata*) treatment. However, the treatment receiving the inoculum of both pathogens (*M. incognita* + *C. fimbriata*), showed a highest reduction in plant growth parameters in comparison with other treatments.

12.3.2.5. Spread and Survival

- The fungus survived in thick walled Aleurioconidia of *C. fim*briata, in soil, host and plant debris and served as a source of primary inoculum. The fungus can survive in infected wood for several years.
- Infected planting material or the soil used for raising the nursery planting material generally are the primary source of infection for wilt organisms.
- Pruning wounds are common entry points for *C. fimbriata*, and the fungus can also be carried on pruning tools and through air.
- Spread to adjacent trees can take place via root grafts generally in plantation at close spacing.
- Flood irrigation/rain water also spread the pathogens from the infected plant/s to healthy plants.
- Wounds in the roots made by insects, nematodes and rodents also predispose the plant to wilt pathogens.
- Infected seedlings, irrigation water, root contact, implements, rain water, budding knife, secateurs and wind are the major sources for the spread of disease (Somasekhara *et al.*, 2009).

12.3.2.6. Management

i) **Biological Control:** Biological formulations if used should be reliable, fresh and used during rest period when no other fungicides/bactericides are used. The soil application of *Bacillus subtilis, Paecilomyces lilacinus, Trichoderma harzianum, Aspergillus niger* at 10-15g/plant along with well-decomposed farm yard manure around the trunk of pomegranate trees helps to prevent wilt infections. Neem cake at 2-3 kg/ plant effectively checks incidence of wilt complex.

Biofertilizer–Kalisena SA having *Aspergillus niger* at 1 kg/acre + Mychorrhizal preparation Josh at 5 kg/acre or Josh ultra 1kg/acre gives effective control of wilt if used from beginning or before the incidence of disease. These two bioagents should be applied twice a year along with sufficient organic matter for effective wilt management. This treatment controls several soil pathogens, improves nutrient uptake and gives disease resistance and also improves yields.

ii) **Integrated Management:** The following practices are recommended for the management of wilt disease complex in pomegranate:

- It is advisable to take cuttings/air layers from disease free orchards and raise saplings in sterilized soil.
- The planting material (sapling as well as soil in which it is planted) should be free from all wilt organisms.

- Soil solarization or sterilization with formalin (50 ml/l) should be done before planting
- Avoid water stagnation and create proper drainage.
- Follow recommended spacing of 4.5 m × 3.0 m in the orchard.
- The nursery beds or soil used for potting mixtures should be sterilized by soil solarization for 6 weeks using 50-100μ thick linear low density polyethylene (LLDPE) sheet during hot summer months or by using chemical sterilants at 2.5-5.0% formalin. If formalin is used ensure that the soil is free from any formalin fumes before transplanting in the bags. Soil solarization is beneficial as it kills harmful pests and pathogens and also increases population of beneficial microrganisms which are present in the soil and are thermo tolerant. However, formalin treatment kills both harmful and beneficial organisms.
- The above formalin treatment can be also be used for sterlizing soil after removing infected dead plant.
- On observing first symptoms of wilt due to fungal pathogens in the orchard, drench soil immediately with propiconazole 25EC (2 ml/l) + chlorpyriphos 20EC (2 ml/l) or carbendazim 50WP (2.0 g/l) + chlorpyriphos 20EC (2 ml/l). Use 5-10 litres solution/plant depending on plant growth to drench soil up to 12 inches depth. Drench at least 3-4 healthy plants all around the infected plant/s, and repeat drenching 3-4 times at 20-25 days interval. Drenching with Ridomil, Metalaxyl or Dithane M-45 (2g/l) will be beneficial if *Phytophthora* is involved. Propiconazole was significantly superior than remaining treatments. Propiconazole treated plants were completely recovered from wilting after 105th day, followed by difenconazole the plants were recovered after 120th day. The control showed maximum percentage of wilting (97.61%).
- For controlling shot hole borer (*Xyleborus* spp.) which is associated with wilt disease, apply 10 litre solution containing red soil (4kg) + Chlorpyriphos 20EC (20ml) + Copper oxychloride (25 g) at plant base up to 2 ft. diameter from second year onwards. To control stem borer, inject in the holes on the trunk with DDVP 2-3 ml and plug the holes with mud.
- Wilt due to root-knot nematodes can be managed with soil application of Phorate 10G at 25g/plant or Carbofuran 3G at 40g/plant or other suitable nematicide in the plant basin, in a ring near root zone and cover it with soil. Drenching with Azadirachtin (1%) at 2ml/l is also recommended. Application of neem cake at 1-3kg/plant depending on age is advisable twice a year. Plant *Tagetes erecta* (African marigold varieties best followed by French marigold) in between row space or in a ring, on the border of plant basin. For effective results, these should be grown for more than 4 to 5 months and incorporated in the soil. Crops like onion, tomato, chilli, potato, bell pepper, gram, legumes, cucurbits, gerbera, gladiolus *etc.* aggravate nematode infestation and hence should be avoided as intercrop. Green manuring with sesbania is beneficial.

iii) General Precautions

- Once disease is detected in the orchard, dig about 3-4 feet long trench between the wilted and healthy plant/s. The partially wilt affected plant/s should be treated with a suitable soil application depending on pathogen involved.

- Dead plants should be removed and burnt, they should not be kept dumped in the orchard for firewood. While removing the wilted plants from the orchard for burning, protect the entire root zone with cover (fertilizer bag *etc.*), so that pathogens in soil or root do not spread within the orchard.
- The soil in the pit from which dead plant has been removed, should be sterilized with drenching of 2.5-5% formalin at 10 litres solution per pit. It should be covered with polyethylene sheet for one week. After 1 week remove polyethylene sheet and rake the soil daily up to 10-15 days, so as to allow escape of gas. Plant new sapling once there is no smell of formalin in soil.
- Pruning tools should be disinfected and cut ends painted with fungicidal oil based paints. Pruning should be avoided during spring to summer and done in winter months. Partially affected plants within the buffer zone should be treated with a suitable chemicals; neighbouring asymptomatic apparently healthy plants should also be treated with appropriate systemic fungicide/insecticide. Plants with more than 30% canopy loss should be uprooted and burnt, soil treated with formalin and new plant grown.

12.3.2.7. Future Lines of Work

- The survey and surveillance of the disease may be extended to other states to know the regional severity of the wilt and mapping of the disease.
- Broad based molecular biological studies are essential to establish pathogenic variation to device better disease management strategies.
- Detailed studies on the influence of various ecological factors on the distribution and incidence of disease complex need to be undertaken.
- To understand the genetic basis of mechanism of interaction, physiological and histological changes in the disease complex may be investigated.
- Studies should be carried out regarding the compatibility of chemicals, antagonists and plant extracts for integrated disease management approach.
- Integrated disease management approaches involving resistant and moderately resistant cultivars, bio-control agents, fungitoxicant components should be developed.

12.4. MULBERRY, *Morus* spp.

Plant parasitic nematodes play an important role in reducing herbage yield and quality of leaves besides the life span of mulberry plants ultimately leading to poor silkworm cocoon production. Among mulberry nematodes, root-knot nematode, *Meloidogyne incognita* is economically important as it affect the crop quantitatively and qualitatively.

12.4.1. Root-knot Nematodes, *Meloidogyne incognita*

The root-knot nematode, *Meloidogyne incognita* is a major constraint in mulberry cultivation and plays an important role in reducing herbage yield and quality of leaves besides the life span of mulberry plants. It is responsible for 10-12% herbage yield loss (Govindaiah *et al.*, 1991)

12.4.1.1. Symptoms:

The severely affected plants showed stunted growth with low water moisture in leaves, and later marginal necrosis and yellowing of leaves. The underground symptoms include the formation of characteristic knots/galls on the root (Plate 12.10). Galls are spherical and vary in size; young galls are too small and yellowish-white in colour, old galls are big and pale brown. The nematode damaged roots do not utilize water and fertilizers as effectively as uninjured roots. Thus, as a result of poor plant growth, about 10 -12 % leaf yield is lost in addition to affecting the leaf quality (Govindaiah *et al.*, 1991).

12.4.1.2. Predisposing Factors

- Nematode spreads primarily through contaminated soil, farm implements and run-off irrigation.
- Planting of infected saplings along with other susceptible crops increases the nematode intensity, some susceptible weeds in and around the mulberry gardens act as the secondary sources of infection.
- Temperature between 27-30 °C, soil moisture of less than 40 % and pH of 5 to 7 are favourable for the development of the root-knot nematode.
- The disease outbreak is throughout the year and more common in sandy soils under irrigated conditions.

12.4.1.3. Management

i) **Physical Methods:** Hot water treatment of mulberry seedlings is effective at 50°C for 15 min. Hot water immersion must be followed up with a cold water treatment.

ii) **Cultural Methods:** Intercropping of mulberry with marigold at wider spacing of 30 to 45 cm controlled *M. incognita*, increased leaf moisture and leaf yield of mulberry (Govindaiah *et al.*, 1990).

Marigold and sun hemp plants at 10 per square meter can be grown in between mulberry rows. Intercropping of these plants and mulching of the same in soil after vegetative growth provides an additional organic matter to the soil which enriches soil fertility in addition to control of the disease.

The root-knot disease can be easily and effectively controlled by the use of amaranth plants as trap crop for root-knot nematodes intercropped with mulberry plants in the naturally root-knot infested field increasing silk production. As intercrop amaranth could be harvested at frequent intervals to keep the nematode population to a minimum level. This way amaranth could serve as highly effective and beneficial to the farmers in protecting other crops from invading nematode larvae as well as by controlling root-knot nematodes in the naturally infected sericultural field, and also through buying and selling of the edible amaranth plant regularly from the intercropped sericultural field (Datta, 2005; Datta and Datta, 2007).

In mulberry, application of neem oil-cake at1ton/ha/year in four equal split dosages was found effective. There is no residual toxicity of neem oil cake on the silkworm.

iii) **Biological control:** Combined soil application of *P. fluorescens* + *T. viride* each at 10 g/plant as soil application was effective to check the root-knot nematode disease and to improve growth of mulberry with increased leaf yield and reduced nematode population (Table 12.8) (Muthulakshmi *et al.*, 2010a).

Table 12.8. Effect of biocontrol agents on plant growth characters of mulberry in *Meloidogyne incognita* infested field (Pooled data from two experiments) (Muthulakshmi et *al.*, 2010a)

Treatment	Plant height (cm)	Leaf yield (kg/plant)	Gall index	Number of egg masses/ g root	Number of eggs/egg mass	Soil antagonistic population (cfu/g soil)
Pseudomonas fluorescens @10g/ plant	270.45 (+25.42)	1.568	2	13.44 (-48.58)	230.51 (-29.67)	2.2×10^6
Trichoderma viride @ 10g/ plant	265.65 (+23.19)	1.546	2	12.93 (-50.53)	256.67 (-21.69)	1.2×10^4
P. fluorescens @ 10g/plant + *T. viride* @10g/plant	290.27 (+34.61)	2.252	2	11.54 (-55.85)	186.98 (-42.95)	3.1×10^6* 4.6×10^4**
P. fluorescens @ 5g/plant + *T. viride* @ 5g/plant	275.48 (+27.75)	1.821	2	12.56 (-51.95)	203.94 (-37.78)	2.2×10^6* 4.1×10^4**
Carbofuran 3G @ 1 kg a.i./ha	240.58 (+11.57)	0.958	3	14.10 (-46.06)	216.43 (-33.97)	—
Control	215 .63	0.810	5	26.14	327 .76	—
CD (P = 0.05)	**0.833**	—		**0.163**	**1.532**	—

Figures in parentheses are per cent increase (+) or decrease (-) over control
* *Pseudomonas fluroscens* in soil
** *Trichoderma viride* in soil

iv) **Integrated Methods:** Combined inoculation of AMF and *Pseudomonas fluorescens* had positive effect on root-knot nematode control on mulberry (Kumutha, 2001).

Soil application of 24 g of neem cake + 200 g FYM + 1 g of bionema + 100 g of AMF mixed in 20 kg of FYM applied in three split doses /year gave maximum reduction of root galling (85.30) (19.9 galls/plant in treated as compared to 135.8 galls/plant in control) and egg mass production (86.45%) (5.00 egg masses/plant in treated as compared to 35.7 egg masses/plant in control) in mulberry cv. Tr 10 (Vijaya Kumari and Sujathamma, 2014) (Plate 12.11 and Table 12.9).

Table 12.9. Effect of Integrated control in the management of root knot nematode *Melodogyne incognita* in different mulberry varieties

Mulberry Variety	No. of galls / plant			No. of egg masses / g root weight		
	Control (infested)	Treated	Reduction over control (%)	Control (infested)	Treated	Reduction over control (%)
Tr10	135.80	19.90	83.30	36.70	5.00	86.45
DD	166.36	28.1	83.10	47.66	7.40	84.47
V1	178.30	30.73	82.75	56.60	9.20	83.63
RFS175	185.00	35.5	80.80	78.80	14.50	81.55
K2	195.53	40.33	79.37	86.30	16.80	80.36
MS8	205.90	48.66	76.30	101.66	22.00	78.32

12.4.2. Root-knot Nematode, *Meloidogyne incognita* and Root Rot, *Macrophamina phaseolina* Disease Complex

Interaction Studies: The root-knot nematode, *Meloidogyne incognita* and the root rot fungus, *Macrophomina phaseolina* plays an important role in reducing herbage yield and quality of leaves besides the life span of mulberry plants. Disease complex reduces the nutritive value of mulberry leaves and adversely affect the growth and development of silkworm ultimately leading to poor cocoon production. Besides causing direct damage, nematodes are also involved in disease complex with other microorganisms and provide easier entrance to weak pathogenic fungi and break the host plant resistance.

Inoculation of *M. phaseolina* prior to *M. incognita* significantly reduced the nematode population. Inoculation of *M. phaseolina* simultaneously with *M. incognita* and 15 days after nematode inoculation caused disease incidence and significant reduction in plant growth compared to the inoculation of fungus or nematode alone and fungus inoculation prior to nematode (Muthulakshmi *et al.*, 2010b).

12.5. CITRUS, *Citrus* spp.

12.5.1. Root-knot Nematode, *Meloidogyne indica*

Whitehead (1968) reported *M. indica* on citrus from India. The root-knot nematode, *M. indica* on citrus has also been reported from Andhra Pradesh and Gujarat. *M. indica* infects *C. aurantium, C. sinensis* (Whitehead 1968) and *C. aurantifolia* (Parvatha Reddy *et al.*, 1981). Kagzi lime orchards in North Gujarat, particularly in Banaskantha district were observed to be heavily infected by the root-knot nematode *M. indica.* The citrus root-knot nematode, *M. indica* infection was observed in Banaskantha, Mahesana and Anand districts of Gujarat.

12.5.1.1. Symptoms:

M. indica infected trees exhibited severe stunting, yellowing, poor in vigour, unthrifty in appearance, twig blight leading to citrus decline symptoms. The plants were showing yellowing of foliage in initial stage and were completely drying at later stage. On uprooting such infected plants, numerous small to very big sized galls/knots were noticed on the roots, resulting in very poorly developed root system (Patel *et al.*, 1999). They fail to flower and produce fruits even after several years of planting. The roots have conspicuous galls on pioneer and fibrous roots (Plate 12.12). In advanced stage, large cavities can be observed in place of galls. Egg masses can be seen as thin films spread over the root surface.

12.5.1.2. Host Range:

Cowpea and bottle gourd were recorded to be very good hosts, while cotton was poor host. Piludi, a weed was found to be severely galled by the nematode and was considered a very good host.

12.5.1.3. Management

i) **Chemical Methods:** The initial seedling dip treatment with carbosulfan at 500 ppm for 6 hr recorded more height and effectively controlled root-knot nematode.

ii) **Organic Amendments:** Among the soil applications, mustard cake, farm yard manure and poultry manure at 2.5 kg/plant were found more effective in increasing the plant growth and reducing root-knot index.

iii) **Host Resistance:** Citrus rootstocks such as Rangpur lime 8784, Sour Orange Tirupati, Citrumello 4479, Rangpur 8748, Acid Lime PDKV, Sai Sarbati and Galgal were found resistant against *M. indica.* Rangpur lime Chethalli, Trifoliate Orange Chethalli, Nasnaran, Hazara Australia, Rangpur lime Kirumakki, Pramalini and Anand Selection were moderately resistant.

12.6. CONCLUSIONS AND RECOMMENDATIONS

The incidences of different nematode pests on several fruit crops have shown an increasing trend. There were outbreaks of root-knot nematode, *Meloidogyne enterolobii* and decline disease complex, *Meloidogyne enterolobii, Fusarium solani* on guava; root-knot nematode, *Meloidogyne incognita* and wilt disease complex caused by fungi (*Ceratocystis fimbriata, Fusarium oxysporum, Rhizoctonia solani*), root-knot nematode (*Meloidogyne incognita*) and shot hole borers (*Xyleborus fornicates* and *X. perforans*) on pomegranate; root-knot nematode, *Meloidogyne incognita* and root-knot nematode, *Meloidogyne incognita* and root rot, *Macrophamina phaseolina* disease complex on mulberry; and root-knot nematode, *Meloidogyne indica* on citrus. The emerging and future nematode problems have to be tackled with INM approach as a part of a sustainable crop production technology based on the following lines:

- Surveys should invariably include data on nematodes and other soil-borne disease pathogens to record concomitant occurrences.
- A network project to study basic aspects like etiology of disease complexes involving nematodes (*Meloidogyne enterolobii* in guava and *M. incognita* in pomegranate) and other soil-borne disease pathogens and effect of other ecological parameters may be developed for external funding.
- Management of disease complexes should be addressed in collaboration with plant pathologists.
- Production of nematode-free planting material in virgin soils may be popularized along with policy recommendations for implementation of such strategies.

- Development Officers/field functionaries of State Government Horticulture Departments may be imparted training in diagnosing nematode infected planting materials, handling nematodes/disease complex problems.
- Whenever root-knot nematode in guava is intercepted, the samples may be sent to State Agricultural Universities/ ICAR Research Institutes for accurate identification.
- Dual action chemicals may be developed by corporate sector for managing disease complexes in guava and pomegranate.
- Role of drip irrigation/mulching/UHDP systems etc. in accentuating nematode problems may be investigated for rapid nematode population build-up in such systems.
- Quarantine agencies may be sensitized on dissemination of nematode infected planting material domestically.

13

Emerging Nematode Problems in Vegetable Crops

13.1. INTRODUCTION

Potato cyst nematodes (*Globodera rostochiensis* and *G. pallida*), which were hither to restricted to South Indian states such as Tamil Nadu (Nilgiris Hills) and Kerala (Kodai Hills), have been detected in Northern Indian state Himachal Pradesh (Shimla, Mandi, Kullu, Chamba and Sirmaur districts), which is a serious concern with ramification on export and quarantine issues.

There has been an outbreak of serious root-knot nematode problems in the recent years in vegetable crops grown under protected cultivation systems such as tomato, bell pepper, cucumber and okra.

13.2. POTATO, *Solanum tuberosum*

13.2.1. Cyst Nematodes, *Globodera rostochiensis, G. pallida*

The potato cyst nematode (PCN) is the most important quarantine pest that is causing concern to potato production and trade throughout the world. In India, the potato cyst nematode, *Globodera rostochiensis* was first reported by Jones (1961) who detected the cysts from the roots of potato plants which exhibited the symptoms of yellowing of leaves, in Nilgiris Hills of Tamil Nadu state. This report initiated a great awareness amongst administrators in the country resulting in establishment of Nematology Unit in Tamil Nadu as well as initiating an Indo-German Collaborative Programme of intensive chemical application (Dasanit) in the infested area to avoid further nematode spread. The potential damage of this nematode to potato cultivation was such that the Destructive Insect Pest Act 1919 was amended by the Tamil Nadu Government in 1971 to ensure strict prohibition of movement of seed potatoes from infested area to any other parts of the state or the country.

The later surveys have indicated the presence of both species of potato cyst nematode (*Globodera rostochiensis* and *G. pallida*) restricted to potato growing regions of Nilgiris and Kodaikanal Hills of Tamil Nadu (Krishna Prasad, 1986) and Kerala (Ramana and Mohandas, 1988).

During 2010, a report on the occurrence of *Globodera rostochiensis* and *G. pallida* has been reported from the Potato breeding Farms of Central Potato Research Institute located in Kufri, Shimla District of Himachal Pradesh by Ganguly *et al.* (2010). The potato fields at Lister House Farm and Main Farms were found to be heavily infested, supporting 58–548 cysts per 100 g soil. The reported occurrence of potato cyst nematodes in Himachal Pradesh, confirming the earlier report of *G. pallida* in this state by EPPO (European and Mediterranean Plant Protection Organisation), is a cause of serious concern with ramifications on export and quarantine issues.

The cysts of *Globodera* spp. were recorded from *kharif* potato only. Out of 16 potato growing localities surveyed, the nematode was recorded from 12 localities in five districts of Himachal Pradesh (Shimla, Mandi, Kullu, Chamba and Sirmaur). The frequency of occurrence in these localities varied from 41.13 to 100%. The highest number of 915 cysts per 100 cc soil was recovered from Kharapathar in Shimla district. The presence of cysts was also noticed in 6 out of 8 Potato Development Stations (PDS) of the State Department of Agriculture surveyed, who supply potato seed material, which forms the source of nematode spread to the farmers' fields.

Contrary to the myth that PCN is restricted to Nilgiri Hills (Tamil Nadu) in India, PCN seems to be widespread in Himachal Pradesh also. So far, incidence of PCN is observed only above 2000 m above MSL. There is now an urgent need to take suitable steps for not only managing these species but also strengthening of strict domestic quarantine regulations to check their further spread to newer areas.

13.2.1.1. Symptoms:

The disease usually occurs in patches (Plate 13.1). The infested plants exhibit typical symptoms of patchy growth of weak and stunted plants. The patches increase in time with continuous potato cropping. Under conditions of severe infestation, the plant growth is stunted (Plate 13.1) and wilting occurs during hot part of the day. Plants show tufting of leaves at the top as the outer leaves turn yellow and die. Root system is smothered, secondary roots are induced at the collar region and the plants can be easily pulled out. Tubers formed are less in number and reduced in size. In extreme cases, tuber formation is arrested. The total photosynthesis per plant is also significantly reduced as a result of reduced leaf area and this is reflected in reduced potato yield.

13.2.1.2. Life Cycle:

G. rostochiensis is amphimictic, with sessile, globose females and motile, vermiform males preceded by four larval stages. The second stage larva is the resistant, dormant and infective stage remaining inside the egg within the cyst formed by tanning of the female's cuticle. The cysts remain in the soil after death of the host, each containing about 500 eggs. After hatching, the second stage larva enters the root of a host plant, just behind the root tip or at the site of a new lateral root and moves through the root away from the tip before starting to feed on a group of pericycle, cortex or endodermis cells which are modified to form a large syncytial transfer cells on which the nematode feeds throughout the rest of its development to the adult stage. Sex is determined by nutrition, a poor food supply resulting in the formation of a male, a good supply, a female. Thus with heavy infections males predominate, the reverse with light infections.

The adult females swells greatly, rupturing the root cortex and protrudes from the root, exposing the vulval region while the head and neck remain inserted in the root (Plate 13.2). As they develop, males become coiled within the earlier larval cuticles and on maturity they rupture them, then leave the root and are attracted to females by a secretion. Both sexes mate several times. Males do not feed but remain active for up to 10 days, females normally mate soon after rupturing the root cortex and within 50 days of larval invasion.

13.2.1.3. Host Range:

All known varieties of potato, tomato and brinjal are severely attacked by the cyst nematode. A number of other Solanaceae are recorded as hosts, but the majority of hosts are species and varieties of tuberous *Solanum*. The only susceptible plant outside the family is snap dragon, *Antirrhinum majus*.

13.2.1.4. Ecology:

Well-drained, aerated sands, silts and peat soils with a moisture content of 50-75% water capacity are suitable for survival and movement of the free-living stages of the potato cyst nematode. A minimum soil temperature of 10°C is necessary for larval activity.

13.2.1.5. Spread and Survival:

The cysts can be spread by the movement of soil through seed material for a very long distance. They can also be spread through field tools, implements, seedlings grown in infested soil for transplanting elsewhere, labours' feet, etc. from one field to another. The cysts can remain viable in the soil without a host for 30 years.

13.2.1.6. Physiological Races:

At present, five pathotypes in *G. rostochiensis* and four pathotypes in *G. pallida* are differentiated using the following differentials (Table 13.1):

Table 13.1. Pathotypes of *Globodera rostochiensis* and *G. pallida*

Differential host	Nematode pathotypes*				
	1	2	3	4	5
	Globodera rostochiensis				
Solanum tuberosum	+	+	+	+	+
ex *andigena* CPC 1673	-	+	+	-	+
ex *kurtzianum* 60.21.19	-	-	+	+	+
ex *vernei* 58.1642/4	-	-	-	+	+
ex *vernei* 62.33/3	-	-	-	-	+
ex *vernei* 65.346/19	-	-	-	-	-
	Globodera pallida				
Solanum tuberosum	+	+	+	+	
ex *multidissectum* P 55/7	-	+	+	+	
ex *vernei* 62.33/3	-	-	+	+	
ex *andigena* x *multidissectum* D 47/1	-	-	-	+	
ex *andigena* x *multidissectum* D 47/2	-	-	-	-	

* (+) indicates ability to produce females, (-)indicates inability to produce females

In India, potato cyst nematodes are restricted to Nilgiri and Kodai hills in South India, where 5 pathotypes – Ro 1, Ro 2 of *G. rostochiensis*, and Pa 1, Pa 2 and Pa 3 of *G. pallida* have been detected (Amalraj and Rao, 1992).

13.2.1.7. Challenges in Management:

Fundamentally, potato growers and their advisers are likely to face one or more of three basic challenges in managing PCN:

- Keeping potato land 'free' of PCN.
- Preventing lightly infested land from becoming heavily infested.
- Bringing high populations back to manageable levels.

The fundamental approaches recommended for each of these situations, based on the information available are set out below.

i) **Keeping Potato Land 'Free' of PCN:** It is almost impossible to detect very small infestations of PCN and to be certain that land is free of infestation. If PCN is present on a farm, then all fields probably have had PCN introduced into them. Management should therefore aim to prevent even small infestations from increasing. To do this:

- Use intervals between potato crops of no less than seven years.
- If grown, alternate resistant cultivars with susceptible cultivars.
- To avoid introducing PCN, clean all machinery, livestock and feed (e.g. potato tubers, carrots), all of which can carry cysts of PCN, unless from other 'PCN-free' land.
- Do not return potentially contaminated soil (e.g. tailings from grading operations) to potato fields.
- Check any non-certified seed potatoes for PCN at source, by testing soil collected from the harvested tubers.
- Collect and test soil from the harvested potatoes to detect PCN infestations in individual fields as early as possible.
- Test large soil samples - at least 200 g for mineral soils, and preferably 500g.

ii) **Preventing Lightly Infested Land from becoming Heavily Infested:** The aim should be to prevent populations from increasing to damaging levels. To do this:

- Rotation is the key to successful PCN management. On many farms, growing potatoes on short rotations (one year in four or less) does not prevent PCN from increasing to damaging levels, even when nematicide is used.
- Sampling is the key to determining whether management is working. Fields should be sampled both before and after each potato crop to monitor population changes. However, note that soil sampling can give variable results unless infestations are relatively uniformly distributed and the soil has been well mixed by cultivations.
- Sampling units should be as small as possible and not more than 2 ha. If possible, use navigational techniques such as the Global Positioning System (GPS) or other approaches to identify specific points for repeated sampling. Bear in mind that infestations will not be uniform immediately

after harvest, so post-cropping sampling should be done after cultivation operations have redistributed cysts.

- Determine the species of PCN so that appropriate resistant cultivars can be used.
- Consider using granular nematicides in an integrated programme to prevent increase even of small, non-damaging infestations, especially if they are of *G. pallida.*

iii) **Reducing Damaging Levels of Pallida and Keeping them Low:** The aim should be, first, to decrease populations to levels where the crop is not damaged when protected with a granular nematicide and, secondly, to reduce populations to low levels where integrated control is sustainable. To do this:

- On very heavily infested land it will be necessary either to have a very long rotation, or to use a fumigant nematicide or other method of partial soil sterilisation, to reduce the soil infestation to a level that can be managed as a lightly infested field.
- Trap cropping should only be considered where PCN infestations are very large and need to be reduced quickly. Recent research has demonstrated that trap cropping can reduce PCN populations by 80% or more and is effective on a commercial scale on a range of soil types. In this regard, *Solanum sysimbrifolium* acts as a very good trap crop.
- Once populations have been reduced, sustainable management of *G. pallida* depends on the integration of all the available control options, particularly rotation, resistant cultivars and nematicides.
- When using fumigant or granular nematicides they must be applied and incorporated as effectively as possible. About 95% kill must be achieved in order to prevent PCN population increase.
- Potato cultivars with high levels of resistance to *G. pallida* are not yet available; if partially resistant cultivars are used they should be alternated with susceptible cultivars to prevent the selection of virulent PCN populations.
- The effects of partially resistant cultivars and granular nematicides on PCN populations are additive. So they can be used in combination to help reduce PCN populations. Nematicides are also needed to protect partially resistant cultivars.

13.2.1.8. Management:

A number of management practices aim to keep PCN numbers in soil below the damage threshold.

i) **Quarantine and Seed Certification**

a) **Quarantine:** The introduction of potatoes (for consumption or for seed propagation) or soil into nematode-free areas should be strictly regulated. It is illegal to move potatoes, soil and associated packaging or machinery into nematode-free areas unless certified as having met specific import conditions.

b) **Seed Certification:** Potato seed and all other potato tubers used for planting of crops pose the highest risk of spreading PCN. All seed producers and growers selling seed from commercial crops should to be registered under certified schemes in all states. Officially PCN certified Seed tubers are:

- From an officially verified, maintained PCN free area.
- Have an officially verified negative test result for PCN.

ii) **Preventive Measures:** Regulations concerning the movement of potatoes and materials associated with PCN risk should be observed at all times. Preventative measures that can be carried out on individual properties include:

- Identifying the property as PCN free and informing visitors that biosecurity practices are in place.
- Restricted access to production areas.
- Identifying people, equipment or machinery in a PCN risk area before coming onto the property.
- Thorough cleaning of machinery and equipment before moving onto the property.
- Purchasing propagative material from reputable suppliers.

iii) **Cultural Methods**

a) **Farm Hygiene:** In most cases the greatest PCN inoculum is moved through seed potatoes, machinery, bags, bins, livestock, people and contaminated soil. PCN is also spread through the return of tare soil from packing sheds and processing factories to agricultural land, usually to different paddocks.

Potato growers should always follow good practices to prevent the spread of pests and diseases, and this is especially important where PCN is known to be present. Machinery, equipment and footwear should be cleaned after working in the infested area or handling tubers from that unit, and care should be taken to avoid any cross-contamination between those tubers and any seed potatoes, including those which may be retained as farm saved seed.

All waste, soil and by-products from potato harvesting, grading and processing operations should be disposed of in line with the Plant Health Code of Practice on Management of Agricultural and Horticultural Waste, and therefore no additional waste disposal requirements apply to material from fields officially recorded as infested. In particular, the following approaches can be used to reduce the amount of waste sent to landfill:

- Excess soil should be left in the field if possible, or where a grower grades his own potatoes on-farm, the soil may be returned to the field of origin.
- Soil and potato waste may be returned to non-agricultural land, or to sacrificial land that is unsuitable for growing potatoes for other reasons.
- Unwashed stock feed should only be used on farms which do not grow potatoes.
- Washed stock feed carries a very low risk and can be used on any farm.

b) **Crop Rotation:** A long rotation period is the most reliable way of reducing PCN populations. There is a significant reduction in viable eggs in the first six years without host plants, therefore this is recommended as the minimum length of rotation. The rate of decline depends on a number of environmental factors, particularly soil type, and *G. pallida* generally declines more slowly than *G. rostochiensis.*

Volunteer potatoes enable PCN to multiply and should be controlled as early as possible each year.

Crop rotation with non-host crops is an effective strategy to reduce nematode numbers in infested soil by breaking the lifecycle. Crop rotation cannot completely eradicate PCN as nematode cysts remain dormant in soil in the absence of host plants. Dormancy is commonly 7-10 years but may exceed 20 years.

Cyst nematode in Tamil Nadu hills is managed by crop rotation with vegetables, particularly cabbage and carrot.

c) **Intercropping:** Cyst nematode in Tamil Nadu hills is managed by intercropping potato with beans or wheat.

d) **Trap Cropping:** A 'trap crop' is a PCN host crop that is grown temporarily before the commercial crop. The trap crop encourages existing eggs in the soil to hatch and the larvae penetrate the roots. Before the nematodes mature and have a chance to increase the population by laying eggs, the trap crop is removed and destroyed (usually 6 to 7 weeks after planting).

Recent research has demonstrated that trap cropping can reduce PCN populations by 80% or more and is effective on a commercial scale on a range of soil types (Table 13.2).

Table 13.2. Effect on *Globodera pallida* populations of trap cropping at three sites

Site	Initial population (eggs /g)	Final population (eggs/g)	Reduction (%)
Site I	160	20	87
Site 2	257	68	74
Site 3	200	55	73

Trap cropping should only be considered where PCN infestations are very large and need to be reduced quickly. The trap crop *Solanum sisymbriifolium* generates high mortality in PCN but requires a full growing cycle to be effective. The technique is costly (at least as costly as soil fumigation) and needs very careful management.

e) **Biofumigation:** The most commercially viable method for management of PCN seem to be biofumigation, a biocidal green manure approach which utilises temperate climate Brassicacae plants, which would most easily integrate into rotations without having detrimental impacts on growing of cash crops (Lord *et al.*, 2011, Vazquez 2012, Tobin *et al.*, 2008).

Several studies have confirmed that PCN mortality is increased after exposure to isothiocyanates (ITC's) with the glucosinolate (GSL) derivatives 2-propenyl-isothiocyanate and 2-phenylethyl-isothiocyanate being suggested to be particularily nematicidal (Lord *et al.*, 2011, Watts,

2012). Lord *et al.* (2011) has shown that high 2-propenyl lines of *B. juncea* such as ISCI99 have the potential to reduce encysted *G. pallida* egg viability by 95% in polyethene covered soil, and by 75% from 2-phenylethyl lines of *E. sativa* in glasshouse experiment. Watts (2012) produced potentially more realistic results at a recorded 42% mortality incurred upon PCN in field. The results were very highly significant illustrating a clear biofumigation effect on PCN viability (Plate 13.3). 100% mortality in PCN is induced at 1mg/ml GSL concentration.

Overall, biofumigation is shown to be a good method of controlling PCN, but further work needs to be conducted to better quantify in-field PCN cyst viability decline caused by ITC gases.

iv) Host Resistance

a) Resistant Cultivars: Growing a cultivar which has resistance to PCN is a more effective way to reduce PCN soil populations compared to just using a non-host crop rotation (Plate 13.4).

Potato varieties resistant to PCN prevent most nematodes developing into adults and therefore reduce the increase in population normally caused by planting host plants. Use of varieties with high resistance can achieve suppression with shorter rotations or less use of chemical treatments than would otherwise be required. However, a very small proportion of cysts do reach maturity on even highly resistant varieties. Care should be taken when using resistant varieties as the resistance to the two species of PCN is rarely similar. Therefore, planting a variety with high resistance to *G. rostochiensis* may enable *G. pallida* to multiply (and *vice versa*) as they may still be present at levels too low to be detected.

The level of resistance of potato varieties, to both *G. rostochiensis* and *G. pallida*, is assessed as part of National List testing. Many varieties score highly in terms of resistance to *G. rostochiensis*, but relatively few have significant resistance to *G. pallida* (Table 13.3).

Table 13.3. The resistance status of the ware potato varieties grown in Great Britain in 2015 (adapted from Swales, 2015)

Potato Variety	Resistance status against *G. pallida* Pa2/3,1 (rating)	Resistance status against *G. rostochiensis* (Ro1) (rating)
Maris Piper	Susceptible (2)	Resistant (9)
Markies	Susceptible (2)	Resistant (9)
Lady Rosetta	Susceptible (2)	Resistant (9)
Melody	Susceptible (2)	Resistant (9)
Harmony	Partially resistant (4)	Partially resistant (4)
Panther	Resistant (8-9)	Partially resistant (4-6)
Maritiema	Partially resistant (5)	Resistant (8)

b) Tolerant Cultivars: Varieties described as "tolerant" to PCN can produce good yields despite the presence of PCN. However, they have little or no impact on the nematodes' development and can in fact lead to an increase in the PCN population. For this reason, the planting of tolerant varieties should only be included in a control programme in combination with other measures which are sufficient to achieve suppression.

v) **Integrated Management:** Three strategies of resistance, rotation, and non-fumigant granular nematicide treatment have been effectively combined in managing PCN in potato production (Alphey *et al.*, 1988; Trudgill *et al.*, 1987; Whitehead, 1986).

Cyst nematode in Tamil Nadu hills is managed by crop rotation with vegetables, particularly cabbage and carrot, intercropping potato with beans or wheat, alternating nematode resistant potato variety 'Kufri Swarna' and application of 2 kg a.i. /ha of Carbofuran 3% G at planting. A three-year adoption of INM practice for cyst nematodes gave efficient and economical production system.

The combined use of soil treatment with nematicides (aldicarb, ethoprophos, fosthiazate and oxamyl) and partially resistant potato cultivars can be adopted to manage PCN when it reaches threshold level. Treatment of soils at planting with granular nematicides to avoid invasion damage, and use of partially resistant cultivars to reduce the rate of population increase.

Double trap crops combined with ethoprop treatment were highly effective in France for seed potato production, providing *G. pallida* population reduction of 98.5%, a level that would have required 21 years of continuous cropping with non-hosts to achieve (Whitehead, 1986).

A successful INM program results from combinations of resistance and tolerance, low rate of nematicide, and rotation alternating resistant and susceptible potato crops with non-host rotation breaks between potato crops.

Integration of bioagents (*Pochonia chlamydosporia, Paecilomyces lilacinus*) with granular nematicides, crop rotation and partially resistant cultivars to provided practical levels of nematode control.

Potato cultivars (e.g., Maris Piper) with the major gene H1 conferring a high level of resistance to *G. rostochiensis* but not to *G. pallida* were incorporated into non-host rotations with considerable success. Three crops over 9 years decreased the nematode Pi in soil from 100 to < 1 egg/g. The first two resistant potato crops required protection with an at-plant application of non-fumigant nematicide (aldicarb or oxamyl) to sustain acceptable yield (Trudgill *et al.* 1987). In related studies, susceptible and resistant (Maris Piper) potato cultivars were alternated in the rotation. Although the susceptible crop required protection with nematicide treatment, which was limited to every second potato crop. An additional benefit was in reducing selection pressure on the nematode population for the virulent *G. pallida* that was present in the field and that increased when Maris Piper was the only potato crop in rotation (Trudgill *et al.*, 1987).

Additional potato cyst nematode management strategies include the early harvesting of potatoes in the less frost-prone areas to prevent the completion of the nematode life cycle (Trudgill *et al.*, 1987) and trap cropping to reduce nematode populations (Mugniery and Balandras, 1984). Other studies evaluated the effects on potato yield and nematode population dynamics of cultivars with different levels of resistance or tolerance in various combinations, when exposed to full, half, or quarter rates of standard aldicarb treatment (Trudgill *et al.*, 1987). These data demonstrated that half or even quarter rates of aldicarb sustained yield and prevented nematode population increase on partially resistant, tolerant cultivars. Field experiments demonstrated that increasing input of compound fertilizer (14:14:22 N-P-K) reduced potato yield loss from *G. rostochiensis*; less aldicarb was required with increased fertilizer rates to protect an intolerant cultivar, and aldicarb was unnecessary to protect a tolerant cultivar (Trudgill, 1987). However, it was necessary to prevent increased final populations and problems for subsequent crops in the rotation.

In summary, a successful integrated management programme with prolonged lives of resistance and nematicide results from combinations of resistance and tolerance, low rate of nematicide, and rotation alternating resistant and susceptible potato crops with non-host rotation breaks between potato crops. Trap cropping and planting and harvest time manipulation are additional options in some regions. With such an integrated management framework in place, opportunity exists to develop and implement additional component strategies, such as a reasonably effective biological control agent or other cultural manipulation. The shift in prevalence from *G. rostochiensis* to *G. pallida* in some areas can be achieved by early use of cultivars resistant only to *G. rostochiensis*.

a) **Integration of Biofumigation and Biological Control:** The biofumigation and nematophagous fungi treatments could be more than compatible, they could have a synergism. *Paecilomyces lilacinus* has a saprophytic stage to its lifecycle which provided biofumigation effects wear off by fungal application at potato planting could take advantage of biofumigant residues causing hyphae proliferation and therefore an increased nematicidal effect (Cannayane and Sivakumar, 2001) might also be of interest and could have some role in enhancing applied nematophagous proliferation. Even if the biofumigation effect does not fully wear away, the fungus has shown to be resilient to fumigant nematicide effects (Stirling, 1991).

   Biofumigation with *Sinapis alba* cv. Architect was found to reduce cyst viability by 48% when compared to the untreated control treatment. However, a trend toward biofumigant and biological control agent combination treatments giving greater nematode control than other treatments seems apparent. A 48% viability reduction to the initial nematode population would result in 14 viable eggs g^{-1} soil at the intermediate stage, whilst a 55% reduction would result in 12 viable eggs g^{-1} soil at the final nematode population stage. The dry fraction of fresh biomass was equal to 47% and that only 298 g^{-1} of dry material was available for incorporation per m^2 from this biofumigant crop. The primary glucosinolate in *Sinapis alba* cv. Architect is glucosinalbin which accounts for 86% of the total glucosinolate profile. The *Sinapis alba* cv. Architect biofumigation potential at incorporation was approximately 63 nmol g^{-1} soil, 86% of which was glucosinalbin at approximately 54 nmol g^{-1} soil.

   Significant mortality (48-55%) of *G. pallida* cysts could be induced by using both individual and combination biofumigant (*Sinapis alba* cv. Architect), biocontrol fungi and bait treatments when compared to an untreated control treatment. As a result of treatment effects on nematode viability, treated Estima potato plants were found to yield more than untreated, an increase in yield by a factor of five in the *P. lilacinus* + *Panagrelus* sp. bait treatment being recorded when compared to the untreated control.

Growers could use these treatments in the event of imminent nematicide withdrawal. Based upon this research, rotations could be reduced by three years which could have substantial economic implications for growers. These treatments could even be more cost effective than currently used synthetic nematicides (*S. alba* cv Architect seed cost £70/ha which is £230 less expensive than fosthiazate which delivers comparable efficacy) (Tobin *et al.*, 2008).

b) Managing potato cyst nematode through maximising natural decline and population suppression

- Identification of the key factors that affect PCN-decline and to explore the manipulation of fungal parasites to increase egg mortality as an alternative to the application of chemical nematicides.
- An inverse relationship was observed between the number of eggs per cyst and the level of egg infection by fungi in some soils.
- *Solanum sisymbriifolium* significantly increased PCN hatch in laboratory tests and could therefore be used as a trap crop in PCN management.
- Granular nematicides (aldicarb, oxamyl) were successful in controlling PCN populations at different sites and in several trials biological control fungi were also able to control PCN multiplication rates. However, the biological control agent produced much less reliable effects.
- The initial PCN population density in soil is critical in the ability of biological control fungi to succeed.
- In general, applications of biological control agents require integration with other control measures to provide practical levels of nematode control and they have been successfully used with granular nematicides, crop rotation and partially resistant cultivars.
- The addition of a commercially produced fungal bioagent developed for the control of PCN will instigate or promote population suppression to PCN.
- Fungal parasites of PCN significantly reduce the survival of eggs of the nematode in soil and establish an equilibrium population, which would allow reductions in the long rotations especially needed to manage *G. pallida.*
- The three PCN parasitic fungi, *Pochonia chlamydosporia, Paecilomyces lilacinus* and *Monographella cucumerina* have a preference for PCN as a nutrient source and will translocate nutrients across the mycelial web as far as efficiency allows so that low levels of PCN will keep the fungi in parasitic rather than saprophytic phase.
- Manage the nematode throughout the crop rotation, which includes a trap crop and the application of a selected isolate of *P. chlamydosporia.*

c) *G. pallida* is More Difficult to Control than *G. rostochiensis*

- *G. pallida* is more difficult to control with the standard treatments of aldicarb or oxamyl, perhaps because of the prolonged period of egg hatch that extends well into the growing season, beyond the time that toxic concentrations of nematicides occur.
- Under crop rotation, *G. pallida* generally declines more slowly than *G. rostochiensis.*

- Many varieties score highly in terms of resistance to *G. rostochiensis*, but relatively few have significant resistance to *G. pallida.*

13.2.1.9. Future Lines of Work

- Enforce strict domestic quarantine.
- Production of certified potato seed material free from PCN both in private and public sectors.
- Survey and surveillance for PCN in potato growing regions which are above 2000 m MSL, particularly Uttarakhand, Jammu & Kashmir and Kodai hills.
- More work on resistance/tolerance to both species.
- Development of suitable INM strategies.
- Biofumigation – more work needed.
- Microplot experiments established with PCN infested soil to develop methods to manage the nematode throughout the crop rotation, which includes a trap crop and the application of a selected isolate of *P. chlamydosporia*:
  - Continue the studies on the spread of the *P. chlamydosporia* in soil using irrigation, to access distribution of the fungus after application.
  - The effects of rotation on the decline of PCN (cereal–oilseed rape-potato, or *S. sisymbriifolium*-potato or other break crops).
  - Test the new isolates of *P. chlamydosporia* obtained in the project and assess their potential as biological control agents.
- Development of a potential quantitative diagnostic kit for PCN using antibodies

Recommendations

- Extensive surveys may be conducted to document endemic areas for the infestation of potato cyst nematode (PCN) in Northern and Eastern hilly areas.
- Enforcement of domestic quarantine effectively to prevent further spread of PCN.
- Ensuring nematode-free seed potato by cleaning the seed material thoroughly with water to remove soil clods.
- Studies on species/pathotypes of PCN in northern parts of the country should be undertaken.

13.3. NEMATODE PROBLEMS ON POLYHOUSE VEGETABLES

In 1980s, the demand for high-quality export oriented horticultural products and need for the availability of horticultural crop produce throughout the year, especially during off-season, compelled growers to cultivate selected vegetable crops under protected cultivation. Protected cultivation of high value vegetable crops under controlled environmental conditions has gained popularity among farmers due to high productivity, better quality of produce and manifold increase in yield per unit area. Vegetable farming (tomato, capsicum, cucumbers, French bean, crucifers, gherkins, muskmelon, watermelon, etc.) inside modern greenhouses is promoted in order to improve earning potential of farmers especially during harsh winters. The nematode incidence under protected cultivation became severe and led to complete crop losses because of congenial conditions of higher temperature, humidity and use of high agronomic inputs like fertilizers and plant growth promoters in polyhouses. The proliferation rates of nematodes in polyhouse cultivation reached up to 10–30-fold more than in the open field cultivation. Polyhouses were observed to be severely infested with major plant parasitic nematode genera viz., *Meloidogyne*, *Pratylenchus*, *Rotylenchulus*, *Helicotylenchus*, *Mesocriconema* and *Tylenchorhynchus*. The most economically important root-knot nematode causes characteristic galls on the roots resulting in damages and dislocation of xylem vessels which impair the movement of water and minerals. The characteristic symptoms are stunted growth and formation of small and chlorotic leaves. They feed on the roots devitalizing the plants ultimately affecting both qualitative and quantitative production. Nematode problems on all these crops under protected conditions have assumed alarming proportions leading to huge losses (up to 80%) in select crops. The nematode infestations exacerbate severity of fungal diseases leading to complete crop failure. In view of damage potential of these tiny hidden enemies of crop plants, an urgent need is felt for developing nematode management stratagies.

13.3.1. Main Causes

In most cases, problems arise from contaminated soil or soil mixture used as a component of the growing medium, mono-cropping and infested planting materials. Commonly followed nematode management practices such as summer ploughing, fallowing and crop rotation which are meant for nematode suppression in field crops are not practicable in polyhouses. Therefore, nematode management here must be considered primarily as exclusion or avoidance. Once nematodes are introduced it is difficult to manage them.

13.3.2. Management

The nematodes may be managed by the following integrated methods:

- Growing cowpea (trap crop for root-knot nematode) closely before the main vegetable crop and removing the plants out of polyhouse at 45 days after sowing.
- Raising marigold (antagonistic to nematodes) and incorporating *in situ* after plucking the flowers.
- If the beds are highly infested with the nematodes, bring down the population of nematodes by applying neem/pongamia cake at 500 g + Carbofuran at 50 g/ sq. m.
- Treat the seeds with biopesticide-*Pseudomonas fluorescens* at 10 g/kg seed.
- Treatment of seed (20 g biopesticide per kg seeds) and substrates with FYM, vermicompost, neem/pongamia cakes enriched with biopesticides (2 kg each of *Paecilomyces lilacinus*, *Trichoderma harzianum* and *P. fluorescens* per 2000 kg of FYM/500 kg of vermicompost/200 kg of neem/pongamia cakes).
- Raise the seedling in the coco-peat or any other substrate by treating with *P. fluorescens* at 1kg + *T. harzianum* at 1kg + neem or pongamia cake at 50 kg/ton.

- Treat the bed with neem or pongamia cake at 50g, *T. harzianum* at 25 g or *Pochonia chlamydosporia* + *P. lilacinus* at 25 g/sq. m at an interval of 2 months.
- Apply 2 tons of farmyard manure enriched with bio-pesticide-*T. harzianum* per acre before planting, along with 100-200 kg of neem or pongamia cake.
- Apply Carbofuran 3 G at 1kg ai/ha at planting and repeat after 45 days.

13.3.3. Cropping Sequences

The cropping sequence has a major role to play in the management of phytonematode problems in vegetable crops under a sustainable production system. Tomato based cropping sequences such as tomato-maize-tomato and tomato-mustard-tomato were found to have maximum suppressive effect on *M. incognita* (Table 13.4) (Singh and Khanna, 2015). Mojtahedi *et al.* (1993) also found suppression of nematodes in soil by using mustard and other *Brassica* crops in cropping sequence. Siddiqui and Alam (2001) proved that fallow-cauliflower-sorghum-coriander was beneficial in reducing the *M. incognita* populations.

Table 13.4. Tomato cropping sequences for management of root-knot nematodes

Rotation	Population range and % incidence of *M. incognita* in 200cc soil
Tomato-Fallow-Tomato	450-2000 (72.3%)
Tomato-Mustard-Tomato	26-98 (44.2%)
Tomato-Maize-Tomato	32-55 (58%)

13.3.4. Recommendations

- Soil testing for nematode infestation should be made mandatory for the establishment of the new polyhouses.
- Region-wise dedicated referral labs for analyzing samples for nematodes be established at State Agricultural Universities/ ICAR Research Institutes.
- Sites for erecting new polyhouses should be properly treated with recommended chemicals and followed by supplementation with biological control agent enriched farm yard manure/oil cakes/ vermicompost.
- There is an urgent need to grant provisional registration for chemicals having proven efficacy against root-knot and other nematodes.
- In case of heavy population of plant parasitic nematodes species, soil fumigation with permissible chemicals should be done.
- Use of quality neem cake, castor cake, coco peat, FYM, or vermicompost enriched with bio-control agents.
- Evaluation of rootstock of crop genotypes against target nematode species (even up to races) should be carried out. The resistant rootstock should be grafted and the same rootstock should be preserved for further studies.
- Efforts on breeding vegetable varieties suitable for polyhouse conditions should be taken up.
- Cluster approach may be followed to tackle the emergent nematode and other pest problems in polyhouse conditions.

13.4. TOMATO, *Solanum lycopersicum*

Productivity of tomato in greenhouse (Plate 13.5) is increased manifold in comparison to open field. Improved flavour and shelf life are the driving force for the increased demand for greenhouse grown tomatoes. Consumers love the taste of a fresh tomato, and they are willing to pay a premium price, especially in the winter. Unlike field-grown tomatoes which require harvesting and marketing over a short period of time (3-4 weeks), greenhouse tomatoes are harvested and marketed over a 30-week period.

13.4.1. Root-knot Nematodes, *Meloidogyne incognita, M. javanica*

13.4.1.1. Symptoms:

Root-knot nematodes cause root galls on the feeder roots and sometimes affect the entire root system showing heavy galling (Plate 13.6). This affects the uptake of nutrients and water and the plants show wilting during warmer part of the day. This causes stunted plants with yellow foliage resulting in yield reduction. In tomato, dynamics of root-knot nematode showed enhanced population build-up from 1 to 30 juveniles per g soil within a period of 6–12 months, which is comparatively higher in contrast to the open cultivation.

13.4.1.2. Management

- Use nematode resistant variety/hybrid if available.
- Follow crop rotation with marigold, wherever possible.
- Seed treatment with biopesticides- *P. fluorescens* at 10g/kg seed.
- Nursery bed treatment with *T. harzianum* at 50 g/sq. m.
- Main field treatment with 2 tons of farmyard manure enriched with *T. harzianum* and *P. lilacinus* per acre before planting, along with 100-200 kg of neem or pongamia cake.
- Apply Carbofuran 3G at 1 kg ai/ha at transplanting.

i) **Soil Solarization:** Soil solarization to control root-knot nematode is simple and economic in its application, and leaves no toxic residues for humans, animals and plants. These attributes make solarization suitable for application in a wide variety of production system. Soil solarization should be done during the hottest period of the year. It offers a satisfactory and environmentally friendly solution for the control of root-knot nematode. This method is easy to use in organic, conventional and integrated growing system and it is more economical than chemical treatments.

Maximum soil temperatures achieved under solarisation were 51°C in the alluvial soil and 52.4°C in the heavy clay soil type at a depth of 20 cm. In the non-solarized unheated greenhouses, the soil temperature achieved is 39°C (Plate 13.7). The effect of soil solarization treatments on the population densities of root knot nematode *M.incognita* was very clear. The larval density in soil was markedly reduced compared with the untreated control (Plate 13.7). The soil solarization method had a significant influence to reduce the severity of root damage. Root galling index is the most important symptom for the root knot nematode. The gall index is dramatically reduced in both solarized greenhouses compared with untreated control where the highest root galling index was found (Plate 13.8). Tomato fruit yields in solarized greenhouses were four fold higher than those obtained in untreated control (Plate 13.8).

ii) **Chemical Methods:** Addition of Dazomet at 1 kg a i /ha to the nursery beds gave effective control of root-knot nematodes. Further, addition of enriched FYM/compost/ vermicompost (with *Trichoderma harzianum/T. viride* + *Paecilomyces lilacinus* + *Pseudomonas fluorescens* each at 2 kg/ton of organic amedment) at 5 t/ha gave effective management of root-knot nematodes.

After 15 to 30 days of planting, addition of enriched neem cake (with *Trichoderma harzianum/T. viride* + *Paecilomyces lilacinus* + *Pseudomonas fluorescens* each at 2 kg/ton) at 50 g/m^2 gave effective management of nematodes under polyhouse conditions. Further, drip application of neem cake suspension enriched with bioagents (20 kg neem cake + *Trichoderma harzianum/T. viride* + *Paecilomyces lilacinus* + *Pseudomonas fluorescens* each at 2 kg) helped in keeping the polyhouse beds free from root-knot nematodes.

13.4.2. Root-knot Nematode, *Meloidogyne incognita* and Wilt, *Fusarium oxysporum* f. sp. *lycopersici/F. solani* Disease Complex

13.4.2.1. Interaction Studies:

The interaction among soil-borne pathogens is well documented in tomatoes on old production land when *Fusarium* wilt disease and root-knot nematode (*Meloidogyne* spp.) are both present (Plate 13.9). Young plants are very susceptible to the combination of pathogens, collapsing prior to harvest. The nematode, by impairing water and nutrient availability, disrupts root function and plant growth processes. These effects combined with the vascular blocking due to the wilt fungi can be particularly severe, and if widespread, result in total crop failure. The root-knot nematode, by causing the development of root galls, provides a nutrient-rich food source on which the fungi colonize rapidly. Root-knot nematodes can thus significantly enhance disease development and yield loss, elevating primary or secondary pathogens to major pest status even though population levels or pathogenic potential of the pathogenic fungi were initially very low and yield losses would have been minimal in the absence of the nematode.

13.4.2.2. Management

i) **Biomanagement**

a) **Antagonistic Fungi:** Soil application of *T. harzianum* at 50 kg/ha (2×10^8 cfu/g) was effective in improving the plant growth parameters and fruit weight, while *P. lilacinus* at 50 kg/ha (2×10^8 cfu/g) was effective in controlling *M. incognita-F. solani* disease complex in tomato (Haseeb *et al.*, 2006).

ii) **Integrated Management**

a) **Bioagents and Botanicals:** Tomato roots that received *P. lilacinus* along with *T. harzianum* and neem cake were free from root knot nematodes (*M. incognita*) and did not wilt due to *F. o.* f. sp. *lycopersici* till harvest. The roots were also free from *Fusarium* infection. The above treatment also reduced the percentage of wilt (10% compared to 90% in control), root-knot index (1.8 compared to 4.4 in control) and increased root colonization with bioagents (74% compared to 0% in control), parasitization of egg masses

(56% compared to 0% in control) and eggs/egg mass (44% compared to 0% in control) (Nagesh *et al.*, 2006 (Table 13.5).

Table 13.5. Effect of integration neem cake and bioagents on the management of *Meloidogyne incognita* infecting tomato

Treatment	Healthy plants (%)	Root-knot index	Colonization by bioagents		
			Root Colonization (%)	Colonization of egg masses/eggs (%)	
				egg masses	eggs
Control	10	4.4	—	—	—
P. lilacinus	30	3.6	48	48	46
T. harzianum	40	4.0	50	44	32
Neem cake	10	4.2	—	—	—
P. lilacinus + Neem cake	60	2.8	62	62	58
T. harzianum + Neem cake	70	3.8	68	36	32
P. lilacinus + *T. harzianum* + Neem cake	90	1.8	74	56	44
CD (P = 0.05)	**8.87**	**1.78**	**3.66**	**4.11**	**3.22**

b) **Bioagents, Cultural Methods and Host Resistance:** Deep ploughing and exposing soil to hot sun in summer, removal and burning of crop debris, soil application of *T. viride* and *P. lilacinus*, use of wilt resistant varieties like Utkal Pallavi, Utkal Deepti, Utkal Kumari, Utkal Urbasi, etc. helped in controlling the disease complex.

c) **Arbuscular Mycorrhizal Fungi and Botanicals:** Bhagawati *et al.* (2000) demonstrated that although the mustard cake and AM fungus, *Glomus etunicatum* were effective in reducing the damage caused by *M. incognita* and *F. o.* f. sp. *lycopersici* on tomato, the performance of concomitant application of both (bioagent and botanical) was much better than the individual application.

13.5. BELL PEPPER, *Capsicum annuum*

Greenhouses can be a means to economically maintain a warm environment during cool season, to protect bell pepper plants from rain, wind, and high solar radiation, and to retain pollinators and beneficial insects while excluding unwanted insect pests. In greenhouses, capsicum fruits are harvested with full maturation colour (Plate 13.10) and fruit yields are greater of high quality and usually produced at a time of the year when production in the field is not possible and market prices for capsicum are highest.

13.5.1. Root-knot Nematode, *Meloidogyne incognita*

13.5.1.1. Symptoms:

Root-knot nematodes cause root galls on the feeder roots which sometimes affect the entire root system showing heavy galling. Root galls caused by root-knot nematodes on sweet pepper are frequently small (Plate 13.11). This affects the uptake of nutrition and water and the plants also show wilting during warmer part of the day. Later, stunting of plants with yellow foliage and reduction in yield is very common. In capsicum a pathogenic bacteria *Ralstonia solanacearum* gets entry into the roots infested by root-knot nematode and together they produce wilting disease that reduces yield by 60–70%.

13.5.1.2. Management

i) **Physical Methods:** Burning of paddy husk or saw dust on infested nursery proved to be reasonably effective.

ii) **Cultural Methods:** Several cultural methods like selection of nematode-free nursery sites; destruction of infested roots after crop harvest; crop rotation with antagonistic crop (marigold); or intercropping with marigold, onion and garlic; flooding; fallowing; deep summer ploughing; either alone or in combination proved to be reasonably effective and economical to check multiplication of root-knot nematodes on capsicum. Ploughing not only leads to disturbance and instability in nematode community but also causes their mortality by exposing them to solar heat and desiccation. Use of plastic sheets for covering soil in greenhouse further enhances nematode reduction. Such an approach also helps in reducing the intensity of weeds, fungi and bacteria in the soil.

iii) **Chemical Methods:** Bare-root dip treatment of capsicum seedlings in 0.1% Carbosulfan/Monocrotophos for 6 hr eliminates root-knot infection.

iv) **Biological Methods**

a) **Antagonistic Bacteria:** Application of a *B. thuringiensis* strain to capsicum in Puerto Rico reduced galling in roots due to *M. incognita* and increased yield significantly (Zuckermann *et al.*, 1993).

b) **Antagonistic Fungi:** Application of 'Royal 350' (*Arthrobotrys irregularis* cultured on oat seed medium) at 140 g/m^2 a month before transplantation of capsicum resulted in good protection against root-knot nematodes. Among the egg parasites, efficacy of *Paecilomyces lilacinus* (commercially formulated as 'Biocon' in Philippines) has been found to be comparatively higher in suppressing the population of *Meloidogyne* spp on capsicum.

Treatment of the greenhouse beds with the formulation of *Pochonia chlamydosporia* at 50g/m^2 was significantly effective in reducing the galling index, number of nematodes in roots and soil, increasing the per cent parasitization of eggs by bioagent and enhancing yields. Seed treatment with *P. fluorescens* alone and the nursery bed treatment with *P. chlamydosporia* alone were also effective (Naik, 2004).

v) **Host Resistance:** Capsicum cvs Mississipi-68, Santanka, Anaheim Chile and Italian Pickling are reported to be resistant to root-knot nematodes. Black Indica, Naharia and Pant C1 were reported to be resistant to *M. incognita*, while All Big as moderately resistant. California Wonder and Naharia were resistant to *M. javanica*, while Early California Wonder as moderately resistant. Naharia was reported to be resistant to *M. arenaria.*

The southern root-knot nematode (*M. incognita*) resistant 'Carolina Wonder' and 'Charleston Belle' are recommended for use by greenhouse growers.

vi) **Integrated Methods:** Integration of *P. fluorescens* and *P. chlamydosporia* in the greenhouse bed has proved significantly effective in reducing the root-galling index (*M. incognita*), number of nematodes in the roots and soil and increasing the yield of the capsicum crop (Naik, 2004).

In capsicum, the seedling stand was good where the combination of neem based *P. fluorescens* and *T. harzianum* was used, followed by seed treatment with *P. fluorescens* + soil application of *T. harzianum* (Naik, 2004).

13.5.2. Root-knot Nematode, *Meloidogyne incognita* and Bacterial Wilt, *Ralstonia solanacearum* Disease Complex

13.5.2.1. Interaction Studies:

Capsicum is prone to many soil borne diseases among which the bacterial wilt (*R. solanacearum*) in combination with root-knot nematode (*M. incognita*) takes heavy toll every year all over the world (Naik, 2004).

13.5.2.2. Management

i) **Integrated Management**

a) **Two Bioagents:** Combined application of neem based formulations of *P. fluorescens* and *P. chlamydosporia*/*T. harzianum* at 40 g/m^2 in nursery beds and transplanting these seedlings in the main field resulted in significant reduction in disease index and root-knot index in capsicum to the tune of 70% and increased the crop yield by 37% (Rao *et al.*, 2002) (Table 13.6). Combinations of the bioagents did not affect the colonization of the individual bioagents on the roots and hence the transplants carried the bioagents to the main field. This has resulted in the effective management of the pathogens involved in the disease complex.

Table 13.6. Effect of integration of neem-based bioagents on plant growth and management of root-knot and bacterial wilt disease complex and yield of capsicum.

Treatment	Seedling weight (g)	Root-knot index (1-10)	Disease index (1-9)	Yield in kg/4 m^2
Seed treat. with *P. fluorescens*	421	5.6	6.4	4.3
Seed treat. with neem-based *P. fluorescens*	428	5.2	6.7	4.7
Nursery treat. with *P. fluorescens*	435	4.6	5.4	4.8
Nursery treat. with *P. chlamydosporia*	364	4.4	7.3	3.2
Nursery treat. with *T. harzianum*	374	4.8	7.0	3.4
Nursery treat. with *P. fluorescens* + *P. chlamydosporia*	463	4.1	5.2	4.0
Nursery treat. with *P. fluorescens* + *T. harzianum*	493	3.8	3.5	5.1
Control	340	8.7	8.2	2.6
CD at 5%	**27.20**	**0.49**	**0.38**	**0.25**

Naik (2004) reported that combination of *T. harzianum* along with *P. fluorescens* increased the yield (3.320 kg/3m² plot) followed by combination of *T. harzianum* and *P. chlamydosporia* (2.960 kg/3m²). The least gall index was present where the combination of *T. harzianum* and *P. fluorescens* was used (1.5) followed by *T. harzianum* + *P. chlamydosporia* (1.7) and *P. fluorescens* + *P. chlamydosporia* (1.7). But among combination treatment of bio-agents, all the treatments were on par (*P. fluorescens* + *T. harzianum*, *P. fluorescens* + *P. chlamydosporia*, *T. harzianum* + *P. chlamydosporia*). Results related to the per cent bacterial wilt disease (*R. solanacearum*) incidence also showed similar trend in which *T. harzianum* + *P. fluorescens* performed very well and the per cent incidence was 16.90 followed by *T. harzianum* + *P. chlamydosporia* (17.20).

Table 13.7. Effect of integration of bioagents on plant growth and management of disease complex in bell pepper in nursery beds

Treatment	Seedling length (cm)	Seedling weight (g)	No. of galls /10 plants	Root colonization (cfu/g)	
				P. lilacinus	*B. pumilis*
Nursery bed treatment with *P. lilacinus* (25 g/ m²)	12.63	3.8	52	13,342	—
Nursery bed treatment with *P. lilacinus* (50 g/ m²)	15.53	4.0	50	16,539	—
T3 - Nursery bed treatment with *B. pumilis*	15.17	4.2	55	—	14,431
T1+ T3	18.43	4.4	44	12,984	13,964
T2 + T3	19.25	4.6	40	16,839	14,291
Control	10.53	3.3	76	—	—
CD (P = 0.05)	**1.29**	**0.37**	**6.28**	**1087.95**	**976.75**

Table 13.8. Effect of integration of bioagents on root galling, wilt incidence, root colonization, propagule density and yield of bell pepper under field conditions

Treatment*	Root-knot index	Mort-ality (%)	Yield kg/6m²	Root colonization (cfu/g)		Propagule density (cfu/g soil)	
				P. lilacinus	*B. pumilis*	*P. lilacinus*	*B. pumilis*
T1	6.5	68.90	4.8	30,653	—	21,346	—
T2	5.3	52.45	5.4	34,289	—	23,467	—
T3	7.0	34.62	7.5	---	25,875	—	17,321
T4	4.3	25.86	8.2	31,764	21,459	20,457	15,347
T5	4.1	17.69	8.5	34,762	20,764	22,689	14,569
T6	8.4	91.45	0.5	—	—	—	—
CD (P=0.05)	**0.58**	**8.94**	**0.35**	**1176.52**	**1087.54**	**1234.87**	**897.34**

* T1 – Nursery bed treatment with *P. lilacinus* (25 g/ m²), T2 - Nursery bed treatment with *P. lilacinus* (50 g/ m²), T3 - Nursery bed treatment with *B. pumilis*, T4 – T1+ T3, T5 – T2 + T3, T6 – Control.

Naik *et al.* (2003) reported that soil application of organically developed *P. lilacinus* (50 g/m^2 of formulated product containing 10^6 spores/g) along with pure culture of *B. pumilis* (10^8 cfu/ml) to the nursery beds of bell pepper increased plant growth, root galling and root colonization by the bioagents (Table 13.7). The above seedlings when transplanted in the main field significantly reduced root galling and wilt incidence, and increased root colonization, propagule density in soil, egg parasitization and fruit yield in the main field (Table 13.8).

13.6. CUCUMBER, *Cucumis sativus*

Cucumber and gherkins are being extensively grown in greenhouses in many countries. There are two types of cucumbers grown under protected cultivation. The Continental or burpless cucumber is the main type of cucumber and has long fruit (Plate 13.12), whereas the Lebanese (mini) cucumber has shorter fruit (parthenocarpic). Compared with field cucumbers, these two types are seedless and do not require pollination to produce fruits. Compared with field crops, protected cultivated cucumbers have flesh that has less total acids (0.11 per cent versus 0.17 per cent) and more sugars (3 per cent versus 2.8 per cent). They are smoother, greener and have a finer texture. The skin is more tender, not peeled and are easier to digest. The crops produced in greenhouses fetch higher income due to good quality produce and higher productivity per unit area as also due to being available in off season. The cropping period gets extended, so regular supply could be maintained for a long time.

13.6.1. *Root-knot* Nematodes, *Meloidogyne incognita*

13.6.1.1. Symptoms:

Root-knot nematodes *(Meloidogyne* spp., especially *M. incognita)* cause swellings or galls on the root system (Plate 13.13). The disease is more common and serious on sandy soils than on clay soils; crop injury is more serious during dry seasons than seasons with adequate rainfall. Root-knot is one of the most common and serious diseases of cucumbers and all growers should practice control measures.

13.6.1.2. Management

i) **Host Resistance:**

   Cucumber cv. Vertina was moderately resistant to *M. incognita* (11-30 galls per root systems) at harvest.

ii) **Biological Control:** *Paecilomyces lilacinus* was found most effective in increasing plant growth parameters like plant height, fresh and dry shoot weight, and root growth. The reason for increased plant growth and other parameters observed could be attributed to the release of growth promoting substances by bio-agents (Baker *et al.*, 1986) or by production of toxic metabolites which inhibited nematodes and other deleterious microorganisms.

   P. lilacinus caused maximum reduction in soil nematode population, root galling and egg mass production (Plates. 3.14 and 13.15) and was superior over *P. fluorescens, Pochonia chlamydosporia,* and marigold intercropping. The reason for reduction in nematode population might be due to parasitic activity of *P. lilacinus* on eggs and all stages of nematodes.

13.7. CONCLUSIONS

In recent years, with changes in the cropping systems and climate, and introduction of high input intensive high yielding varieties/hybrids, a shift in pest status has been observed. There were outbreaks of cyst nematodes, *Globodera rostochiensis* and *G. pallida* on potato in Himachal Pradesh. There has been an outbreak of serious root-knot nematode problems in the recent years in vegetable crops grown under protected cultivation systems such as root-knot nematodes, *Meloidogyne incognita, M. javanica* and root-knot nematode, *M. incognita* and wilt, *Fusarium oxysporum* f. sp. *lycopersici*/*F. solani* disease complex on tomato; root-knot nematode, *M. incognita* and root-knot nematode, *M. incognita* and bacterial wilt, *Ralstonia solanacearum* disease complex on bell pepper, and root-knot nematode, *M. incognita* on cucumber. Hence, there is an urgent need for rapid and accurate detection, diagnosis, and development of management recommendations for these new nematode diseases.

14

Emerging Nematode Problems in Ornamental and Medicinal Crops

14.1. INTRODUCTION

Foliar nematode, *Aphelenchoides besseyi* is of great concern for profitable cultivation of tuberose (*Polianthes tuberosa*). It induces a typical 'floral malady' symptoms which are a prime threat for growing quality flowers particularly in major tuberose growing areas of West Bengal and Odisha of India. The widespread detection of white tip disease of rice in the southern and eastern states of India was reported and the foliar nematode problem is restricted to tuberose in WB and Odisha. The foliar nematode is responsible for 38% loss in spike yield and 59% loss in loose flower yield.

Crops such as carnations and gerbera are being grown throughout India under protected cultivation (in polyhouses/greenhouses/shade nets) which are seriously infested with nematodes such as *Meloidogyne incognita*, *M. javanica* (root-knot nematodes) and *Rotylenchulus reniformis* (reniform nematode). Nematode problems on all these crops under protected conditions have assumed alarming proportions leading to huge losses (up to 80%) in select crops. The nematode infestations exacerbate severity of fungal diseases leading to complete crop losses. *M. incognita* infection makes the plants highly susceptible to *Fusarium oxysporum* f. sp. *dianthi* attack. *Phytophthora parasitica* + *M. incognita* interact to produce a disease complex in gerbera leading to reduction in the yield around 40–60%.

Root-knot disease due to *Meloidogyne incognita* and the disease complex it causes in conjunction with other soil-borne pathogens like *Fusarium* spp. are upcoming diseases in respect of medicinal crops like coleus (*Coleus forskohlii*) and aswagandha (*Withania somnifera*).

14.2. ORNAMENTAL CROPS

14.2.1. Tuberose, *Polianthes tuberosa*

14.2.1.1. Foliar Nematode, *Aphelenchoides besseyi*

The infestation of foliar nematode, *Aphelenchoides besseyi* has emrged as a serious potential threat of great concern for profitable cultivation of tuberose (*Polianthes tuberosa*) in West Bengal, India. This nematode was reported for the first time from the Hawaii Islands (Holtzmann, 1968) on the leaves of tuberose. Later, the occurrence of 'floral malady' caused by *A. besseyi* on tuberose was found in the

Ranaghat areas of the Nadia district of West Bengal, India (Chakraborti and Ghosh, 1993). The nematode has recently got spread to Odisha presumably through bulbs. *A. besseyi* is nowadays a major limiting factor for cultivation of tuberose in Ranaghat, Haringhata and Panskura areas of West Bengal. The 'Calcutta single' cultivar of tuberose is more vulnerable to *A. besseyi* than the 'Calcutta double' cultivar. Research results confirmed that *A. besseyi* is the primary causal agent for malformed flowers and population of *A. besseyi* causing white tip disease in rice is the same population infecting tuberose causing floral disease (Khan, 2001). The nematode species is widespread in rice in West Bengal (Das and Khan, 2007). The foliar nematode problem is restricted to tuberose in WB and Odisha. The presence of nematode species in the cut flower stalk is one of the main constraint to export the flowers to other countries.

i) **Symptoms:** Ectoparasitic feeding of *A. besseyi* on the scape (floral stalk) leads to the formation of prickle like rough, irregular structure. Scape becomes curled and twisted. Scapes of the infected plants are many times smaller than healthy and normal scapes. Endoparasitic feeding of *A. besseyi* was observed in leaves, bracts and petals where brownish to black necrotic spots of varying shapes and sizes were developed (Plate 14.1). The nematode forms 'nematode wool' on dark brown spots. The ovary contains large number of nematodes. This nematode is generally more serious during rainy season from June to September and cent per cent loss of the second year crop of the 'Calcutta single' cultivar of tuberose was encountered. However, in 'Calcutta double' cultivar 30% to 40% flower stalk renders unsaleable and individual flower stalk occur to the extent of 59% (Pathak and Khan, 2009).

ii) **Host Range:** Host tests indicated that the *A. besseyi* population from tuberose induces symptoms in oat (cv. OS-6), rice (cv. IET-4094), and teosinte (cv. TL-1) but not on onion, maize, *Sorghum halepense*, gladiolus and chrysanthemum.

iii) **Survival and Spread:** Studies on various parts of tuberose plant for understanding their role in dissemination of *A. besseyi* revealed that flowers, stalk, leaf, scales and bulbs of 'Calcutta single' cultivar were the major sources of nematode infection. Further, *A. besseyi* was found infecting tuberose in field naturally either through infected bulbs or infested plants parts.

Infested bulbs harbour nematode in coiled anhydrobiotic condition (quiescent pre-adult and adult stages) in the scaly leaves outside the bulbs. The nematode can also survive in the dried scaly leaves, stems and flowers more than 25 months; however, they cannot survive in soil for long time (Khan, 2004; 2006a, b).

iv) **Management:** Out of the six tuberose cultivars viz. Calcutta single, Calcutta double, Suvasini, Swarna rekha, Rajat rekha and Shringar, only Suvasini was immune to attack by *A. besseyi.*

The treatments viz., pre-soaking of bulbs in plain water for overnight + hot-water treatment at 50°C for 30 min. + three sprayings with monocrotophos 36 SL at 500 ppm at 15 days interval (repeated the same spraying in second year from March onwards) had low nematode infestation (26%) and PDI value (25%), reduction (53%) of nematode population and higher spike yield (mean 286/plot - 1.5 × 1.8 m^2) as compared to untreated plots (mean 207/plot) (Table 14.1) (Pathak and Khan, 2009).

Table 14.1. Effect of treatments on yield losses, nematode infestation, per cent disease index (PDI), nematode population and cost: benefit (C: B) ratio in tuberose.

Treatment	Nematode population density	% decrease	% infestation	PDI	Spike Yield/ Plot (kg)	Spike yield loss	Loose flower Yield/ plot (kg)	Yield loss	C:B ratio
Treated	1516	52.55	25.89	25.21	286	38.15	36.1	59.03	1:19
Untreated	3152		42.84	42.01	207		22.7		
t calculated (P= 0.05)			29.57	26.80	5.92		4.62		

With the adoption of nematode management practice, spike yield could be saved up to 38% and that could be to the extent of 59% when loose flower yield/plot was taken into consideration.

In conclusion, the following practices are recommended for the management of foliar nematode on tuberose:

- Select healthy nematode free bulbs for planting.
- Pre-soaking of tuberose bulbs in plain water or in neem seed kernel extract (NSKE 4%) for overnight followed by hot water treatment at 50°C for 20 min (Khan *et al.*, 2005, 2006) or dipping in monocrotophos 36SL at 500-700 ppm for 4-6 hours (Khan *et al.*, 2008).
- After sprouting, three to four sprayings with carbosulfan 25 EC or monocrotophos 36SL each at 500 ppm at 15 to 20 days interval is needed.
- During peak flowering stage additional spraying with NSKE 5% at 15 days interval may be advocated.
- Run-off or irrigation water from infested field should not be allowed to enter the uninfested tuberose fields (Khan, 2006).
- Infested plant parts should be burnt immediately on inspection.
- In the second and third year crop, spraying for 3-4 times with monocrotophos 36SL at 500 ppm at 15 to 20 days interval should start from April-May onwards to reduce the nematode infestation.
- Field sanitation is most essential for reducing infestation. Infested plant parts (plants and inflorescence) should be burnt or buried into the soil.
- Grow nematode tolerant tuberose cultivars, Phule Rajani, Prajwal and Shringar (Khan and Ghosh, 2007).
- Growing tuberose away from rice field is encouraged to avoid contamination of tuberose field.
- Proper spacing should be followed.

v) **Conclusions:** In West Bengal, infestation of foliar nematodes i.e. *Aphelenchoides besseyi* has become a potential threat to the tuberose crop. Identification of nematode infestation at early stage of infestation to advocate suitable management strategy is an integral part of foliar nematode management in tuberose crops. Use of resistant cultivars and hot water or nematicidal treatments is important preventive measure to defend the menacing infestation of *A. besseyi.* They are more abundant in rice-tuberose cultivation system in this region and in monocotyledonous plant in particular. Considering the host diversity of this nematode species, it may attack many more plants in the state with the passage of time. Hence, a close vigilance on their infestation is an urgent need of the day. Government and non- governmental institutions should come forward to take initiative to carry forward the research work towards managing this quarantined nematode pest which may hamper export of tuberose in near future.

14.3. POLYHOUSE FLOWER CROPS

In 1980s, the demand for high quality, export-oriented horticultural products and need for the availability of horticultural crop produce round the year especially in off season, compelled the growers to cultivate selected flower crops under protected cultivation. As a result people started cultivation of flower crops under protected conditions in several states of India. Soon the nematode incidence particularly under protected cultivation became severe and led to total crop losses because of congenial conditions of higher temperature, humidity and use of high agronomic inputs like fertilizers and plant growth promoters in polyhouses.

An overall average annual yield loss in major ornamental crops due to nematodes goes up to 60% under protected cultivation. Rapid spread of nematode infestation through soil, crop residues and indiscriminate use of agro-chemicals in horticultural ecosystems is a major concern for crop protection specialists and policy makers.

Symptoms such as chlorosis and stunting appear after sufficient damage is inflicted. The proliferation rates of nematodes in polyhouse cultivation reached up to 10 to 30 folds more than in the open field cultivation. As yet farmers continue to incur losses in crops under protected cultivation without appropriate solutions to problems posed by nematodes. The population build up is very rapid in the polyhouses and nematode population reaches 5 - 6 times of threshold levels within 18 - 24 months, making the polyhouse cultivation a wasteful exercise.

Crops such as carnations, roses, gerbera and anthuriums are being grown under protected cultivation (in polyhouses/ greenhouses/ shade nets). These crops which are grown throughout India are seriously infested with nematodes such as *Meloidogyne incognita, M. javanica* (root-knot nematodes) and *Rotylenchulus reniformis* (reniform nematode). Nematode problems on all these crops under protected conditions have assumed alarming proportions leading to huge crop losses (up to 80%) in select crops. The nematode infestations exacerbate severity of fungal diseases leading to total crop loss.

M. incognita infection makes the plants highly susceptible for the attack of *Fusarium oxsporum* f. sp. *dianthi*. *Phytophthora parasitica* + *M. incognita* interact to produce a disease complex in gerbera leading to reduction in the yield around 40 to 60 %.

Management practices adopted by farmers include continual use of chemical nematicides, often at higher than recommended rates resulting in build-up of resistance. In addition, biomagnification and environment deterioration due to hazardous chemicals has rendered several cultivated ecosystems unstable and non-profitable.

Successful management strategies for nematodes and other disease complexes have been evolved using bio-pesticides like *Paecilomyces lilacinus*, *Pochonia chlamydosporia*, *Trichoderma harzianum*, *T. viride* and *Pseudomonas fluorescens*. Farmers who adopted these technologies reduced the use of agro-chemicals to the extent of 40 to 45% and obtained 30 to 35% increased yields in gerbera and carnations.

14.3.1. Preparation of Beds

- Raised Beds are to be prepared and the soil should be brought to fine tilth. The bed size can vary according to the requirement and the type of crop grown.
- Add recommended doses of fertilizers, carbofuran or phorate at 50g/sq.m + 200g neem/ pongamia/ mahua cake enriched by bio-pesticides per sq. m.
- Further incorporate bio-pesticide enriched FYM at 2kg/sq. m or bio-pesticides enriched vermicompost at 500g/sq.m in top 18 cm of soil in the beds.
- Water the beds for 7-10 days for proper decomposition of these organic materials.

14.3.2. Enrichment of Organic Amendments (FYM/ Neem/ Pongamia/ Mahua Cake/ Vermicompost)

1 ton of farm yard manure/ neem/ pongamia/ mahua cake/ vermicompost has to be enriched by mixing with 2 kg each of *Pseudomonas fluorescens, Trichoderma harzianum* and *Paecilomyces lilacinus*. It has to be covered with plastic mulch and optimum moisture of 25 - 30% has to be maintained by regular watering for a period of 15 days (Plate 14.2). Once in a week thoroughly mix the mixture for maximum multiplication and uniform spread of microorganisms in the entire lot of substrates.

14.3.3. Application of Enriched Organic Amendments

14.3.3.1. Soil Application:

Apply 100 g of neem / pongamia / mahua cake or 250 g of vermicompost enriched with *Pseudomonas fluorescens* + *Trichoderma harzianum* + *Paecilomyces lilacinus* to 1 sq. m. beds in the rhizosphere region of plants.

14.3.3.2. Spray Application

- The organic formulation containing *Pseudomonas fluorescens* and *Trichoderma harzianum* has to be sprayed on the plants at regular intervals of 20 days at a dosage of 5g/ lit or 5ml/ lit.
- Alternately, take 20 kg of neem/ pongamia/ mahua cake enriched in the above mentioned manner and mix it in 200 litres of water, leave it for a period of 2-3 days. Filter this suspension and use it for spraying by mixing 250 ml of suspension in 1 lit. of water and spray at regular interval of 20 days.

14.3.3.3. Application through Drip Irrigation System

- The organic formulation of biological control agents has to be given through drip/ by drenching at 5g/ lit or 5ml/ lit at regular interval of 20 days.

- Alternately, take 20 kg of neem/ pongamia/ mahua cake enriched in the above mentioned manner and mix it in 200 litres of water, leave it for a period of 2-3 days. Filter this suspension and use it for drenching at regular interval of 20 days.

By following the above methods, farmers can get significant increase in the yield of flower crops with the cost: benefit ratio of 1: 3 and above.

14.3.4. Recommendations

- Soil testing for nematode infestation should be made mandatory for the establishment of the new polyhouses.
- Region-wise dedicated referral labs for analyzing samples for nematodes be established at State Agricultural Universities/ ICAR Research Institutes.
- Sites for erecting new polyhouses should be properly treated with recommended chemicals and followed by supplementation with biological control agent enriched farm yard manure/oil cakes/ vermicompost.
- There is an urgent need to grant provisional registration for chemicals having proven efficacy against root-knot and other nematodes.
- In case of heavy population of plant parasitic nematodes species, soil fumigation with permissible chemicals should be done.
- Use of quality neem cake, castor cake, coco peat, FYM, or vermicompost enriched with bio-control agents.
- Evaluation of root-stock of crop genotypes against target nematode species (even up to races) should be carried out. The resistant root-stock should be grafted and the same root stock should be preserved for further studies.
- Efforts on breeding flower varieties suitable for polyhouse conditions should be taken up.
- Cluster approach may be followed to tackle the emergent nematode and other pest problems in polyhouse conditions.

14.4. CARNATION, *Dianthus caryophyllus*

14.4.1. Root-Knot Nematode, *Meloidogyne incognita*

The root-knot nematode, *M incognita* is one of the serious limiting factors in commercial cultivation of carnation and gerbera under polyhouse conditions. Nagesh and Parvatha Reddy (2000) reported that *M incognita* was responsible for 27 % loss in flower yield of carnation.

14.4.1.1. Economic Importance and Losses:

Most of the highly fetching exotic cultivars of Carnation from Europe have shown 40 to 60% mortality in polyhouse beds due to root-knot nematode infection in and around Bangalore (Nagesh and Parvatha Reddy, 2005).

14.4.1.2. Symptoms:

Root knot nematode is a serious pest of carnations. Infected plants usually appear stunted and tend to wilt on warmer days. When such plants are dug, the root galls are generally conspicuous and be easily observed (Plate 14.3).

14.4.1.3. Management

i) **Chemical Methods:** Pre-plant treatment of beds with Chlorpyriphos and Carbofuran (each applied twice in 6 months) significantly reduced nematode populations in roots and soil. However, there was a build-up of nematode populations in beds treated with these two chemicals after 1 year.

ii) **Biological Methods:** Under polyhouse conditions, carnation plants treated with bio-agents such as *Paecilomyces lilacinus* and *Pochonia chlamydosporia* gave significantly higher reduction of root galling as compared to *T. harzianum* (Shylaja, 2005).

The organic formulation of biological control agents (*Pseudomonas fluorescens* + *Trichoderma harzianum* + *Paecilomyces lilacinus*) has to applied through drip/ drenching at 5g/ lit or 5ml/ lit at regular interval of 20 days for effective management of root-knot nematodes on carnation.

iii) **Host Resistance:** Carnation cvs. Desco, Castelaro, Kappa, Rara, Logupink, Target and Antalia were highly resistant to *M. incognita.*

Screening of the four commercially available varieties of carnation for their resistance to root-knot nematode, *M. incognita,* revealed that *cv.* Ivonne was the most susceptible while, Randez was moderately tolerant. The other two varieties, Dark Randez and Sunshine were observed to be moderately susceptible (Shylaja, 2005).

iv) **Integrated Methods**

a) **Bioagents and Botanicals:** Integration of neem cake at 0.5 kg with *T. viride* at 100 g per m^2 was found effective for the management of root-knot nematodes and increased carnation flower yields.

Integration of *P. lilacinus*/*T. harzianum* at 0.5 l per m^2 (aqueous spore suspension containing 2 x 10^4 spores per ml) with neem cake at 0.5 kg per m^2 or fenamiphos at 2 g a.i. per m^2 increased plant growth parameters and flower yield of carnation. The above treatments also increased root-knot egg parasiti-zation by the parasitic fungi (Nagesh and Parvatha Reddy, 2005) (Plate 14.4).

Three antagonistic fungi viz., *T. harzianum*, *V. lecanii* and *P. lilacinus* at 2 × 10^4 spores/ml in combination with neem cake reduced *M. incognita* population in both soil and roots of carnation.

Application of *P. lilacinus* at 0.5 g/kg of soil along with neem cake at 1.0 t/ha efficiently suppressed the nematode population and checked its build up, enhancing the plant growth parameters resulting in better production with increased flower stalk length and flower diameter in carnation. The plants also came to flower early (Johnson, 2000).

b) **Arbuscular Mycorrhizal Fungi (AMF) and Bioagents:** Combined inoculation of AMF and *Pseudomonas fluorescens* had positive effect on root-knot nematode control on carnation (Anusuya and Vadivelu, 1999).

c) **Bioagents, Botanicals and Chemicals:** Pre-plant treatment of beds with dazomet (at 40 g/m^2) followed by the application of neem cake (at 1 kg/m^2 15 days later) along with antagonistic fungi, *P. chlamydosporia*/ *P. lilacinus* (at 2 x 10^{12} spores/ m^2) significantly reduced root-knot nematode population (*M. incognita*), mortality of plants and suppressed the nematode infection for 2 years. The antagonistic fungi established better in the beds treated with dazomet. The above treatment also reduced root galling, nematode multiplication rate and plant mortality, and increased spike/stem length, flower yield and root colonization with the bioagents (Nagesh and Parvatha Reddy, 2005). On a long term basis, soil management with pre-plant treatment of Dazomet, followed by the application of oil cakes plus antagonistic fungi, was found more effective against *M. incognita* than post-plant treatment with Carbofuran, Carbosulfan and Chlorpyriphos on carnation grown in polyhouses.

14.4.2. Root-knot Nematode, *Meloidogyne incognita* and Wilt, *Fusarium oxysporum* f. sp. *dianthi* Disease Complex

14.4.2.1. Interaction Studies:

M. incognita in association with *F. oxysporum* f. sp. *dianthi* has been reported to cause wilt disease complex in carnation (Plate 14.5). In an experiment carried out to study the role of root-knot nematode, *M. incognita* in predisposing carnation to *Fusarium* wilt, it was observed that the appearance of the wilt symptoms were accelerated when *M. incognita* was inoculated two weeks prior to *F. oxysporum* f. sp. *dianthi*. The rate of wilting was observed to be 2.1 and 4.6, respectively during 12th and 25th week of observation, while the root galling index (RGI) values were recorded to be 4.15 and 4.75, respectively. It was observed that maximum mortality was recorded when *M. incognita* was inoculated 2 weeks prior to *F. oxysporum* f. sp. *dianthi* which was 33.4% and 79.2 %, respectively, at 12th and 25th week of observation. The plant growth was also reduced significantly due to prior inoculation of *M. incognita* (Shylaja, 2005).

14.4.2.2. Integrated Management:

The studies carried out to evaluate combination of bio-agents for the management of wilt (*F. o.* f. sp. *dianthi*) and root-knot nematode (*M. incognita*) disease complex in carnation revealed that a combination of *Pochonia chlamydosporia* + *Paecilomyces lilacinus* each at 20 g/m^2 gave significant increase in plant growth parameters (plant height, root length and root weight) and flower yield (stalk length, stalk weight and flower diameter) (Table 14.2) (Shylaja, 2005).

Table 14.2. Effect of biocontrol agents on the plant growth and flower yield of carnation infected with *M. incognita* and *F. oxysporum* f. sp. *dianthi*

Treatment	Plant height (cm)	Stalk length (cm)	Root length (cm)	Root weight (g)	Stalk weight (g)	Flower diameter (cm)
Formulations of *P. chlamydosporia* and *P. lilacinus* each at 20 g/m^2	54.0	48.8	21.3	20.5	17.24	5.85
Formulations of *P. chlamydosporia* and *T. harzianum* each at 20 g/m^2	52.2	43.4	15.7	16.44	15.90	5.15

Treatment	Plant height (cm)	Stalk length (cm)	Root length (cm)	Root weight (g)	Stalk weight (g)	Flower diameter (cm)
Formulations of *T. harzianum* and *P. lilacinus* each at 20 g/m^2	47.8	39.2	18.0	18.06	14.66	4.99
Control	22.8	17.6	12.8	14.86	10.82	3.88
CD (P = 0.05)	**5.16**	**4.54**	**2.9**	**5.6**	**2.01**	**0.81**

The lowest root galling (1.64), wilting index (2.0) and plant mortality (49.5%) was observed in plants treated with *P. chlamydosporia* and *P. lilacinus*. The lowest number of nematodes in both soil and roots (44.4 J2 per 100 cm^3 soil and 10 nematodes/5g roots) were also recorded in plants treated with *P. chlamydosporia* and *P. lilacinus* (Table 14.3.) (Shylaja, 2005).

Table 14.3. Effect of biocontrol agents on root galling and wilt disease index in carnation infected with *M. incognita* and *F. oxysporum* f. sp. *dianthi*

Treatment	Root galling index	Wilt disease index	% plant mortality	No. of nemas/100 cm^3 soil	No. of nemas/5g roots
Formulations of *P. chlamydosporia* and *P. lilacinus* each at 20 g/m^2	1.64	2	49.5	44.4	10.0
Formulations of *P. chlamydosporia* and *T. harzianum* each at 20 g/m^2	2.70	3	60.4	73.6	13.6
Formulations of *T. harzianum* and *P. lilacinus* each at 20 g/m^2	2.17	3	63.9	62.4	21.0
Control	3.66	5	95.8	125.8	42.2
CD (P = 0.05)	**0.82**	**0.53**	**10.46**	**5.20**	**3.12**

14.5. GERBERA, *Gerbera* spp.

The root-knot nematode is recognized as the major limiting factor in successful production of gerbera crop.

14.5.1. Root-knot Nematode, *Meloidogyne incognita*

The root-knot nematode, *M incognita* is one of the serious limiting factors in commercial cultivation of gerbera under polyhouse conditions. Nagesh and Parvatha Reddy (2000) reported that *M incognita* was responsible for 31% loss in flower yield of gerbera.

4.5.1.1. Economic Importance and Losses:

Most of the highly fetching exotic cultivars of Gerbera from Europe have shown 40 to 60% mortality in polyhouse beds due to root-knot nematode infection in and around Bangalore (Nagesh and Parvatha Reddy, 2000).

14.5.1.2. Symptoms:

Symptoms include stunting of plants, yellowing of leaves (Plate 14.6), wilting of plants, and heavy galling on roots.

14.5.1.3. Management:

Soil application of *Bacillus subtilis* (BbV 57) or *Pseudomonas fluorescens* at 2.5 kg/ha at the time of planting is effective for the management of root knot nematode.

Integration of *P. lilacinus*/*Trichoderma harzianum* at 0.5 l per m^2 (aqueous spore suspension containing 2 x 10^4 spores per ml) with neem cake at 0.5 kg per m^2 or fenamiphos at 2 g a.i. per m^2 increased plant growth parameters and flower yield of gerbera. The above treatments also increased root-knot egg parasitization by the parasitic fungi (Nagesh and Parvatha Reddy, 2005).

14.5.2. Root-knot Nematode, *Meloidogyne incognita* and Foot Rot, *Phytophthora parasitica* Disease Complex

14.5.2.1. Interaction Studies:

Sustainable production of gerbera is seriously hampered by the disease complex caused by *M. incognita* and *P. parasitica*. These two pathogens reduce the productivity of gerbera significantly to the tune of 40-60%.

14.5.2.2. Integrated Management:

Combined application of neem cake enriched with either *Trichoderma harzianum* or *Pseudomonas fluorescens* [mixing 50 g of *T. harzianum* (2 x 10^6 cfu/g) or *P. fluorescens* (2 x 10^8 cfu/g) in 1 kg of neem cake] applied at 25 g/m^2 was found effective for the management of disease complex and increased the flower yield by 26% in gerbera cv. Debora (Manoj Kumar *et al.*, 2010).

14.6. MEDICINAL CROPS

14.6.1. Coleus, *Coleus forskohlii*

14.6.1.1. Root-knot Nematode, *Meloidogyne incognita*

The root-knot nematode infestation was reported on coleus from Kerala and Orissa. The dry weight of the tubers was reduced by 20% due to root-knot nematodes. The percentage of starch on fresh weight basis showed drastic reduction (16%) in the infested tubers. *M. incognita* has been reported to cause a yield reduction of up to 86% (Senthamarai *et al.*, 2006b).

i) **Symptoms:** The newly inoculated *C. forskohlii* plant exhibited stunted growth, yellow patches and severe galling of roots due to *M. incognita*. Tuber formation was found commensurate with the degree of galling and yield reduction was 65 per cent under field conditions. The pathogenicity study which was carried out in controlled condition revealed that the effective damage to the plant was caused by 1000 larvae/plant.

    The galls on coleus roots are very big and pronounced (Plate 14.7). The root-knot nematode damage often leads to crop failure. The infested tubers swell in size with irregular surface and cracking of the skin. Rotting sets in even before harvest, if the infestation is severe. Infested tubers rot after harvest and rarely reach market.

ii) Management

a) **Cultural Methods:** Sree Bhadra, a high yielding variety of sweet potato released by Central Tuber Crops Research Institute, Trivandrum, is identified as a resistant trap crop for the root-knot nematode. The nematodes were able to penetrate the root but giant cell formation was not induced and as a result the nematodes die inside the roots. Planting this variety in root-knot nematode infested field helped in clearing the field free of nematodes and giving good tuber yield. Subsequently growing of susceptible crop like coleus escaped nematode damage (Mohandas, 2001).

b) **Chemical Methods:** Chemical agents such as methyl isothiocynate mixtures, oxamyl, thionazin and carbofuran have been found effective for the management of nematodes. Application of carbofuran 3% granules at the rate of 20 kg per hectare under wet conditions near the root zone was also found effective.

Treatment with fensulfothion reduced damage to 5.7% as against 21% in control. The above treatment also increased the production of healthy tubers to the extent of 47% (Pillai, 1976).

c) **Biological Methods:** Soil application of *Trichoderma harzianum* and *Pseudomonas fluorescens* each at 2.5 kg/ha recorded increased plant growth and reduced *M. incognita* population (Senthamarai *et al.*, 2006a).

Soil application with bio-agents like *T. viride* and *P. fluorescens* significantly reduced the nematode population in soil and root and increased the growth and yield of *C. forskohlii* crop (Senthamarai *et al.,* 2008).

Increase in plant height, number of branches, fresh and dry weight of the plant as well as less root-knot index was observed due to PGPR treatments (fluorescent *Pseudomonas* sp.) as compared to control. Application of PGPR strains (RB50 and RB31) recorded maximum plant height, number of branches, fresh and dry weight of the plants with less root-knot index (Table 14.4) (Mallesh, 2008).

Table 14.4. Biocontrol potentiality of PGPR strains in coleus against root-knot nematodes

PGPR Strains	Plant height (cm)	No. of branches	Fresh weight (g)	Dry weight (g)	Root-knot index
RB 01	31.5	4.5	94.0	15.00	2.0
RB 10	27.5	4.0	92.5	14.25	3.5
RB 13	32.0	5.0	101.0	16.50	2.0
RB 22	26.0	4.0	93.0	14.25	3.0
RB 31	34.0	5.5	107.0	16.90	1.5
RB 43	28.0	4.5	93.5	14.50	2.5
RB 50	35.0	6.0	110.5	17.50	1.5
Control	22.0	3.5	69.5	11.12	4.0

d) **Integrated Methods:** Integration of soil solarization in the nursery for 15 days with 150 gauge LDPE film and application of *Paecilomyces lilacinus* + neem cake or *P. lilacinus* + *Bacillus macerans* in the main field are the best treatments in increasing plant height (64.3 and 60.3 cm compared to 40.0 cm in control), number of leaves (593.3 and 583.3 compared to 310.0 in control), weight of tubers/plant (560.0 and 546.6 g compared to 350.0 g in control), yield (11.5 and 11.3 kg/plot compared to 6.9 kg/plot in control) and in reducing root galls (0.3 and 1.0 compared to 50.6 in control), nematode population in soil (25.0 and 30.0/100 ml soil compared to 196.6/100 ml soil in control) and roots (1.0 and 1.6/5 g roots compared to 79.0/5 g roots in control) (Nisha and Sheela, 2006).

Dipping of stem cuttings in *P. fluorescens* + soil application of neem cake at 400 kg/ha + growing marigold as intercrop followed by their biomass incorporation during earthing up significantly increased the yield (22.7-30.0%) and reduced the root-knot nematode population (71.2-73.8%), followed by the integration of *P. fluorescens* + marigold intercrop, which were almost equally effective (Seenivisan and Devrajan, 2008).

14.6.1.2. Root-knot Nematode (*Meloidogyne incognita*) and Wilt Disease (*Fusarium chlamydosporum*) Complex

i) **Interaction Studies:** Root-knot disease due to Meloidogyne incognita and the disease complex it causes in conjunction with other soil-borne pathogen like *Fusarium chlamydosporum* is an upcoming disease which appeared on severe form (Kumar, 2008) in coleus.

In the interaction studies, *M. incognita* was the most aggressive pathogen compared to *F. chlamydosporum.* However simultaneous inoculation of *M. incognita* and *F. chlamydosporum* caused greater reduction in plant growth as well as nematode multiplication (Plate 14.8). However, with respect to individual pathogens, *M. incognita* caused greater reduction in plant growth compared to *F. chlamydosporum.* In case of sequential inoculation of *M. incognita* 7 days prior to *F. chlamydosporum* caused reduction in plant growth parameters.

The effect of simultaneous inoculation of *M. incognita* with *F. chlamydosporum* on coleus was additive in nature. However, when *M. incognita* was inoculated with *F. chlamydosporum* the resultant effect was almost equal to sum of individual effect. These results suggest that the nematode can predispose the coleus to infection by fungus namely *F. chlamydosporum* and can aggravate the disease. Reduction in number of galls per plant and final nematode population was observed in simultaneous inoculation.

ii) **Management:** The talc-based bioformulations of rhizobacterial strains RB50 and RB31 decreased root-knot index and the incidence of disease complex while increasing plant height, number of branches, fresh and dry biomass; and tuber yield in coleus (Table 14.5 and Plate 14.9) (Mallesh, 2008).

Table 14.5. Efficacy of talc formulations of PGPR strains on fresh tuber yield parameters and the disease incidence in coleus under field conditions

PGPR strain	Tubers/ plant	Total biomass(g)		Root-knot index	Disease incidence (%)	Decrease over control (%)	Tuber yield (kg/plot)
		Fresh	Dry				
RB 01	15.25	964.44	127.61	1.13	16.67	41.05	4.18
RB 10	12.87	823.13	99.55	1.60	25.00	11.75	3.25
RB 13	16.63	1016.74	169.85	0.80	20.00	29.40	4.60
RB 22	11.87	811.84	88.60	1.53	23.33	17.64	3.36
RB 31	16.50	1094.54	190.89	0.73	13.33	52.94	5.45
RB 43	14.92	841.83	112.50	1.33	18.33	41.05	3.95
RB 50	18.00	1243.02	212.27	0.66	10.00	64.70	6.74
Control	10.70	730.44	80.00	2.20	28.33	—	3.00
CD at 5%	**1.52**	**14.66**	**6.55**	**0.65**	**5.02**	**—**	**0.79**

Combined application of plant products (Neem seed kernel powder at 5 g/kg of soil) with biocontrol agents (*Paecilomyces lilacinus, Trichoderma viride* + *Pseudomonas fluorescens* at 10 g/kg of soil) was found effective in reducing the number of galls, nematode population, number of egg masses, root-knot index, root rot index and improving the plant growth parameters as compared to inoculated control.

Studies on screening of different genotypes of coleus against *M. incognita* and *F. chlamydosporum* disease complex showed that the genotypes viz., Yellow Tubers and Rabakavi Local showed resistant reaction. Orange tubers and Nimbanur Local genotype showed moderately resistant reaction.

14.6.1.3. Root-knot Nematode (*Meloidogyne incognita*) and Collar Rot (*Rhizoctonia bataticola, Sclerotium rolfsii, Fusarium chlamydosporum*) Disease Complex

i) **Interaction Studies:** *Rhizoctonia bataticola, Sclerotium rolfsii, F. chlamydosporum* and *M. incognita* were found to be the most commonly associated fungi and nematode with collar rot disease complex (Plate 14.10) (Ramaprasad Shresti, 2005).

ii) **Symptoms:** Wilt symptoms were first recorded at 45 days after inoculation in treatment *M. incognita* inoculated seven days prior to inoculation of all the fungal pathogens simultaneously (*F. chlamydosporum* + *R. bataticola* + *S. rolfsii*) (Plate 14.11) (Ramaprasad Shresti, 2005).

iii) **Management:** The wilt incidence, nematode population, number of galls and colony forming units of *F. chlamydosporum* and *R. bataticola* were significantly minimum in the plots where *T. viride* (10 ml/plant spore suspension) combined with neemto (500 g/5 m^2) were applied compared to other treatments (Ramaprasad Shresti, 2005).

The treatment *T. viride* + Neemto recorded significantly lowest per cent wilt incidence (12.76) over control (35.52) followed by *T. harzianum* (18.87) and *P. fluorescens* (19.98). The number of galls/5 g of root was significantly lowest in treatment *T. viride* + Neemto (10.13), followed by carbofuran (14.93) which was on par with each other. Colony forming units of *Fusarium chlamydosporum* was significantly highest in the control (19.60) followed by FYM. Colony forming units of *Rhizoctonia bataticola* was found to be significantly highest in the treatment control (19.60) followed by FYM (Table 14.6) (Ramaprasad Shresti, 2005).

Table 14.6. Management of collar rot disease complex of *Coleus forskohlii* using different biocontrol agents, organic amendments and chemicals (Observations recorded at harvest - 150 days after planting)

Treatment	Wilt incidence (%)*	No. of galls/5 g roots	Cfu**	
			Fusarium chlamydosporium	*Rhizoctonia bataticola*
Trichoderma viride @ 10 ml/plant (8 x 10^3 cfu/ml)	21.09 (27.33)	21.13	7.60	12.20
T. harzianum @ 10 ml/plant (8 x 10^3 cfu/ml)	18.87 (25.74)	19.53	8.00	12.60
Pseudomonas fluorescens @ 10 ml/plant (24 x 10^5 cfu/ml)	19.98 (26.51)	18.27	8.00	14.20
Pronto @ 5% soil drench	23.31 (28.84)	17.33	10.60	15.60
Neemto @ 500 g/5 m^2	21.09 (27.24)	16.07	12.60	16.40
Carbofuran 3G @ 15 g a.i./5 m^2	24.42 (29.57)	14.93	16.20	17.60
Farm yard manure @ 5 kg/5 m^2	25.53 (30.38)	25.67	15.20	18.80
T. viride @ 10 ml/plant (8 x 10^3 cfu/ml) + Neemto @ 500 g/5 m^2	12.76 (20.93)	10.13	6.20	9.60
Carbendazim @ 0.1% soil drench	21.19 (27.33)	23.33	3.60	6.80
Propiconazole @ 0.1% soil drench	23.31 (28.84)	23.00	3.80	7.40
Control	35.52 (36.59)	28.40	19.60	21.60
CD (P= 0.05)	3.48	5.38	2.49	2.72

* Figures in parentheses are arc sin angular transformed values.
** Cfu – Colony forming units x 10^{-3}/g of soil (Average of 5 replicates).

Greenhouse studies using the plant growth promoting rhizobacterial (PGPR) fluorescent pseudomonads (bioformulations containing RB 50 and RB 31 strains) showed significant increase in seedling biomass besides reduction in coleus wilt complex due to combination of pathogens. RB 50 and RB 31 strains decreased root-knot index and incidence of disease complex while increasing biomass and tuber yield. Biochemical analyses in the above treatments showed elevated expression of defense enzymes (peroxidase, polyphenol oxidase and phenylalanine ammonialyase) and higher accumulation of phenolic compounds (activity being highest in respect of RB 50 and RB 31 treated plants) compared to respective inoculated checks. The suppression of disease complex in coleus is largely due to aforementioned biocontrol mechanism as well as induction of systemic resistance by efficient PGPR strains (Lingaraju and Mallesh, 2010).

14.6.1.4. Root-knot Nematode, *Meloidogyne incognita* and Root Rot *Macrophomina phaseolina* disease complex

The productivity of coleus has been hampered by its susceptibility to nematode, and root rot disease. Pathogens associated with root rot are *Macrophomina phaseolina* and the root knot nematode *Meloidogyne incognita.* Severely affected areas were Salem, Attur and Rasipuram in Tamil Nadu. Due to this disease complex the yield loss was ranged up to 50 to 60 per cent. Therefore, it is necessary to take effective control measures to save the crop from complete devastation.

i) **Symptoms:** Simultaneous inoculation of *M. incognita* and *M. phaseolina* as well as nematode inoculation followed by fungus 15 days later caused significant reduction in tuber yield and 100% root rot disease in medicinal coleus. The nematode multiplication was adversely affected when fungus was inoculated prior to nematode (Table 14.7) (Senthamarai *et al.*, 2008).

Table 14.7. Effect of *Meloidogyne incognita* and *Macrophomina phaseolina* on root galling and yield of coleus

Treatment	Tuber yield / plant (g)	No. of galls/plant	% disease incidence
M. incognita (1 J_2/g soil)	44.00	768	0
M. phaseolina (5 g/kg soil)	61.50	0	50
M. incognita (prior) + *M. phaseolina* (15 days later)	34.50	373	100
M. phaseolina (prior) + *M. incognita* (15 days later)	52.50	110	50
M. incognita + *M. phaseolina* (simultaneously)	13.00	316	100
Uninoculated control	84.38	0	0
CD (P = 0.05)	**10.07**	**2.08**	**43.99**

ii) **Integrated Management:** Integrated nematode management strategy (INMS) includes dipping of stem cuttings in *Psuedomonas fluorescens* (strain Pf1) + growing marigold (*Tagetes erecta*) as intercrop and their biomass incorporation during earthing up. Biointensive disease management strategy (BDMS) include soil drenching with *P. fluorescens* (strain PfC6) at 2.5 kg/ha at planting, 30, 60, 90 and 120 days after planting. INMS along with BDMS treatment increased the tuber yield by 40.6% and reduced nematode infestation in terms of mean number of juveniles per 100 ml soil (73.2%), mean number of adult females per g of root (82.4%), mean number of egg masses/g root (85.9%), mean number of eggs per g of root (87.9%) with least gall index (1.6). This treatment also decreased the incidence of *Macrophomina phaseolina* root rot disease up to 50.4%. Effect of this treatment on *M. incognita, M. phaseolina* and root tuber yield of medicinal coleus was almost equal to that of the treatment of INMS. But the latter is highly economical with lesser BC ratio than other treatments including standard chemical check i.e. carbofuran 3G at 1 kg a.i./ha + soil drenching with carbendazim 0.1%. Hence it could be concluded that treatment of INMS i.e. dipping of stem cuttings in 0.1% *P. fluorescens* + marigold intercropping can be commercially exploited for the management of *M. incognita* and *M. phaseolina* disease complex in medicinal coleus (Seenivsan, 2010).

The following practices have been recommended for the management of disease complex in coleus:

- Select the coleus cuttings from disease free plants.
- Coleus cuttings should be treated with carbendazim solution (0.1 per cent).
- Soil drenching with carbendazim (0.1 per cent) or propiconazole (0.1 per cent).
- Soil application of FYM at 12.5 ton/ha + *Trichoderma viride* at 2.5 kg/ha before planting.
- Application of carbofuran 3G at 1 kg a.i./ha before planting.
- Stem cuttings to be dipped in 0.1 per cent *Pseudomonas fluorescens* at the time of planting.
- Use drip irrigation to minimize the spread of pathogens from infected plants to healthy plants.

14.6.2. Aswagandha, *Withania somnifera*

14.6.2.1. Root-knot Nematode, *Meloidogyne incognita*

The root-knot nematode, *M. incognita* (race-2) is an important limiting factor in the successful cultivation of ashwagandha (Pandey and Kalra, 2003). The nematode is very widespread affecting more than 80% plants in India.

i) **Symptoms:** The nematode infected plants typically show chlorotic, stunted, less branched with fewer and smaller leaves (Plate 14.12) and poor response to fertilizer and irrigation. Such symptoms usually are not noticeable until severe damage to root system has been done by the nematodes. Roots of such plant were severely galled (Plate 14.12) with reduced alkaloids. When stem touches the soil it was also found to be infested with root-knot nematode. The root-knot nematode infected plants are more likely to be killed early than healthy non-infested plants.

ii) **Management**

a) **Botanicals:** Neem compound, *Artemisia annua*, *Mentha* and *Murraya koenigii* distillates were found highly useful to suppress the root-knot development and *M. incognita* population on *W. somnifera.*

b) **Biological Methods:** The efficiency of *T. harzianum* was found to be comparable to that of carbofuran (RKI=2), followed by *P. lilacinus*, *Arthrobotryis oligospora* and neem compound. Besides reducing nematode infestation, the biocontrol agents also enhanced the growth of the plant (Table 14.8) (Sharma and Pandey, 2009).

Table 14.8. Effect of bioagents and botanicals on plant growth parameters and root galling in *Withania somnifera*

Treatments*	Fresh Weight (g)		Dry Weight (g)		Length (cm)		RKI
	Shoot	Root	Shoot	Root	Shoot	Root	
T. harzianum	38.66	8.50	10.60	2.40	30.50	18.07	2.00
P. lilacinus	46.00	10.60	10.63	3.17	36.33	17.37	2.33
A. oligospora	34.73	6.57	8.70	1.93	32.67	9.47**	2.66
Neem compound	33.66	7.47	8.00	1.87	26.87	12.33	2.66
Carbofuran	26.00	8.40	3.97**	2.20	17.33**	15.23	2.00
Controls							
Medium + NI	12.33	3.17	2.90	0.77	12.30	6.47	4.00
Medium + NU	23.66	6.77	5.13	1.83	26.37	13.27	—
No medium + NI	10.66	2.77	1.67	0.57	12.10	7.60	3.66
No medium + NU	18.50	4.93	4.90	1.33	14.83	10.53	—
CD (P=0.05)	**9.14**	**1.87**	**1.65**	**0.80**	**9.10**	**3.51**	**0.66**

* NI = Nematode inoculated, NU = Nematode uninoculated.
** Non-significant,

c) **Integrated Methods:** Integration of neem cake/ vermicompost/ cow urine with *T. harzianum* considerably reduced the root-knot nematode development and enhanced plant growth and yield. Maximum root-knot suppression was noticed in vermicompost with *T. harzianum* followed by mentha distillate with *Glomus aggregatum*. Highest increase in plant yield was recorded when the soil was amended with mentha/curry leaf distillates along with *T. harzianum*/*G. aggregatum* (Pandey and Kalra, 2005) (Table 14.9).

Table 14.9. Effect of organic materials and bioagents on growth and root-knot nematodes on Ashwagandha

Treatment	Dry plant weight (g)	Root-knot index (RKI)	Total nema. popn. (soil + roots)
Untreated – Uninoculated	8.5	—	—
Untreated – Inoculated	5.2 (-38.8)	4.00	6460
Carbofuran	8.0 (-5.9)	1.66	3012
Neem compost + Davana	23.8 (+62.4)	1.66	2820
Neem compost + Curry leaf	7.4 (-12.9)	1.33	2261
Neem compost + Vermicompost	12.0 (+41.2)	1.66	2649
Neem compost + *T. harzianum*	12.8 (50.6)	1.66	2674
Neem compost + *G. aggregatum*	14.3 (+68.2)	3.00	2964
Neem compost + *Mentha* distillate	14.2 (67.1)	1.66	2409
Davana + Curry leaf	10.8(+27.1)	1.33	2030
Davana + Vermicompost	14.1 (65.9)	3.00	2594
Davana + *T. harzianum*	9.6 (+12.9)	1.66	2110
Davana + *G. aggregatum*	10.3 (+21.2)	1.33	2050

Treatment	Dry plant weight (g)	Root-knot index (RKI)	Total nema. popn. (soil + roots)
Davana + *Mentha* distillate	13.2 (+55.3)	1.33	2150
Curry leaf + Vermicompost	14.0 (+64.7)	1.66	2440
Curry leaf + *T. harzianum*	15.0 (+76.5)	3.33	3252
Curry leaf + *G. aggregatum*	13.4 (+57.6)	3.00	2635
Curry leaf + *Mentha* distillate	15.3 (+80.0)	1.66	2404
Vermicompost + *T. harzianum*	14.4 (+69.4)	0.66	1400
Vermicompost + *G. aggregatum*	13.6 (+60.0)	1.99	2480
Vermicompost + *Mentha* distillate	13.9 (+63.5)	1.33	2000
G. aggregatum + *Mentha* distillate	15.6 (+83.5)	1.00	1784
CD (P = 0.05)	**0.71**	**0.01**	**465.70**

Root-knot indices were much lower in the treated plots indicating the effectiveness of the treatments in inhibiting the development of *M. incognita.* Substantial reduction in root galling was noticed in neem oil seed cake + *T. harzianum*, vermicompost + *T. harzianum*, cow urine + *T. harzianum* treated plots as compared to untreated control (Table 14.10). A significant and marked improvement in plant growth and yield was also noticed in plots treated with vermicompost + *T. harzianum*, cow urine + *T. harzianum*, neem oil seed cake + *T. harzianum* with maximum root yield obtained from vermicompost + *T. harzianum* treated plots (Pandey *et al.*, 2011).

Table 14.10. Influence of organic and biological amendments on plant growth, root yield and root galling in *Withania somnifera*

Treatments	Shoot dry weight (kg/m^2)	Root dry weight (kg/m^2)	Root-knot Index (RKI)
Control (zero fertilizers)	1.3f	0.15h	3.33a
Farmyard manure	1.8e	0.20g	1.66b
Trichoderma harzianum	2.3d	0.25e	0.66cd
Cow urine	2.7b	0.28d	0.83cd
Vermicompost	2.3d	0.29bc	1.33bc
Neem oil seed cake	2.5c	0.23f	1.16bc
Farmyard manure + *T. harzianum*	2.5c	0.23f	1.16bc
Cow urine + *T. harzianum*	2.8ab	0.30b	0.33d
Vermicompost + *T. harzianum*	2.9a	0.32a	0.66cd
Neem oil seed cake + *T. harzianum*	2.8ab	0.29bc	0.33d

Mean in each column followed by same letters do not differ significantly ($P \leq 0.05$) according to Duncan's multiple range test.

It is concluded that *T. harzianum* strain Thu when inoculated with vermicompost can be an environment friendly strategy to reduce the population of *M. incognita* on *W. somnifera* and to obtain higher root yields. Besides providing a complete nutrition to the plant, vermicompost supports the growth of *T. harzianum* in the rhizosphere and thus substantially restricting the nematode population build-up.

14.6.2.2. Root-knot, *Meloidogyne incognita* and Root Rot, *Fusarium solani* Disease Complex

Root-knot disease due to *Meloidogyne incognita* and the disease complex it causes in conjunction with other soil-borne pathogen like *Fusarium solani* is an upcoming disease which appeared in severe form on ashwagandha.

Gupta *et al.* (2004) reported occurrence of severe form of root rot disease complex in ashwagandha nurseries and regular crop under Lucknow (Uttar Pradesh, India) conditions. The disease was found to cause 30-50 per cent plant mortality. Further, investigations to characterize the infecting fungus led to identification of *Fusarium solani* as the causative organism and it was a new report of root rot disease complex in *Withania somnifera* from Uttar Pradesh.

In Karnataka, Mallesh (2008) noticed maximum disease incidence of 30.33% in Dharwad district followed by Bijapur district (21.55%). The pathogens, *viz.* *F. solani* and *M. incognita* were isolated from disease complex plots.

i) **Symptoms:** Pots treated with *F. solani* took 25 days for expression of typical wilt/root rot symptoms and *M. incognita* took still more days for infection and for gall formation.

   Infected plants exhibited loss of turgidity, withering and drooping of leaves and wilting of plants. Death and decaying of roots, brown to black discoloration of roots with pulpiness were observed (Plate 14.13). White cottony growth of the fungus was observed near collar region.

   Inoculation of root-knot nematode alone did not produce root-rot in ashwagandha. Maximum root-rot index (5.0) and more disease severity (4.0) was observed in case of ashwagandha in treatment *Meloidogyne* + *Fusarium* followed by *Fusarium* alone *i.e.*, the expression of symptoms was early and disease severity, root rot index was more. Plants infected with root-knot and root rot disease complex showed rotting, decaying and disintegration of roots.

ii) **Management:** Application of Talc-based PGPR strains RB50 and RB31 decreased root knot index and incidence of disease complex while increasing plant height, number of branches, fresh and dry biomass; and tuber yield in ashwagandha (Table 14.11and Plate 14.14) (Mallesh, 2008).

Table 14.11. Efficacy of talc formulations of PGPR strains on yield parameters and disease incidence in ashwagandha under field conditions

PGPR strains	Total Biomass (g)		Gall index	% disease incidence	% decrease over control	Fresh berry wt. (g/plant)	Fresh root yield (g/plant)
	Fresh	Dry					
RB 01	166.65	63.84	1.33	13.13	58.54	73.56	279.67
RB 10	144.00	51.30	1.93	26.67	15.78	38.19	210.00
RB 13	186.15	66.40	1.00	16.67	47.36	61.32	319.60
RB 22	135.83	50.83	1.46	23.33	26.33	40.00	214.00
RB 31	200.20	83.78	0.80	11.67	63.15	87.38	354.60
RB 43	158.47	69.75	1.40	18.33	42.12	63.44	252.80
RB 50	225.32	87.45	0.75	08.33	73.69	97.00	373.00
Control	111.86	46.92	3.06	31.67	—	35.13	185.60
CD 5%	**19.62**	**8.91**	**0.97**	**4.19**	—	**9.14**	**17.92**

Biochemical analyses in the above treatments showed elevated expression of defense enzymes (peroxidase, polyphenol oxidase and phenylalanine ammonialyase) and higher accumulation of phenolic compounds (activity being highest in respect of RB 50 treated plants) compared to respective inoculated checks. The suppression of disease complex in ashwagandha is largely due to aforementioned biocontrol mechanism as well as induction of systemic resistance by efficient PGPR strains (Lingaraju and Mallesh, 2010).

14.7. CONCLUSIONS

The incidences of different nematode pests on several ornamental and medicinal crops have shown an increasing trend. There were outbreaks of foliar nematode, *Aphelenchoides besseyi* on tuberose. There has been an outbreak of serious root-knot nematode problems in the recent years in flower crops grown under protected cultivation systems such as root-knot nematode, *Meloidogyne incognita* and *M. incognita* and wilt, *Fusarium oxysporum* f. sp. *dianthi* disease complex on carnation; and *M incognita* and *M. incognita* and foot rot, *Phytophthora parasitica* disease complex on gerbera. The emerging nematode pests among medicinal crops include *M. incognita*, *M. incognita* and wilt disease, *Fusarium chlamydosporum* complex, *M. incognita* and collar rot, *Rhizoctonia bataticola, Sclerotium rolfsii, Fusarium chlamydosporum* disease complex, and *M. incognita* and root rot, *Macrophomina phaseolina* disease complex on coleus; and *M. incognita* and *M. incognita* and *Fusarium solani* disease complex on ashwagandha. Hence, there is an urgent need for rapid and accurate detection, diagnosis, and development of management recommendations for these new nematode pests.

Section V
FUTURE THRUSTS

15

Pests likely to become Serious Threats in Future

15.1. INTRODUCTION

Pests can assume serious proportions in future due to changes in the ecosystems and habitats (Table 15.1). Recently, Tabashnik *et al.* (2008) has reported the occurrence of resistance in Bt cotton to *Helicoverpa zea* in Arkansas and Mississippi states of USA. There are also indications of development of resistance in *H. armigera* in parts of China (Wang *et al.*, 2009) and occasional reports of survival of *H. armigera* on Bt cotton in Australia (Downes *et al.*, 2007) and India (Kranthi *et al.*, 2009). Therefore, *H. armigera* may again reemerge as major pest of cotton if the resistance in Bt cotton breaks down. In addition, some alien and new pests could become invasive and some secondary pests could assume serious proportions. Some of the examples include sugarcane woolly aphid, *C. lanigera*; coconut eriophyiid mite, *Aceria guerreronis*; coffee berry borer, *Hypothenemus hampei*; sapota seed borer, *Trymalitis margarias*; spiralling whitefly, *Aleurodicus disperses* and serpentine leaf miner, *Liriomyza trifolii* (Puri and Ramamurthy, 2009). The recent epidemic of *S. litura* on soybean in Vidarbha region of Maharashtra in August 2008 and outbreak of brown plant-hopper, *Nilaparvata lugens* on basmati rice in Haryana and Western Uttar Pradesh during September- October 2008, were mainly due to favourable prevailing weather conditions (Bambawale *et al.*, 2009).

Table 15.1. List of insect pests likely to become serious in future due to changes in ecosystems and habitats

Insect pest	Crop(s)
American bollworm, *Helicoverpa armigera*	Cotton, chickpea, pigeon pea, sunflower, tomato
Whitefly, *Bemisia tabaci*	Cotton, tobacco
Brown plant hopper, *Nilaparvata lugens*	Rice
Green leaf hopper, *Nephotettix* spp.	Rice
Serpentine leaf miner, *Liriomyza trifolii*	Cotton, tomato, cucurbits, several other vegetables
Fruit fly, *Bactrocera* spp.	Fruits and vegetables
Mealy bugs (Several species)	Several field and horticultural crops

Insect pest	Crop(s)
Thrips (Several species)	Groundnut, cotton, chillies, roses, grapes, citrus and pomegranate
Wheat aphid, *Macrosiphum miscanthi*	Wheat, barley, oats
Pink stem borer, *Sesamia inferens*	Wheat
Gall midge, *Orseolia oryzae*	Rice
Gall midge (Several species)	Fruit crops
Diamondback moth, *Plutella xylostella*	Cabbage, cauliflower
Hoppers (Several species)	Rice and mango
Pyrilla, *Pyrilla perpusilla*	Sugarcane or rice at times
Polyphagous pests like termites, white grubs, hairy caterpillars and tobacco caterpillar (several species)	Many agro-ecosystems

Source: Modified from Puri and Ramamurthy (2009)

15.2. POTENTIAL IMPACT OF CLIMATE CHANGE ON EMERGING PESTS IN FUTURE

15.2.1. Increased CO_2 Concentration

- May induce plant growth and enhance the availability of food for the insects associated with their host plants.
- Under high growth rate of plants they may produce denser canopies with high humidity thus providing favourable microclimate.
- Expected to cause changes in physical or chemical defence, and chemical composition of their tissues, decrease in protein content and increase in C: N ratio may reduce the nutritive quality of food.
- Slow rate of plant material decomposition is expected to leave plant residue undecomposed for longer period which could favour the survival rate of hibernating insects.

15.2.2. Rising Temperature

- Host exposure to more number of pest generations per growing season due to high ambient temperature could trigger deactivation of host plant resistant gene.
- Increase in food quality due to increase in nitrogen under stress i.e. increased C: N ratio resulting in increased feeding of herbivores to derive more amino acids and sudden emergence of pests.
- Increase in temperature shorten the length of crop growing season for tropical and subtropical crops will reduce the developmental period and increase the number of generations resulting in increased pest load.
- Some cultivars could become more vulnerable to insect pests as a result of stress caused by higher temperature or other factors due to climate change.

- Some natural control mechanisms could 'decouple' insect pests and their natural enemies as they react differently to changes in temperatures. e. g. brown plant hoppers are 17 times more tolerant to 40 ^{0}C than its predator.
- New biotypes could evolve with greater possibility in multi-voltine insect pests.

15.2.3. Precipitation

- The water stress changes the plant process and its thermal environment in such a way that the stressed plants become more susceptible and suitable to certain phytophagous insects.
- The stressed plant become more susceptible through erosion of their defence system and may become more suitable due to accumulation of certain constituents, which lead to faster growth, higher survival and more reproduction of insects.
- The insect outbreaks are often preceded by the periods of unusual drought conditions.
- There could be population explosions in species that respond favourably to high rainfall or drought by increasing their feeding behaviour and/or breeding behaviour.

15.3. ANTICIPATED EFFECT OF CLIMATE CHANGE ON PESTS

Direct effects of climate change on herbivorous insects can be divided into the following two main categories:

- Geographical distribution range shift.
  - Natural expansion due to global warming.
  - Accidental introduction caused by the increase of world trade, resulting in the establishment of herbivorous insects outside of their natural distribution range.
- Increase of the number of generations per year.

Examples of plant pests whose geographical distribution and pest incidence and/or intensity will be substantially influenced by climate change (i.e. change in ecological conditions will substantially widen the area where the plant pest could be introduced and/or could substantially increase its potential economic importance) and of which economic (including environmental) consequences will be substantial. Some examples of insect and mite pests, disease pathogens, nematodes and weeds are given below:

15.3.1. Insect Pests

15.3.1.1. Geographical Distribution Range Shift

i) Colorado Potato Beetle, *Leptinotarsa decemlineata*

a) **Present Distribution:** North America, Europe (except Denmark, Finland, Norway, Sweden and the UK), and much of Asia, to the Pacific Ocean.

b) **Anticipated Effect of Climate Change:** Baker *et al.* (1998) estimated that for an average warming of 2.3°C, the Colorado beetle, *Leptinotarsa decemlineata*, could expand its potential range in the UK by 120%. This would extend the potential northern limit of the species by 400 km, thus posing a potential risk to more than 99% of potato producing areas.

ii) **Western Corn Rootworm, *Diabrotica virgifera virgifera***

a) **Present Distribution:** North America and spreading throughout Europe.

b) **Anticipated Effect of Climate Change:** *D. virgifera virgifera* was introduced into Europe near Belgrade in 1992, and has since been spreading eastwards and northwards. This major pest has not yet spread to its potential limits under existing climatic conditions, but global warming has the potential for increase its range considerably northwards. Baker *et al.* (2003) showed that under current climate conditions, *D. virgifera virgifera*, appears to be at the edge of its range in the UK, but by 2050, under global warming, a large area of SE England will be much more suitable for this species.

iii) **South American Leaf Miner, *Liriomyza huidobrensis***

a) **Present Distribution:** Cosmopolitan, but absent from a number of countries in Europe including UK, Denmark, Finland, Germany, Ireland and Sweden.

b) **Anticipated Effect of Climate Change:** Warmer winter temperatures are likely to increase the probability of the non-indigenous, South American leaf miner (*Liriomyza huidobrensis*) overwintering outdoors in the UK (Baker *et al.*, 1996).

iv) **Diamondback Moth, *Plutella xylostella***

a) **Present Distribution:** Probably of European origin, now cosmopolitan in distribution.

b) **Anticipated Effect of Climate Change:** At the northern extremity of its range, this species generally occurs at relatively low population densities and it does not usually overwinter. If however, as a result of global warming, overwintering occurred more frequently, the pest status of this species would increase dramatically (Dosdall, 1994).

v) **European Corn Borer, *Ostrinia nubilalis***

a) **Present Distribution:** Central and southern Europe, North America, northern India and western China.

b) **Anticipated Effect of Climate Change:** The effect of climate change for multivoltine species such as the European corn borer will be that it may be able to produce additional generations, relative to current conditions, in a given locale, with a potentially greater impact on their host plants. For example, *O. nubilalis* is predicted to become bivoltine – i.e. to produce two generations per season rather than one – in the Czech Republic as a result of predicted increases in temperatures during the period 2025-50 (Trnka *et al.*, 2007). This species is at present near the northern limit of its range in southern England. A 3°C increase in mean annual temperature would advance the limit for maize (*Zea mays*) providing an opportunity for a substantial northward shift in the distribution of *O. nubilalis* (Porter, 1995).

vi) **Old World Bollworm, *Helicoverpa armigera***

a) **Present Distribution:** Widespread throughout the tropics, subtropics and warmer regions of the Old World, including southern Europe and the Mediterranean.

b) **Anticipated Effect of Climate Change:** Migrant moths, such as *H. armigera* have also shown a "phenomenal increase" over the period 1969-2004, penetrating inland more frequently (Parsons and Davey, 2007), although this has not (yet) led to an increase in its pest status in the UK. However, there have been outbreaks in areas at the northern edge of its range in Europe, including the first finding in 2003, of *H. armigera* in Austria; and a damaging outbreak, also in 2003, in Baden-Württemberg, Germany, on a great diversity of plants grown outdoors (EPPO Reporting Service, 2004/012 and 2004/013).

vii) Cottony Cushion Scale, *Icerya purchasi*

a) **Present Distribution:** Cosmopolitan throughout the warmer regions of the world

b) **Anticipated Effect of Climate Change:** Outdoor populations appear to be naturally spreading northwards perhaps as a consequence of global warming. In 1999, a severe infestation was reported in the Jardin des Plantes, Paris and it has also recently been found breeding outdoors in London (Watson and Malumphy, 2004).

viii) The Cottony Camellia Scale, *Pulvinaria* (= *Chloropulvinaria*) *floccifera*

a) **Present Distribution:** Cosmopolitan

b) **Anticipated Effect of Climate Change:** *P. floccifera* has become much more common in the UK, extending its range northwards in England and increased its host range in the last decade or so. This is almost certainly a response to climate change. In Sweden, *P. floccifera* was previously only known as a greenhouse species, but is now established as an outdoor species (Gertsson, 2005).

ix) Laurel Scale, *Aonidia lauri*

a) **Present Distribution:** Mediterranean

b) **Anticipated Effect of Climate Change:** *Aonidia lauri* has been found breeding outdoors sporadically in the London area since about 1996 (Malumphy, 1997). It is likely to respond to climate change and become more common.

x) Cabbage White Butterfly, *Pieris brassicae*

a) **Present Distribution:** Europe, NW Africa and Asia. Introduced into N. America and Australia.

b) **Anticipated Effect of Climate Change:** Most butterflies are predicted to increase in diversity, range and abundance in the UK under a warmer climate, but according to Roy *et al.* (2001) *P. brassicae* would likely decline.

xi) Cabbage Root Fly, *Delia radicum*

a) **Present Distribution:** Widespread throughout Europe, N. Africa, W. Asia, and N. America

b) **Anticipated Effect of Climate Change:** A 3°C increase in mean daily temperatures would cause cabbage root fly (*Delia radicum*) populations to become active about a month earlier in the UK (Collier *et al.*, 1991), which may translate into increased pest severity.

xii) Fruit flies (Diptera: Tephritidae): Melon Fly, *Bactrocera cucurbitae*; Oriental Fruit Fly, *Bactrocera dorsalis*; Malayasian Fruit Fly, *Bactrocera latifrons*

a) **Present Distribution:** South of the Japanese Archipelago

b) **Anticipated Effect of Climate Change:** Invasion of the Japanese Archipelago

xiii) Southern Green Stink Bug, *Nezara viridula* (Hemiptera: Pentatomidae); Rice Bug, *Leptocorisa chinensis* (Hemiptera: Alydidae)

a) **Present Distribution:** Tropical and subtropical countries in the world.

b) **Anticipated Effect of Climate Change:** Northward range shift to the temperate zones in the world.

xiv) Asian Rice Gall Midge, *Orseolia oryzae* (Diptera: Cecidomyiidae)

a) **Present Distribution:** South of the Philippines.

b) **Anticipated Effect of Climate Change:** Northward range shift to the subtropical Asia.

xv) *Quadrastichus erythrinae* (Hymenoptera: Eulophidae)

a) **Present Distribution:** Tropical countries in Africa and Asia (found in Australia, Hawaii, Florida, Japan etc. in recent years).

b) **Anticipated Effect of Climate Change:** Expanding its distribution range and causing serious damages to *Erythrina* trees (Fabaceae) by inducing galls on twigs and leaves.

xvi) Sweet Potato Weevil, *Cylas formicarius* (Coleoptera: Curculionidae)

a) **Present Distribution:** Tropical and subtropical countries in the world

b) **Anticipated Effect of Climate Change:** Accidental introduction to the temperate zones and establishment there under warmer conditions.

xvii) West Indian Sweet Potato Weevil, *Euscepes postfasciatus* (Coleoptera: Curculionidae)

a) **Present Distribution:** Tropical and subtropical countries in the world.

b) **Anticipated Effect of Climate Change:** Accidental introduction to the temperate zones and establishment there under warmer conditions.

xviii) Blossom Midge, *Contarinia maculipennis* (Diptera: Cecidomyiidae)

a) **Present Distribution:** Southeast Asia (accidentally introduced to Hawaii, Florida, and Japan).

b) **Anticipated Effect of Climate Change:** Accidental introduction and subsequent establishment in green houses, expansion of host range from orchids to various sorts of crops.

xix) Asian Citrus Psyllid, *Diaphorina citri* (Homoptera: Psyllidae)

Transmits citrus greening disease.

a) **Present Distribution:** Tropical and subtropical countries in the world.

b) **Anticipated Effect of Climate Change:** Northward range shift to the temperate zones.

xx) **Plant Hoppers (Hemiptera: Delphacidae): Brown plant hopper, *Nilaparvata lugens*; White-backed plant hopper *Sogatella furcifera***

a) **Present Distribution:** Tropical and subtropical Asia and Australia.

b) **Anticipated Effect of Climate Change:** Northward range shift to the Temperate Zones (long distant migration).

xxi) ***Plautia crossota* (Hemiptera: Pentatomidae), *Glaucias subpunctatus* (Hemiptera: Pentatomiidae)**

a) **Present Distribution:** Tropical, subtropical, and temperate zones of Asia.

b) **Anticipated Effect of Climate Change:** Range shift to further north and overlapping of generations.

15.3.1.2. Increase in the Number of Generations per Year

i) **Aphids (Hemiptera: Aphididae): *Aphis gossypii, Myzus persicae, Aulacorthum solani***

a) **Present Distribution:** Worldwide.

b) **Anticipated Effect of Climate Change:** Increase of the number of generations a year due to lower developmental zero point and low thermal totals required for one generation, resulted in more infestation.

ii) **Lepidopteran Species (Lepidoptera: Noctuidae): *Helicoverpa armigera, Mamestra brassicae, Spodoptera litura, S. exigua***

a) **Present Distribution:** Worldwide.

b) **Anticipated Effect of Climate Change:** Increase of the number of generations a year, resulted in more infestation.

iii) **Mite Species: Particularly Rust Mites**

a) **Present Distribution:** Worldwide.

b) **Anticipated Effect of Climate Change:** Increase of the number of generations a year due to lower developmental zero point and low thermal totals required for one generation, resulted in more infestation.

15.3.1.3. Other Effects

There are a range of non-specific indicators, which could, and are being used as indicators of climate change, i.e. beyond the specific ones of increases in the geographical range or abundance of given species, in particular, phylogenetic changes. For example, the phenology of a range of agricultural and horticultural events in Germany, over the period 1951–2004, was estimated to have advanced on average by 1.1 – 1.3 days, per decade (Estrella *et al.,* 2007). Some of the best evidence for climate change is provided by non-pests, e.g. dragonflies and damselflies, are expanding their flight period, flying both earlier and later in the season (Hassal *et al.,* 2007). The same is true of a large number of butterflies, grasshoppers, bush crickets and wasps. Plants are flowering earlier, an average of 4.5 days earlier in the decade 1991-2000, compared to the previous 35 years (Fitter and Fitter, 2002). This presumably applies to crop plants as well, although the effects could be masked by varietal selection.

i) **Larch Bud Moth, *Zeiraphera diniana* (Lepidoptera: Tortricidae)**

   a) **Present Distribution:** Central European Alps.

   b) **Anticipated Effect of Climate Change:** Disappearance of periodical outbreaks, resulted in continuous outbreaks (Baltensweiler, 1993).

ii) **Pests in Greenhouses, such as Whiteflies, Aphids, Mites, etc.**

   a) **Present Distribution:** In greenhouses.

   b) **Anticipated Effect of Climate Change:** In Japan, previously pests in greenhouses were all domestic species, but in recent years about 50% of them are alien species. They came to Japan and can overwinter in greenhouses. Some of them will be able to go into the field from the greenhouses in future.

iii) **Camellia Scale (*Pulvinaria – Chloropulvinaria – floccifera*)**

   a) **Present Distribution:** In greenhouses.

   b) **Anticipated Effect of Climate Change:** In Sweden this species was previously only known as a greenhouse species, but is now established as an outdoor species.

iv) **Various Herbivorous Insects**

   a) **Anticipated Effect of Climate Change:** Increasing of C/N ratio affects feeding habit and development of herbivorous insects such as leaf hoppers, plant hoppers, etc.

v) **Aphids:** Aphid flight phenology is thought to be a good indicator of climate warming and Fleming and Tatchell (1995) predicted that over the next 50 years, aphids will appear at least eight days earlier in the spring – as measured by their presence in suction traps. This may increase their severity as pests, although the extent to which this will affect their pest status will also depend partly on how the phenologies of their host plants (and natural enemies) change as well (Harrington *et al.,* 2007). The green spruce aphid, *Elatobium abietinum* could be at an advantage, in that a longer period favourable for growth would be available (Straw, 1995). Zhou *et al.* (1996) concluded that climate change-induced changes would increase the severity of aphid outbreaks and extend the period of virus infection further into the growing season in the UK. The Green citrus aphid, *Aphis spiraecola* is a sporadic, transient visitor to the UK, which survives in sheltered conditions and/or in warm years. It is possible that *A. spiraecola* may be expanding its range as a result of climate change.

vi) **Migrant Pest Species:** In many northern European countries, populations of many migrant insect species occur seasonally, as a result of immigration from more favourable regions to the south. Migrants will be able to respond quickly to climate change and will be difficult to exclude if they move naturally. The numbers of migratory butterflies arriving in the south of England has increased steadily over the period 1992-2005 (Sparks *et al.*, 2007), and moths, such as the important pest species, *Helicoverpa armigera* have also shown a "phenomenal increase" over the period 1969-2004, penetrating inland more frequently (Parsons and Davey, 2007).

15.3.2. Disease Pathogens

According to Chakraborty *et al.* (2000), "Climate change could have positive, negative or no impact on individual plant diseases" and more research is needed to obtain base-line information on different disease systems. These authors concluded that despite the significance of weather on plant diseases, a comprehensive analysis of how climate change will influence plant diseases in agricultural systems is presently unavailable. Some discussion of the impact of climate change on both indigenous and non-indigenous diseases in the UK is contained in Hardwick *et al.* (1996).

Increased rainfall correlated with a high incidence of canker (*Leptosphaeria maculans*) in oil seed rape, whilst temperature was an important factor in the development of brown rust (*Puccinia hordei*) in cereals.

There are a number of quarantine diseases, which might become more serious under global warming, including brown rot, *Ralstonia solanacearum* and sudden oak death, *Phytophthora ramorum.* Incidences of brown rot in Europe are primarily caused by R. *solanacearum* race 3 biovar 2A, which is reported to have strains adapted to cooler conditions. There are also other phylotypes of R. *solanacearum,* originating from Asia and South America, which would have a greater likelihood of establishment under predicted changes in climate, thereby threatening a much wider range of host plants. Wright and Woodhall (2006) considered that the impact of P. *ramorum* might increase through climatic changes producing more optimal conditions for the pathogen.

Dickeya dadantii (syn. *Erwinia chrysanthemi*, *Pectobacterium chrysanthemi*) is a phytopathogenic Enterobacteriaceae that can cause soft rot diseases in a wide range of economically important crops. Laurila *et al.* (2008) reported for the first time that *Dickeya*, originally known as a pathogen in tropical and warm climates, may cause diseases in potato in northern Europe.

Waage *et al.* (2006) identified a substantial increase in the rate of new disease reporting over last decade, possibly as a consequence of globalization, although the influence of global climate change could also be a contributing factor. In another of the UK Government's Foresight project studies, 'The Effects of Climate Change on Infectious Diseases of Plants' Chancellor and Kubiriba (2006) concluded that whilst the 'geographical distribution of some plant hosts and disease pathogens will alter as a result of future changes in climate and, especially in Africa, this effect is likely to be strongest in areas of significant water stress'. However, they emphasised that the absence of reliable epidemiological information 'makes predictions hazardous'.

Barker *et al.* (2006) suggest that plant diseases 'will continue to be characterized by the extreme variety of hosts and the extreme variety of pathogens', which they suggest means there will be a need for generic broad-spectrum diagnostic tests. Such devices will need to screen larger numbers of plant products than at present, and be faster and cheaper than current methods.

15.3.2.1. Fusarium Head Blight, *Fusarium graminearum, F. culmorum* and Related Species

i) **Present Distribution:** Northern America, Southern Cone (Southern Brazil, Argentina, Paraguay), China (Yangtse River basin), Northern Europe.

ii) **Anticipated Effect of Climate Change:** More people shall be subjected to hidden mycotoxin infections. Reduction of incomes is the possibility for some countries. With increasing rainfall at critical time of the growing season, more wheat areas will be subjected to mycotoxin infections and this can affect trade since international regulations on authorized mycotoxin contents are becoming more stringent. Climate change expected to affect cropping systems with more maize being grown in some areas and pathogen surviving on a new crop in the rotation. Expansion of conservation agriculture and maize are also likely to change the pathosystem composition. In North Africa, unusual high rainfall is already showing more Fusarium head blight outbreaks. Indian Sub-continent is currently spared from rains and thus of Fusarium head blight during wheat growing season. If 'sporadic' rains occur more frequently with climate changes, as a result of mycotoxin contamination, Fusarium head blight may turn-out as a huge problem for public health (food) and industry (feed). See risks in Bangladesh too. More fungicides will also be used.

15.3.2.2. Wheat Blast, *Magnaporthe grisea*

It emerged in Brazil in the mid-80's. The disease first considered related to rice blast appears originated probably from a grass. Host range is broad and it is seed transmitted.

i) **Present Distribution:** Southern Brazil, Bolivia, and Paraguay.

ii) **Anticipated Effect of Climate Change:** Heavy losses may occur due to absence of resistance. Adaptation of the pathogen in warm environments (i.e. South Asia) would be a disaster. Increased yield losses, infestation in new areas due to ability/possibility of pathogen to survive in unknown alternate hosts.

15.3.2.3. Wheat Spot Blotch, *Cohliobolus sativus*

i) **Present Distribution:** Southern Brazil, Bolivia, Paraguay, and Eastern India.

ii) **Anticipated Effect of Climate Change:** Heavy losses may occur due to insufficiency of satisfactory resistance. Adaptation of the pathogen is likely to occur in warm environments (i.e. South Asia). More areas will be affected in South Asia as 50% of areas considered presently under favourable wheat growing conditions will become more subject to heat and water stress. Observations over 6 years in India, Nepal and Bangladesh have already shown that climate change is taking its toll as a result of the increase of average night temperature penalizing grain filling and increasing the spot blotch severity. This effect is likely to increase in more optimum growing areas.

15.3.2.4. Wheat Yellow rust, *Puccinia striiformis*

i) **Present Distribution:** Cool wheat growing areas of Andean region, North America, western China, Northern Europe, Middle East and Central Asia.

ii) **Anticipated Effect of Climate Change:** New severe epidemics are likely to occur due to change of races. Indications exist that the pathogen has become more tolerant to heat. Heat average increase triggering races adapted to new thresholds of temperature requirements. More areas are affected by epidemics or epidemics have become more frequent.

15.3.2.5. Wheat Leaf Rust, *Puccinia tritici*

i) **Present Distribution:** North and South America, Europe, North Africa, Russia, Middle East, and Indian Sub-continent.

ii) **Anticipated Effect of Climate Change:** New epidemics may occur due to change of races. In south Asia, the disease tends to appear late in the growing season when temperature is rising. Climate change can lead to disease spread in areas where its importance is minor at present, e.g. facultative and winter wheat growing areas of China, parts of Europe, Pacific Northwest of US, winter and facultative wheat areas of Central Asia.

15.3.2.6. Wheat Stem Rust, *Puccinia graminis*

i) **Present Distribution:** North and South America, China, Middle East, Indian Sub-continent, and Australia.

ii) **Anticipated Effect of Climate Change:** New epidemics may appear due to change of races and lack of genetic resistance. Change in heat average and relative humidity may trigger races adapted to new thresholds of environmental requirements. There is a study done in the UK indicating where in Europe Ug99 can become important with global warming.

15.3.2.7. Indian Peanut Clump Virus (IPCV)

i) **Present Distribution:** Restricted to a few areas (i.e. Rajasthan, southern India) where wheat is grown on peanut fields.

ii) **Anticipated Effect of Climate Change:** Yield losses are likely to occur. Transmitted by seed and soil fungi (Polymixa). No resistance in wheat or largely unknown. Climate change is likely to affect cropping systems patterns. Changes in rotations or cropping sequences due to high cereal prices may induce new infections of field or spread of the disease.

15.3.2.8. Barley Yellow Dwarf Virus (BYDV)

i) **Present Distribution:** Northern America, Ecuador, China, Pakistan, and Northern Europe.

ii) **Anticipated Effect of Climate Change:** Yield losses may occur. Not enough resistance in wheat. Resistance on virus diseases in developing countries is presently largely under-funded, delaying to capacity of reaction against the problem. Climate change (drought) will increase the survival and spread of the aphids. In Pakistan it has been observed that some years, dry air favours aphid flights.

15.3.3. Weeds

15.3.3.1. Ragweed, *Ambrosia artemisiifolia*

i) **Present Distribution:** An invasive North American annual (Asteraceae), which occurs in many European countries.

ii) **Anticipated Effect of Climate Change:** Ragweed plants, *Ambrosia artemisiifolia* produce highly allergenic pollen. A doubling of the atmospheric CO_2 concentration stimulated ragweed-pollen production by 61%, which suggests that there may be significant increases in exposure to allergenic pollen under global warming (Wayne *et al.,* 2002). Rogers *et al.* (2006) also showed that ragweed pollen production is expected to increase significantly under predicted future climate conditions.

15.3.4. Nematodes

There is a close correlation between soil temperatures and the distributions of some plant-parasitic species, such as *Longidorus caespiticola* (Boag and Neilson, 1996). A 1° C rise in temperature would allow this species to become established further north in Great Britain. *Meloidogyne incognita* was recently found in The Netherlands, a species that previously was considered as limited to the Mediterranean area.

The burrowing nematode, *Radopholus similis* occurs only below ~1,400 m altitude in the East African Highlands where it is a principal pest of banana and plantain, a regional key starch staple for over 20 million people. A small raise in temperature would result in *R. similis,* which is cold-sensitive, infecting millions more bananas grown at higher altitudes. Similarly, the rice root knot nematode, *Meloidogyne graminicola,* can be maintained under damaging levels through good water management. However, with reduced availability of water following climatic changes and/or competition for urban use, reduced quality of water management, or the introduction of water saving mechanisms such as direct wet seeding is favouring the development of high populations of *M. graminicola,* drastically raising its economic significance as a damaging pest (de Waele and Elsen, 2007).

The pine wood nematode, *Bursaphelenchus xylophilus,* is associated with pine wilt disease but depends on bark beetles (*Monochamus* spp.) to spread from tree to tree. It is native to North America and is thought to have been carried to Japan at the beginning of the twentieth century on timber exports. In Japan, it is causing massive mortality of native pine trees. In 1999, *B. xylophilus* was found in Europe for the first time, in Portugal.

15.4. CONCLUSIONS

Several pests and diseases can assume serious proportions in future due to changes in the ecosystems and habitats. *Helicoverpa armigera* may again reemerge as major pest of cotton if the resistance in *Bt* cotton breaks down. In addition, some alien and new pests could become invasive and some secondary pests could assume serious proportions. There is clear evidence that climate change is altering the distribution, incidence and intensity of plant pests and diseases. Major drivers of emerging pest change include increases in temperature, variability in rainfall intensity and distribution, change in seasonality, drought, CO_2 concentration in the atmosphere and extreme events (e.g. hurricanes, storms), intrinsic pest characteristics (e.g. diapause, number of generations, minimum, maximum and optimum growth temperature of fungi, interaction with the host) and intrinsic ecosystem characteristics (e.g. monoculture, biodiversity). Drivers for the spread of transboundary plant pests and diseases are international trade and traffic. The increase in movement of people, animals, plants, goods and conveyances has accelerated the redistribution of plant pests and diseases and climate change will create new ecological niches allowing for the establishment and spread of pests and diseases into new geographical areas and from one region to another. This expansion will continue to result in huge financial losses and require large eradication programmes and control measures. New vectors, selection and recombination of disease genotypes may occur when plant species and varieties mix or when insect pests and vectors are introduced without their natural enemies. Change in climate resulting in changes in species composition and interactions will augment the emergence of new diseases and pests.

Additional information and research is required on distribution of plant pests and diseases, and on their epidemiology. In particular there is a need for better surveillance methodologies; fast and cheap identification methods, epidemiological knowledge; and information on biological control organisms and mechanisms; resistant crop species. Coordinated research on programmes related to climate change and food security, will be needed to improve the range of options available to countries. Better accessibility and analysis of existing historical data and more detailed data for all regions in relation to different climate change scenarios will improve the baseline studies needed to assess adaptation.

16

The Way Forward

16.1. INTRODUCTION

A major challenge facing agricultural scientists today is the need to secure food for an increasing world population. This growth occurs predominantly in developing, mostly tropical countries, where the majority of hungry people live. Reducing yield losses caused by pests of tropical agricultural crops is one measure that can contribute to increased food production. They cause extensive damage and substantial yield losses. The effects of agricultural, environmental, socioeconomic, and policy changes on the occurrence of crop pests in the tropics and the losses these pathogens cause are largely undocumented. The absence of sufficient data on the impact of pest species on agricultural crops prevents the timely identification of emerging pest problems.

Crop production provides about 84 per cent of global food, feed and fibre needs and virtually all other human endeavours depend on food security. Global crop losses due to trans-boundary pests are estimated at more than 50 per cent of potential crop output. Insect pests are estimated to have caused the destruction of 15 per cent of crops, and 13 per cent each by pathogens and weeds, and post-harvest pest infestations another 10 per cent. Losses of 100 per cent occur in specific circumstances, creating enormous variability in productivity and risks to the livelihoods of farm families. Most threatening are outbreaks of migratory pests because of their sudden emergence, often without or with limited warning, triggered by changing ecological conditions or agricultural practices which favour an explosive increase in the pest population and its rapid spreading into neighbouring areas. The unexpected appearances of such massive pest outbreaks hit the countries which in most cases are ill prepared, with no or only poor capacities to effectively cope with rapidly evolving and fast moving threats. New and re-emerging plant pests have raised fear of their potential impact on livelihoods, food security and global markets. Crop pests, including disease pathogens, insect pests, and weeds should be overcome to meet future food needs.

Emergence of crop pests has been affecting crop production from time immemorial. In fact, they were among the ten plagues of ancient Egypt. As a result of drastic reduction of crop productivity, farmers were periodically confronted with famines, hunger and distress, but because of the extensive character of production systems and relative isolation of production areas and countries at that time, the magnitude of subsequent crises was relatively limited: the spread of pathogens was quite slow.

Nowadays, the movement of plant pests and diseases across boundaries has given rise to global threats to food security and agricultural development and trade and often creates global concern especially when such diseases and pests affect commercial crops. Increased movement of people, plants and products in the globalized economy on the one hand, and the concentration and intensification of production systems on the other, have accelerated and enlarged redistribution of plant pests with a clear tendency to expand to all regions of the globe. In addition, climate change is creating new ecological niches for the (re)emergence and spread of pests and diseases. As a result, the impact of crop pests has considerably increased. There are indications that the poorest countries with vulnerable sanitary regulation and infrastructure carry a larger burden than others. The list of imported exotic diseases and pests has grown while no substantial progress has been made in the control of local entrenched pests. Poor countries and producers may perceive different risks and incentives associated with crop pests, and it is essential to recognize these differences in the design and implementation of pest prevention and control measures. It is in the interest of the international community to avoid creating country or regional reservoirs for crop pests that will maintain the threat at global level.

Plant pests reduce the availability of food and affect quality. Although quantification of losses and potential losses due to trans-boundary pests and diseases is limited, historically, plant pest introduction, establishment, re-emergence and outbreaks have resulted in major food problems either directly through yield reductions of food crops, or indirectly through the reduction of yields of cash crops and loss of consumer confidence, e.g. potato blight or locusts.

16.2. POSSIBLE TECHNICAL AND POLICY RESPONSES

Climate change will result in a higher probability of entry, establishment and spread of pests and diseases of crops, for the following reasons:

- Climate change will create winners and losers. For some plant pests and diseases, the climate will become more conducive and for others the meteorological conditions will become less favourable. This will result in unstable situations with a high probability of entry and establishment in areas that are presently protected by unsuitable conditions.
- Meteorological and related environmental circumstances may change the geographical distribution of host species, putting them in contact with plant pests and diseases of related hosts to which they do not possess resistance.
- New plant pests and diseases may emerge due to evolving selection and adaptation to new situations.

16.2.1. Data for Projecting Risk

Projections of future distribution, severity and incidence can be made for plant pests and diseases that are presently of quarantine significance, using various risk analysis methods and tools. However, the combination of climate change, new environments and new ecological conditions and limited data that pertain to these situations make the assessments of future situations less reliable. Emergence of plant pests and diseases rarely can be foreseen, and lack of reliable data will make projections of the potential spread of such plant pests and diseases highly unreliable. Changes in rainfall, very complex to foresee, will have a major impact on outbreak and plague patterns of migratory plant pest species, in particular on locust species which are totally dependent on moisture and temperature.

16.2.2. Early Warning and Prevention Strategies

The strategy to address trans-boundary plant pests and diseases is prevention, early warning including forecast, early detection, early control and research. Investments in early control and detection mechanisms will be critical, to avoid higher costs of eradication and control. Prevention and early warning requires a reduction of the possibilities of entry and establishment and can be accomplished through better border control and rapid diagnostic tools for better surveillance of plant pests and diseases. To be successful, surveillance systems require monitoring and input from farmers as well as government services. Prevention and early warning also requires cooperation of countries within the same geographic/eco-climatic region to ensure better monitoring of plant health in the region. FAO's Emergency Prevention System for Trans-boundary Plant Pests and Diseases (EMPRES) provides support to governments in all of these areas.

16.2.3. Eradication, Containment, and Impact Reduction

Once plant pests and diseases establish, an immediate decision is required on follow-up action. Where possible and economically feasible, eradication and containment should start as early as possible. Countries need appropriate emergency capacity to take action as well as regional infrastructures that can support and coordinate action among countries. Joint action of countries in the same region is an absolute necessity. Where eradication and containment is judged not feasible, actions should be undertaken that will reduce the impact of the introduced plant pest or disease. These actions can include changes in agronomy, introduction of new varieties, carefully considered introduction of biological control agents and integrated pest management, all of which are within the framework of overall autonomous and planned adaptation as identified in the Fourth Assessment Report of the IPCC (IPCC, 2007: Climate Change 2007).

16.2.4. Information Options

The effects of climate change on migratory pests will most likely require that new areas have to be surveyed in different time periods with control capacity available at different periods of the year and more locations than at present. Addressing changing locust situations will require better surveillance and monitoring, and earlier control.

Additional information is required on distribution of plant pests and diseases, and on their epidemiology. In particular, there is need for better surveillance methodologies; fast and cheap identification methods; epidemiological knowledge; and information on biological control organisms and mechanisms and resistant crops. Coordinated research, including the Consultative Group on International Agricultural Research (CGIAR) programmes related to climate change and food security, will be needed to improve the range of options available to countries. Better accessibility and analysis of existing historical data and more detailed data for all regions in relation to different climate change scenarios will improve the baselines needed to assess adaptation.

16.2.5. Food Trade Industry

The introduction of diseases and pests will result in higher costs to national food industry in relation to inspection, treatment and compliance with obligations of the importing trading partners. Trade disputes in the WTO systems could become more frequent. Information exchange mechanisms exist at global and regional levels. At national level, there are many national databases as well as those maintained by non-governmental organizations (NGOs) and universities. However, data are of varying quality and are often incomplete or out of date. As provision of data on distribution of plant pests and diseases can be perceived as harmful to trade interests, provision of such data needs cooperation and commitment of all parties. To enable risk assessment, prevention, monitoring, early detection and warning, and control, there is a need for an overall global data exchange mechanism covering the distribution of diseases, pests, and correlated ecological conditions including climate. In order to improve information exchange, it will be necessary to increase cooperation among national, regional and global organizations, specifying the data required and the safeguards that should be applied to protect national interests. Government agencies and relevant stakeholders should come together and discuss specifications and sustainable systems for practical use.

16.2.6. Government Constraints

The national plant protection infrastructure, in particular in developing countries, is often unable to execute the range of activities required for prevention, early warning and early control of trans-boundary plant pests and diseases. National systems are often fragmented among agencies and ministries. The high level of uncertainty and the concomitant requirements for better legislation, increased risk analysis, better border control and the increase in requirements for eradication, containment and adaptation are beyond the possibilities of most plant health services, in particular those of developing countries. Learning and sharing lessons from failures as well as from successes is especially important at national level. Governments also need to be aware of the importance of maintaining capacity for dealing with new plant pests and diseases. Those that reduce funding when there is no crisis often suffer later from failure to maintain capacity.

At present most countries have insufficient enabling legislation and resources allocated to:

- Surveillance and monitoring
- Border control and inspections
- Expertise in risk assessment
- Diagnostic tools for early detections
- Expertise in diagnosis (taxonomy)
- Data collection and access to information
- Tools for rapid response to entry, establishment and spread
- Control measures at the source of the produce.

16.2.7. Government Priorities

A top priority for dealing with plant pests and diseases is strengthening national plant health services and plant health systems through capacity building. This includes improvement of infrastructure, border control, better legislation and enforcement, and better surveillance. Other priorities include improving the ability to respond to movements of plant pests and diseases through increasing preparedness, ensuring maintenance of expertise and adopting rapid diagnostic tools and forecasting models. Investment in capacity building will contribute to reduction of emerging plant pests and diseases at source. Basic sciences such as climate change science, taxonomy, modelling, population ecology and epidemiology should be given highest priority by governments.

Resources to address plant health are often distributed among national ministries and agencies. In a number of countries, there is a move to establish "biosecurity" agencies that bring several of these functions together. In view of the additional strains climate change puts on these systems, governments may wish to design and implement national strategies that capture synergies across the agencies and entities responsible for managing plant pests and diseases, and consider moving towards biosecurity approaches.

16.2.8. Ecosystem Processes

Climate affects both local and regional ecosystem processes and production. Many threats are trans-boundary and countries will not be able to address these issues individually. Regional cooperation is a high priority for risk analysis, regional standard setting, exchange of information and coordinated action. Countries should examine and, where appropriate, strengthen their regional organizations and cooperation in plant health. Regional and sub-regional organizations and cooperation exists but coverage, functions and efficiency vary among regions. In such cooperative frameworks, issues such as standard setting, joint risk assessment, joint action and access to information should be carefully considered and, where required, organizations should be strengthened.

16.2.9. Global Frameworks

Global regulatory frameworks are provided by the World Trade Organization (WTO), the FAO International Plant Protection Convention (IPPC) and the Convention on Biological Diversity. OIE and IPPC also provide standard setting mechanisms for plant health. IPPC and OIE have the structures to serve adequately under climate change scenarios, but their resources are limited.

Relevant global organizations should seek further cooperation in appropriate fields through information exchange and capacity building. The Standard and Trade Development Facility, a WTO-hosted joint programme of FAO, OIE and World Health Organization, is a good example.

16.3. POLICY CONSIDERATIONS

Trans-boundary plant pests represent a serious threat to global food security in all of its dimensions – availability, stability, access and utilization (safety). They reduce production and productivity, disrupt local and national economies, threaten human health and exacerbate poverty. They can also result in huge financial losses and require prevention and control measures and eradication programmes.

16.3.1. Disaster Prevention

More cost effective, livelihood saving and ecologically less devastating prevention strategies require timely and coordinated reaction from national, regional and international partners and the ability to mobilize resources at short notice to address both life and livelihoods risks as well as environmental concerns. Emergency prevention can be defined as actions taken in anticipation of an emergency to facilitate rapid and effective response to a threat and aims at:

- Early warning and detection systems providing relevant and timely information and analysis for effective decision-making.
- Governments and relevant partners prepared to prevent, mitigate and manage threats to agriculture, food and health.

16.3.2. Capacity Development

Strengthening the capacities of national plant health services and systems should be a top priority in coping with trans-boundary bio-threats in a more effective manner. This includes development of self reliant training capacities, development of appropriate legal and policy frameworks, introduction of economic and environmentally sound coping technologies, building national surveillance and diagnostic systems, support to adequate infrastructure and tools, and practised contingency plans and field exercises.

16.3.3. Protecting Livelihoods

To address the challenges of large-scale emergencies caused by trans-boundary bio-threats more effectively and to provide better coordinated and timely assistance to the affected countries, several considerations are critical to protect livelihoods, health and nutrition, and food security:

- Disaster risk reduction and prevention should have priority and should ensure cross border multi-stake holder programmes and coordinated approaches.
- The international community and the public need to be made aware that investment in emergency and disaster prevention is economically beneficial and reduces pressure on already vulnerable communities.
- Support should be given to build human capacities, enhance networking among stakeholders, regional and interregional cooperation between affected countries, and to develop self-reliant surveillance systems, knowledge bases, tools and standards to implement trans-boundary management concepts.
- Regional organizations and research institutes should play a more active part to systematically address trans-boundary plant health issues.
- The scientific community and industry should be involved in developing environmentally friendly control technologies and strategies.
- All stakeholders, including producers at all levels and capacities, must be engaged in the design and implementation of disease and pest prevention and control measures to protect their livelihoods and to ensure their participation.

- Partnerships and coordination among relevant international, regional and national organizations as well as financial institutions should be encouraged and fostered.

16.4. ROADMAP AHEAD

- New pest problems are likely to crop up in near future due to diversification in agriculture. Hence, there is a need to maintain close vigil on these problems and studies on pest surveillance, host parasite relationship and crop loss estimation against such pests of economic importance has to be taken up on priority basis.
- In the present regime WTO and of IPR, there is a need to generate epidemiological data on important pests so as to fix their tolerance limit for developing and identifying pest risk analysis and pest free paradigms of such pests. Hence, the first and foremost option for pest management is to prevent the introduction of pest problems in the area where these do not exist.
- There is a need and scope for intensification of researches in areas of organic farming, use of bioagents, plant growth promoting rhizobacteria (PGPR), host resistance and botanical pesticides in developing integrated pest management technologies.
- In the present age of molecular biology and biotechnology, there should be more emphasis on basic researches at cellular/ molecular levels to understand the aspects of host parasite relationship at cellular level and mechanism of host resistance which will ultimately lead to exploitation of genetic diversity in crop improvement and development of pest resistant cultivars.
- The organic content in the soil not only influence the plant health but also has direct bearing on the activities of soil inhabiting pests. The soils rich in organic content support higher saprozoic, predatory and microvorous microbial activities. The higher microbial is known to adversely affect the activities of pests. Hence, it is necessary to investigate the incorporation of organic amendments in the form of farm yard manure (FYM) or green manuring which are invariably recommended for better soil health and also for reducing the populations of crop pests.
- Integrated pest management technologies which are cost effective and eco-friendly against key pests should be transferred to the farmers through field level demonstrations in collaboration with the Krishi Vigyan Kendras.
- There is a need for strengthening of crop pest quarantine facilities.

16.5. CONCLUSIONS

The ever increasing human population require increased amount of food and fibre from decreasing agricultural land. Modern agriculture has largely achieved these objectives but intensive cultivation of high yielding varieties and use of agrochemicals in geometric rate has resulted in environmental deterioration and aggravation of pest problems. Many reported or hitherto unknown species have assumed serious status and some of the serious pests have developed resistance to one or more group of pesticides. On the top of this, we have to increase rather than maintaining production and productivity levels from existing agricultural land. This will put further strain on our fragile agro-ecosystem. The crop protection scientists have to work hand in hand with other agricultural scientists and farmers to meet this need of an hour. The emerging and future pest problems have to be tackled with IPM approach as a part of a sustainable crop production technology.

References

Abd-Elgawad, M.M.M. (2014). Plant-parasitic nematode threats to global food security. *Journal of Nematology* 46(2): 130–260.

Abdel-Momen, S.M. and Starr, J.L. (1997). Damage functions for three *Meloidogyne* species on *Arachis hypogaea* in Texas. *J. Nematol* 29: 478-483.

Abdul Karim, S., Yuen, P.M. and Norlia, Y. (2002). Challenges in the integrated management of the root-knot nematode on guava. *Proceeding Mesyuarat Teknikal Pusat HR Port Dickson*: 27.

Abhary, M.K., Anfoka, G.H., Nakhla, M.K. and Maxwell, D.P. (2006). Post-transcriptional gene silencing in controlling viruses of the Tomato yellow leaf curl virus complex. *Archives of Virology* 151(12): 2349-2363.

Adegbola, M.O.K. (1993). Determination of the most suitable frequency and time of fungicide spraying schedule for the adoption in the control of *Phytophthora* black pod (pod rot) disease of Cocoa. In: *Proceedings of the 11th International Cocoa Research Conference*, pp 18–24. Yamoussoukro, Cote d'Ivoire.

Adejumo, T.O. (1997). *Identification, Incidence, Severity and Methods of Control of the Causal Organism of False Smut Disease of Cowpea* (Vignia unguiculata *L Walp.*). Ph. D. thesis, University of Ibadan, Nigeria, 201 pp.

Aegerter B.J., Nuñez J.J. and Davis R.M. (2003). Environmental Factors affecting rose downy mildew and development of a forecasting model for a nursery production system. *Plant Disease* 87(6): 732-738.

Agrios, G.N. (1997). *Plant Pathology*. Academic Press.

Agyekum, E. (2004). *The Effect of Plant Density and the Relative Time of Planting on Colocasia/Rice Intercropping System*. MSc thesis, Faculty of Agriculture and Renewable Natural Resources, Kwame Nkrumah University of Science and Technology, Kumasi, Ghana.

Ahmed, T., Ansari, M. and Ali, H. (2009). Outbreak of diamondback moth, *Plutella xylostella* in Aligarh, India. *Trends Biosci.* 2(1): 10-12.

Akad, F., Eybishtz, A., Edelbaum D., Gorovits, R., Dar-Issa, O., Iraki, N. and Czosnek, H. (2007). Making a friend from a foe: expressing a GroEL gene from the whitefly Bemisia tabaci in the phloem of tomato plants confers resistance to tomato yellow leaf curl virus. *Archives of Virology* 152(7): 1323-1339.

Alam, S.K.F., Atanu Seni. and Sahoo, A.K. (2011). Biology of the mealybug, *Phenacoccus solenopsis* Tinsley (Pseudococcidae: Hemiptera) on sprouted potato and brinjal plant. *The Journal of Plant Protection Sciences* 3(2): 32-36.

Aleksandrova, I.V. (1981). Decontamination of seed rice from *Aphelenchoides besseyi* Christie, 1942 by combined treatment with chemical and irradiation. *Byulleten' Vsesoyuznogo Instituta Gel'mintologii im. K.I. Skryabina* 31:105.

Ali, M., Kumar, S. and Singh, M. (eds.) (2009). *25 Years of Pulses Research at IIPR 1984-2009.* Indian Institute of Pulses Research, Kanpur, Uttar Pradesh, India.

Almeida, A.M., Gomes, V.M. Souza, R.M. (2011). Greenhouse and field assessment of rhizobacteria to control guava decline. *Bragantia* 70: 837–842.

Almeida, E.J., Santos, J.M. and Martins, A.B.G. (2009). Resistance of guava and araca to *Meloidogyne mayaguensis. Pesq. Agropec. Bras.* 44: 421-423.

Al-Musa, A. (1982). Incidence, economic importance, and control of tomato yellow leaf curl in Jordan. *Plant Disease* 66(7): 561-563.

Alphey, T.J.W., Phillips, M.S. and Trudgill, D.L. (1988). Integrated control of potato cyst nematodes using small amounts of nematicide and potatoes with partial resistance. *Annals of Applied Biology* 113: 545-552.

Amalraj, S.F.A. and Rao, G.R. (1992). Potato cyst nematode pathogens in the southern hills of India. *Indian Journal of Nematology* 22: 82-85.

Amarasinghe, L.D. (2012). An integrated approach to the management of rice root knot nematode, *Meloidogyne graminicola* in Sri Lanka. *Journal of Science of the University of Kelaniya Sri Lanka* 6: 55–63.

Ameta, O.P. and Kumar, A. (2005). Field evaluation of newer insecticides for the management of white fly *Bemisia tabaci* in cotton. *Indian Journal of Applied Entomology* 19(1): 58-60.

Amosa, F. and Wati, P. (1997). Effects of taro/maize intercropping systems on the incidence of and severity of taro leaf blight. In: *Taro Genetic Resources: Conservation and Utilisation*, pp 1–2. 1995 Annual Research Report, The Institute for Research, Extension and Training in Agriculture (IRETA) and the School of Agriculture (SOA), University of the South Pacific, Alafua Campus, Apia, Samoa.

Anderson, P.K. (1994). A conceptual framework of integrated epidemiology for application to emerging diseases. *Ann. NY Acad. Sci.* 740: 439–444.

Anderson, P.K. and Morales, FJ. (1993). The emergence of new plant diseases: the case of insect transmitted plant viruses. In: M.E. Wilson, R. Levins and A. Spielman (eds.) *Disease in Evolution: Global Changes and Emergence of Infectious Diseases,* pp. 181–194. New York Academy of Sciences, New York.

Anita, B. and Samiyappan, R. (2012). Induction of systemic resistance in rice by *Pseudomonas fluorescens* against rice root knot nematode *Meloidogyne graminicola. J Biopest* 5 (Suppl.): 53-59.

Anjos, J.R.N. and Charchar, M.J.A. (2000). Natural infection of barley by *Pyricularia grisea* in Brazil. *Fitopatol. Bras.* 25: 205.

Anjos, J.R.N., Silva, D.B., Charchar, M.J.D. and Rodrigues, G.C. (1996). Ocorrência de brusone (*Pyricularia grisea*) em trigo e centeio na região dos cerrados do Brasil Central. *Pesqui. Agropecu. Bras.* 31: 79-82.

Anon. (1963). *Annual Work Progress Report on Crop Improvement Program of Rice, Sugarcane, Vegetable and Field Crops.* Directorate of Rural Affairs, Vietnam, 266 pp.

Anon. (1980). Rep. ICRISAT for 1979.

Anon. (2003). *Coffee Guide.* Central Coffee Research Institute, Balehonnur, Chikamagalur district, Karnataka, pp. 81-85.

Anon. (2010). *Training Manual on e-Pest Surveillance (Awareness and Surveillance Programme for Management of Major Pests of Paddy) under Rashtriya Krishi Vikas Yojna-2nd Green Revolution.* Directorate of Agriculture & Food Production, Bhubaneswar, Orissa, 54 pp.

Anon. (2011). *Meloidogyne enterolobii.* EPPO Bull. 41: 329-339.

Anon. (2013). *Recovery Plan for Wheat Blast Caused by* Magnaporthe oryzae triticum *Pathotype.* National Plant Disease Recovery System (NPDRS), USDA, 33 pp.

Anon. (2014). *Annual Report- 2013-14.* Indian Institute of Vegetable Research, Varanasi, Uttar Pradesh, India.

Ansar, M., Saleem A. and Iqbal A. (1994). Cause and control of guava decline in Punjab (Pakistan). *Pak. J. Phytopath.* 6: 41-44.

Antignus, Y., Lapidot, M., Hadar, D., Messika, Y. and Cohen, S. (1998). Ultraviolet-absorbing screens serve as optical barriers to protect crops from virus and insect pests. *Journal of Economic Entomology* 91(6): 1401-1405.

Antignus, Y., Vunsh, R., Lachman, O., Pearlsman, M., Maslenin, L., Hananya, U. and Rosner, A. (2004). Truncated Rep gene originated from tomato yellow leaf curl virus-Israel [Mild] confers strain-specific resistance in transgenic tomato. *Annals of Applied Biology*, 144(1): 39-44.

Anusuya, K. and Vadivelu, S. (2002). Mutualistic symbiosis of VAM fungi on plant growth promoting rhizobacteria (PGPR) on micro propagated carnation. *J. Microbiology World* 4: 105-107.

Araki, Y., Sugihara, M., Sawada, H. and Masuko, M. (2005). Monitoring of the sensitivity of *Magnaporthe grisea* to metominostrobin 2001–2003: No emergence of resistant strains and no mutations at codon 143 or 129 of the cytochrome b gene. *Journal of Pesticide Science* 30: 203–208.

Arayarungsarit, L. (1987). Yield ability of rice varieties in fields infested with root-knot nematode. *Int. Rice Res. Notes* 12: 14.

Arif, M.I., Rafiq, M. and Ghaffar, A. (2009). Host plants of cotton mealy bug (*Phenacoccus solenopsis*): A new menace to cotton agro ecosystem of Punjab, Pakistan. *Int. J. Agric. Biol.* 11: 163-167.

Arif, M.J., Hussain, S., Shahid, M.R., Abbas, G. and Gogi, M.D. (2008). *In vivo* and *in vitro* bioassay of potential insecticides against mealy bugs: A cotton worse catastrophe in Pakistan. In: *Proceedings of 2nd Congress on Insect Science-Pest Management in Global Context*, p.130. Punjab Agricultural University, Ludhiana.

Arora, R. and Dhawan, A.K. (2013) Climate change and insect pest management In: AK Dhawan, B. Singh, M.B. Bhullar and R. Arora (eds). *Insect Pest Management,* pp. 44-60. Scientific Publishers, Jodhpur.

Arora, R.K. (2000). Bio-control of potato late blight. In S.M. Paul Khurana, G.S. Shrkhawat, BP Singh and SK Pandey (eds.) *Potato Global Research and Development, Vol. I*, pp. 620-623. Indian Potato Association, Central Potato Research Institute, Shimla, India.

Arora, R.K., Rose de Leema, J. and Paulraj, F. (1992). Phylloplane microflora in biocontrol of potato late blight. *National Symp. on Non-Chemical Methods for the Management of Plant Diseases*, p. 49. Indian Phytopathological Society, Bhavnagar.

Ascencio-Ibáñez, J.T., Díaz-Plaza, R., Méndez-Lozano, J., Monsalve- Fonnegra, Z.I., Argüella-Astroga, G.R. and Rivera-Bustamante, R.F. (1999). First report of *Tomato yellow leaf curl Geminivirus* in Yucatán, Mexico. *Plant Dis.* 83: 1178.

Ashok, A. and Mehrotra, R.S. (1987). Control of *Phytophthora* leaf blight of taro (*Colocasia esculenta*) by fungicides and rouging. *Phytoparasitica* 15(5): 299–305.

Ashok, B. and Saikia, U.N. (1996). Fungicidal management of leaf blight of *Colocasia. Int J Trop Agric* 14(1–4): 231–233.

Atkins, J.G. and Todd, E.H. (1959). White tip disease of rice III. Yield tests and varietal resistance. *Phytopathology* 49: 189-191.

Atkinson, A.H., Heath, R.L., Simpson, R.J., Clarke, A.E. and Anderson, M.A. (1993). Proteinase inhibitors in *Nicotiana alata* stigmas are derived from a precursor protein which is processed into five homologous inhibitors. *Plant Cell* 5: 203-213.

Atwal, A.S. and Dhaliwal, G.S. (2009). *Agricultural Pests of South Asia and Their Management.* Kalyani Publishers, New Delhi.

Aujla, S.S., Grewal, A.S., Gill, K.S. and Sharma, I. (1980). Effect of Karnal bunt on chappati making properties of wheat grains. *Crop Improve.* 7: 147-149.

Aujla, S.S. and Sharma, I. (1990). Evaluation of fungicides against Karnal bunt disease of wheat. *Journal of Research, Punjab Agricultural University* 27(3): 434-436.

Avelar-Mejia, J.J., Teliz-Ortiz, D. and Zavaleta-Mejia, E. (2001). Patógenos asociados con el "declinamiento del guayabo". *Revista Mexicana de Fitopatologia* 19: 223-229.

Averna-Sacca, R. (1912). "Brusone" do arroz. *Bol. Agric.* 13a: 291-302.

Bade, B.A. and Ghorpade, S.A. (2009). Life fecundity tables of sugarcane woolly aphid, *Ceratovacuna lanigera* Zehntner. *J Insect Sci* 22(4): 402-405.

Bajaj, H.K. and Kanwar, R.S. (2005). Parasitization of maize by *Heterodera avenae* and *H. filipjevi*. *Nematol. Medit* 33: 203-207.

Baker, R., Paulitz, T., Windham, M.H. and Elad, Y. (1986). Enhancement of growth of ornamentals by a biological control agent. *Colorado Greenhouse Grow. Assoc. Res. Bull.* 431: 1.

Baker, R.H.A., Cannon, R.J.C. and MacLeod, A. (2003). Predicting the potential distribution of alien pests in the UK under global climate change: *Diabrotica virgifera virgifera. The Proceedings of BCPC International Congress – Crop Science & Technology Volume 2*, pp. 1201–1208. SECC, Glasgow, Scotland, UK.

Baker, R.H.A., Cannon, R.J.C. and Walters, K.F.A. (1996). An assessment of the risks posed by selected nonindigenous pests to UK crops under climate change. In: R.J. Froud-Williams, R. Harrington, T.J. Hocking, H.G. Smith and T.H. Thomas (eds.) *Implications of "Global Environment Change" for crops in Europe. Aspects of Applied Biology* 45: 323-330.

Baker, R.H.A., MacLeod, A., Cannon, R.J.C., Jarvis, C.H., Walters, K.F.A., Barrow E.M. and Hulme, M. (1998). Predicting the impacts of a non-indigenous pest on the UK potato crop under global climate change: reviewing the evidence for the Colorado beetle, *Leptinotarsa decemlineata. Brighton Crop Protection Conference - Pests and Diseases, Vol. III,* pp. 979-984. BCPC, Surrey, UK.

Bale, J.S. and Hayward, S.A.L. (2010). Insect over wintering in a changing climate. *J Exptl Biol* 213: 980-994.

Baltensweiler, W. (1993). Why the larch bud-moth cycle collapsed in the subalpine larch-cembran pine forests in the year 1990 for the first time since 1850. *Oecologia* 94: 62-66.

Bambawale, O.M., Tanwar, R.K. and Chander, S. (2009). IPM strategies and orientation towards changes in pest scenario due to climatic change. In: V.V. Ramamurthy, G.P. Gupta and S.N. Puri (eds.) *Proc. Natn. Symp. IPM Strategies to Combat Emerging Pests in the Current Scenario of Climate Change,* pp. 14-24. CAU, Pasighat, Arunachal Pradesh.

Bansal, R., Singh, D.V. and Joshi, L.M. (1984). Effect of Karnal bunt pathogen (*Neovossia indica* [Mitra] Mundkur) on weight and viability of wheat seed. *Indian J. Agric. Sci.* 54: 663-666.

Barea, G. and Toledo, J. (1996). Identificación y zonificación de piricularia o bruzone (*Pyricularia oryzae*) en el cultivo del trigo en el dpto. de Santa Cruz. In: *CIAT. Informe Técnico.* Proyecto de Investigación Trigo, Santa Cruz, pp. 76–86.

Barker, I., Brownlie, J., Peckham, C., Pickett, J., Stewart, W., Waage, J., Wilson, P. and Woolhouse, M. (2006) *Foresight. Infectious Diseases: preparing for the future. A Vision of Future Detection, Identification and Monitoring Systems.* Office of Science and Technology, London.

Barlow, H.S. and Chew, P.S. (1970). The rhinoceros beetle *Oryctes rhinoceros* in young oil palms replanted after rubber on some estates in Western Malaysia. In: R.L. Wastie and

B.J. Wood (eds.) *Proceedings of the Malaysian Crop Protection Conference,* pp. 133-144. Incorporated Society of Planters, Kuala Lumpur, Malaysia.

Basavaraj, K., Sreenivas, A.G., Prabhuraj, A., Bheemanna, M., Hosmani, A.C. and Nargund, V.B. (2011). Seasonal incidence of chilli gall midge (*Asphondylia capparis* Rubsaaman.) and its parasitoids. *Karnataka J. Agric. Sci.* 24(4): 555-557.

Basavaraju, B.S. Kulapati Hipparagi Chinnamadegowda, C. and Krishnamurthy, N. (2001). Management of giant African snail in betelvine garden. *Current Research* 30(7/8): 116-118.

Basu, A.N. (1995). Bemisia tabaci (Gennadius): *Crop Pests and Principal Whitefly Vector of Plant Viruses.* Oxford and IBH Publishing, New Delhi, 183 pp.

Bedford, G.O. (1980). Biology, ecology and control of palm rhinoceros beetles. *Annual Review of Entomology* 25: 309-339.

Beevi, S.P. and Lyla, K.R. (2001). Host plant mediated natural parasitism by *Encarsia* spp. (Aleyrodidae: Homoptera). In: A Verghese, P Parvatha Reddy (eds.) *Proceedings, 2nd National Symposium on Integrated Pest Management in Horticultural Crops: New Molecules, Biopesticides & Environment,* pp. 208-210. Association for Advancement of Pest Management in Horticultural Ecosystems, Bangalore, India.

Beevi, S.P., Lyla, K.R. and Vidya, P. (1999). Report of *Encarsia* (Hymenoptera: Aphelinidae) on spiralling whitefly *Aleurodicus dispersus* Russell (Homoptera: Aleyrodidae). *Insect Environment* 5: 44.

Begum, N. and Babu, H.N.R. (2013). Impact of coconut eryophyid mite *Aceria guerreroneis* K (Acari; Eryphyidae) in different agro climatic zones of Karnataka. *Environ. Ecol.* 31(2): 659-662.

Ben Moussa, A., Marrakci, M. and Makm, M. (2005). Characterization of Tospovirus in vegetable crop in Tunisia. *Infection, Genetics and Evolution* 5: 321.

Bendahmane, M, and Gronenborn, B. (1997). Engineering resistance against tomato yellow leaf curl virus (TYLCV) using antisense RNA. *Plant Molecular Biology* 33(2): 351-357.

Beniwal, M.S., Chawla, P., Singh R., Chawla, P. and Singh R. (2000). Effect of soaking of bunted seeds in water on teliospore germination of *Neovossia indica. Indian Phytopathology* 53(2): 219-220.

Bergquist, R.R. (1974). Effect of fungicide rate, spray interval, timing of spray application, and precipitation in relation to control of *Phytophthora* leaf blight of taro. *Ann Bot* 38(154): 213–221.

Bhanupriya, M. (2006). *Biological and molecular characterisation of Indian Tospovirus isolates from economically important crops and development of transgenic tomato with nucleocapsid gene.* Ph. D. thesis, Sri Venkateswara University, Tirupathi, 198 pp.

Bhargava, A.K., Sobti, A.K., Ghasolia, R.P. (2003). Studies on guava (*Psidum guajava* L.) drying/wilt disease in orchards of Pushkar Valley. *J. Phytological Res.* 16: 81-84.

Bhat, N.S., Raju, G.T., Manjunatha, M., Nagabhushana, G.G. and Deshpande, V.P. (1988). Chemical control of cowpea pod borer. *Indian Journal of Plant Protection* 16:197-200.

Bhatti, D.S., Dutt, R. and Verma, K.K. (1997). Larval emergence from cysts of *Heterodera avenae* and H. cajani as affected by plant leaf extracts. *Indian Journal of Nematology* 27(1): 63-69.

Bhuiyan, S.I., Sattar, M.A. and Khan, M.A.K. (1995). Improving water use efficiency in rice irrigation through wet seeding. *Irrigation Science* 16: 1-8.

Bianchi, F.A. (1957). Notes and exibhitions. *Proc. Hawaiian Ent. Soc.* 16(2): 184.

Biswas, H. and Rao, Y.S. (1971). Studies on nematodes of rice and rice soils. II. *Meloidogyne graminicola*, on deep water rice (*Oryza sativa* sub sp. *indica*). *Oryza* 8: 101- 102.

Blok, V.C., Wishart, J., Fargette, M., Berthier, K. and Phillips, M.S. (2002). Mitochondrial DNA differences distinguishing *Meloidogyne mayaguensis* from the major species of tropical root-knot nematodes. *Nematology* 4: 773-781.

Boag, B. and Neilson, R. (1996). Effects of potential climate changes on plant-parasitic nematodes. R.J. Froud- Williams, R. Harrington, T.J. Hocking, H.G. Smith and T.H. Thomas (eds.) *Implications of "Global Environment Change" for crops in Europe. Aspects of Applied Biology* 45: 331-334.

Bogantes-Arias, A. and Mora-Newcomer, E. (2010). Evaluation of four rootstocks for graft in guava (*Psidium guajava* L.). *Agronomia Mesoamericana* 21: 103-111.

Bonde, M.R., Peterson, G.L., Schaad, N.W. and Smilanick, J. (1997). Karnal bunt of wheat. *Plant Dis.* 81: 1370–1377.

Booker, R.H. (1965). Pests of cowpea in northern Nigeria. *Bulletin of Entomological Research* 55: 663-672.

Borbon, O. (1990). Pérdidas de café provocadas por la broca del fruto del cafeto en Togo *Hypothenemus hampei* (Ferr). In: PROMECAFE (ed.) *IV Taller Regional sobre la Broca del Fruto del Cafeto.* PROMECAFE, San Salvador (El Salvador).

Borgen, A. (2004). Organic seed treatment to control common bunt (*Tilletia tritici*) in wheat. *Seed Testing International* 128: 12-13.

Borokova, W. (1967). Host parasite relations and host range studies of a new *Meloidogyne* species in Southern USA. *Phytopathology* 55: 1359-1361.

Botermans, M., Verhoeven, J.T.J., Jansen, C.C.C., Roenhorst, J.W., Stijger, C.M.M. and Pham, K.T.K. (2009). First report of *Tomato yellow leaf curl virus* in tomato in the Netherlands. *Plant Dis.* 93: 1073.

Brasier, C.M. (2001). Rapid evolution of introduced plant pathogens via interspecific hybridization. *Bioscience* 51: 123–133.

Bridge, J. and Page, S.L.J. (1982). The rice root-knot nematode, *Meloidogyne graminicola*, on deep water rice (*Oryza sativa* subsp. *indica*). *Revue de Nematologie* 5: 225-232.

Brito, J.A., Stanley, J.D., Mendes, M.L., Cetintas, R. and Dickson, D.W. (2007). Host status of selected cultivars plants to *Meloidogyne mayaguensis* from Florida. *J. Nematol.* 36: 232-240.

Bruno, A.C. and Urashima, A.S. (2001). Inter-relação sexual de *Magnaporthe grisea* do trigo e de outros hospedeiros. *Fitopatol. Bras.* 26: 21-26.

Buangsuwon, D, Tonboon-ek P., Rujirachoon, G., Braun, A.J. and Taylor, A.L. (1971). Nematodes. *Rice diseases and pests of Thailand* (English Edition), pp. 61-67. Rice Protection Research Centre, Ministry of Agriculture, Thailand.

Bucher, E., Lohuis, D., van Poppel, P.M., Geerts-Dimitriadou, C., Goldbach, R. *et al.* (2006). Multiple virus resistance at a high frequency using a single transgene construct. *J Gen Virol* 87: 3697–3701.

Buffo, E., Battisti, A., Stastny, M. and Larsson, S. (2007). Temperature as a predictor of survival of the pine processionary moth in the Italian Alps. *Agricultural and Forest Entomology* 9: 65-72.

Buiel, A.A.M., Dwivedi, S.L., Prasad, M.V.R., Singh, A.B., Dharmaraj, P.S. and Parlevliet, J.E. (1995). Multi-environment testing for reduced incidence of peanut bud necrosis disease in India. In A.A.M. Buiel, J.E. Parlevliet and J.M. Lenne (eds.) *Recent Studies on*

Peanut Bud Necrosis Disease, pp. 47-54. *Proceedings of a Meeting*, ICRISAT Asia Center, Patancheru, Andhra Pradesh, India, 80 pp.

Butani, D.K. and Jotwani, M.G. (1984). *Insects in Vegetables*. Periodical Expert Book Agency, Delhi, pp. 65-66.

Butler, L. (1990). Karnal bunt, quarantine and the international shipment of CIMMYT wheat seed. *Proc. 7th Bien. Workshop on Smut Fungi.*

CABI. (2000). *Distribution Maps of Plant Diseases No. 804.* Meloidogyne mayaguensis. CABI, Wallingford (GB).

CABI/EPPO. (2014). *Distribution Maps of Plant Diseases, Map 1011 (Edition 3) - Iris Yellow Spot Virus.* CABI, Wallingford, UK.

Caciagli, P. (2008). Vegetable viruses. In : B.W.J. Mahy and M.H.V. Van Regenmortel (eds.) *Encyclopedia of Virology*, pp. 282- 290. Academic Press, New York.

Cahill, M., Byrne F.J., Gorman, K., Denholm, I. and Devonshire, A.L. (1995). Pyrethroid and organophosphate resistance in the tobacco whitefly *Bemisia tabaci* (Homoptera: Aleyrodidae). *Bulletin of Entomological Research* 85(2): 181-187.

Callaway, E. (2016). Devastating wheat fungus appears in Asia for first time. *Nature* 532: 421-422.

Cannayane, I. and Sivakumar, C.V. (2001). Nematode egg parasitic fungus I: Paecilomyces lilacinus – a review. *Agricultural Review* 22: 79-86.

Canterbury, G.E. and Blockstein, D.E. (1997). Local changes in a breeding bird community following forest disturbance. *J. Field Ornithol.* 68: 537–546.

Capinera, J.L. (2011). *ENY-512 (IN904).* Entomology and Nematology Department, Florida Cooperative Extension Service, Institute of Food and Agricultural Sciences, University of Florida at http://edis.ifas.ufl.edu.

Cardoso, C.A., De, A., Reis, E.M. and Moeira, E.N. (2008). Desenvolvimento de sistema de previsão de brusone causada por *Pyricularia grisea*. *Summa Phytopathologica* 34: 216–221.

Carneiro R.M.D.G., Castro J.M.C., Carneiro R.G., Gomes C.B., de Freitas V.M. and Mattos J.K. (2012). Major guava nematodes and control prospects using resistance on *Psidium* spp. and non-host crops. In C.A.F. Santos *et al.* (eds.) Proc. 3rd IS on Guava and Other Myrtaceae *Acta Hort.* 959: 41-49.

Carneiro, R.M.D.G., Almeida, M.R.A., Braga, R.S., Almeida, C.A. and Gioria, R. (2006a). Primeiro registro de *Meloidogyne mayaguensis* parasitando plantas de tomate e pimentão resistentes à meloidoginose no estado de São Paulo. *Nemat. Bras.* 30: 81-86.

Carneiro, R.M.D.G., Cirotto, P.A., Silva, D.B. and Carneiro, R.G. (2007). Resistance to *Meloidogyne mayaguensis* in *Psidium* spp. accessions and their grafting compatibility with *P. guajava* cv. 'Paluma'. *Fitopatol. Bras.* 32: 281-284.

Carneiro, R.M.D.G., Mônaco, A.P.A., Moritz, M.P., Nakamura, K.C. and Scherer, A. (2006b). Identificação de *Meloidogyne mayaguensis* em goiabeira e em plantas invasoras, em solo argiloso no Estado do Paraná. *Nemat. Bras.* 30: 293-298.

Carneiro, R.M.D.G., Moreira, W., Almeida, M.R.A. and Gomes, A.C.M.M. (2001). Primeiro registro de *Meloidogyne mayaguensis* em goiabeira no Brasil. *Nemat. Bras.* 25: 223-228.

Casassa, A.M., Crozzoli, R., Matheus, J., Bravo, V. and Marin, M. (1998). Effect of the root-knot nematode, Meloidogyne incognita, on the growth of guava (*Psidium* spp.) in nurseries. *Nematologia Mediterranea* 26: 237-242.

Casassa, A.M., Matheus, J., Crozzoli, R., Bravo, V. and Gonzalez, C. (1997). Response of some selections of guava to *Meloidogyne incognita*, in Mara County, Zulia State, Venezuela. *Fitopatologia Venezolana* 10: 5-8.

Castagnone-Sereno, P. (2012). *Meloidogyne enterolobii* (=*M. mayaguensis*): Profile of an emerging, highly pathogenic, root-knot nematode species. *Nematology* 14: 133-138.

Castle, S.J., Palumbo, J.C., Prabhaker, N., Horowitz, R. and Delholm, I. (2010). Ecological determinants of *Bemisia tabaci* resistance to insecticides. In: P.A. Stansly and S.E. Naranjo (eds.) *Bemisia: Biomics and Management of a Global Pest,* pp. 423-465. Springer, Dordrecht, Heidelberg, The Netherlands, London, New York, Germany, UK, USA.

Catley, A. (1969). The coconut rhinoceros beetle *Oryctes rhinoceros* (L.). *PANS* 15: 18-30.

Cetintas, R., Kaur, R., Brito, J.A., Mendes, M.L., Nyczepir, A.P. and Dickson, D.W. (2007). Pathogenicity and reproductive potential of *Meloidogyne mayaguensis* and *M. floridensis* compared with three common *Meloidogyne* spp. *Nematropica* 37: 21–31.

Chakraborti, H.S. and Ghosh, S.C. (1993). Studies on the floral malady of *Polianthes tuberosa* L. *Journal of Mycopathology Research* 33: 25-28.

Chakraborty, K. and Singh, R.N. (1989). Guava wilt correlation between variation in disease syndrome and edaphic factors. *Indian Phytopath.* 42: 310.

Chakraborty, S., Tiedemann A.V. and Teng, P.S. (2000). Climate change: potential impact on plant diseases. *Environmental Pollution* 108: 317-326.

Chancellor, T. and Kubiriba, J. (2006). T7.2 The Effects of Climate Change on Infectious Diseases of Plants. *Foresight: Infectious Diseases: preparing for the Future.* Office of Science and Innovation. http://www.foresight.gov.uk/Previous_Projects/Detection_and_Identification_of_Infectious_Diseases/Reports_and_Publications/Final_Reports/T/t7_2a.pdf.

Chandrayudu, E., Srinivasan, S. and Rao, V.N. (2008). Evaluation of certain newer insecticides against spotted pod borer, *Maruca vitrata* (Geyer) on cowpea (*Vigna ungiculata* (L) Walp). *Current Biotica* 2: 240-243.

Chattopadhyay, S.B. and Bhattacharjya, S.K. (1968a). Investigations on wilt disease of guava (*Psidum guajava* L.) in West Bengal - I. *Indian J. Agric. Sci.* 38: 65-72.

Chattopadhyay, S.B. and Bhattacharjya, S.K. (1968b). Investigations on wilt disease of guava (*Psidum guajava* L.) in West Bengal - II. *Indian J. Agric. Sci.* 38: 176-183.

Chattopadhyaya, S.B. and Sen Gupta, S.K. (1955). Studies on wilt of *Psidum guajava* in West Bengal. *Indian J. Hort.* 12: 76-79.

Chaudhry, M.H. and Khan, M.A. (1990). Effectiveness of different spray fungicides against Karnal bunt of wheat. *Rachis* 9(2): 5-6.

Cherim, M. (2007). *Living with Pests on a Warming Planet* (http://greenmethods.com).

Chiapello, H., Mallet, L., Guérin, C., Aguileta, G., Amselem, J., Kroj, T., *et al.* (2015). Deciphering genome content and evolutionary relationships of isolates from the fungus *Magnaporthe oryzae* attacking different host plants. *Genome Biol Evol.* 7: 2896–2912.

Chowdappa, P., Nirmal Kumar, B.J., Madhura, S., Mohan Kumar, S.P., Myers, K.L., Fry, W.E. and Cooke D.E.L. (2014). Severe outbreaks of late blight on potato and tomato in South India caused by recent changes in the *Phytophthora infestans* population. *Plant Pathology* 64(1):191-199.

CNN. (1997). *Termites threaten New Orleans' Treasures.* http://www.cnn.com/ EARTH/ 9703/28/formosa.termites/ [acccssed 10 July 2003].

Coffey, M.D. (1991). Strategies for integrated control of soil borne *Phytophthora* species. In: J.A. Lucas, R.C. Shattock, D.S. Shaw and L.R. Cooke (eds.) *Phytophthora.* Cambridge University Press, Cambridge, UK, 447 pp.

Cohen, S. and Antignus, Y. (1994). *Tomato yellow leaf curl virus* (TYLCV), a whitefly-borne *Geminivirus* of tomatoes. *Adv. Dis. Vector Res.* 10: 259-288.

Cohen, S. and Lapidot, M. (2007). Appearance and expansion of TYLCV: A historical point of view. *In:* H. Czosnek, (ed.) *Tomato Yellow Leaf Curl Virus Disease*, pp. 3-12. Springer, Dordrecht, The Netherlands.

Cohen, S., Melamed-Madjar, V. Hameiri, J. (1974). Prevention of the spread of tomato yellow leaf cu,rl virus transmitted by *Bemisia tabaci* (Gennadius) (Homoptera, Aleyrodidae) in Israel. *Bulletin of Entomological Research* 64(2): 193-197.

Collier, R.H., Finch, S., Phelps, K. and Thompson, A.R. (1991). Possible impact of global warming on cabbage root fly (*Delia radicum*) activity in the UK. *Annals of Applied Biology* 118: 261-271.

Córdoba-Sellés, C., Martínez-Priego, L., Munoz-Gómez R. and Jordá-Gutiérrez, C. (2005). Iris yellow spot virus: a new onion disease in Spain. *Plant Disease* 89(11): 1243.

Costa, A.S. (1975). Increase in the population density of *Bemisia tabaci*, a threat of widespread virus infection of legume crops in Brazil. In J Bird and K Maramorosch (eds.) *Tropical Diseases of Legumes*, pp. 27–49. Academic Press.

Costa, S.R., Santos, C.A.F. and Castro, J.M.C. (2012). *Tolerance of* Psidium guajava × P. guineense *Hybrids to* Meloidogyne enterolobii. Série Documentos 247, Petrolina, PE, Brazil, 39 pp.

Cralley, E.M. (1956). A new control measure for white tip. *Arkansas Farm Research* 5:5.

Cralley, E.M. and Adair, C.R. (1949). Rice diseases in Arkansas in 1948. *Plant Disease Reporter* 33: 257-259.

Cranshaw, W.S. (2006). *Colorado Insecticide Trials for Control of Thrips on Onion, 1995-2006.* Colorado State Univeristy Agricultural Experiment Station Report TB06-01, Fort Collins, USA.

Cruz, M.F.A., Diniz, A.P.C., Rodrigues, F.A. and Barros, E.G. (2011). Foliar application of products on the reduction of blast severity on wheat. *Trop. Plant Pathol.* 36: 424-428.

Cruz, M.F.A., Prestes, A.M., Maciel, J.L.N. and Scheeren, P.L. (2010). Resistência parcial à brusone de genótipos de trigo comum e sintético nos estádios de planta jovem e de planta adulta. *Trop. Plant Pathol.* 35: 24-31.

Cuadra, R. and Quincosa, A. (1982). Potential of different *Psidium gujava* species as sources for resistance of guava to *Meloidogyne. Ciencias-de-la-Agricultura* 13: 19-26.

Cunningham, A.A., Daszak, P. and Rodnguez, J.P. (2003). Pathogen pollution: defining a parasitological threat to biodiversity conservation. *J. Parasitol.* 89: S78–S83.

Czosnek, H. and Laterrot, H. (1997). A world-wide survey of tomato yellow leaf curl viruses. *Arch. Virol.* 142: 1391-1406.

Dabur, K.R., Taya, A.S. and Rana, B.P. (2010). Effect of rice root nematode, Meloidogyne graminicola on wheat crop. National Conf. on Innovations in Nematological Research for Agricultural Sustainability – Challenges and A Roadmap Ahead, Tamil Nadu Agri. Univ., Coimbatore.

Dahal, G., Lecoq, H. and Albrechtsen, S.E. (1997). Occurrence of papaya ringspot potyvirus and cucurbit viruses in Nepal. *Annals of Applied Biology* 130(3): 491.

Dale, D. (1994). Insect pests of rice plant-Their biology and ecology, In E.A. Heinrichs (ed.) *Biology and Management of Rice Insects,* pp. 364-485. Wiley Eastern Limited, London, 779 pp.

Das Gupta, M.K. and Ghoshal B.K. (1977). Is it possible to control guava wilt through oil cake amendments? *Sci. Cult.* 43: 131-133.

Das Gupta, S.N. and Rai, J.N. (1947). Wilt disease of guava (*Psidum guajava* L.). *Curr. Sci.* 16: 256-258.

Das T.K. and Khan, M.R. (2007). Occurrence and distribution of white tip nematode, *Aphelenchoides besseyi* in West Bengal, India. *Indian Journal of Nematology* 27: 94–97.

Das, B.K. and Talukdar, B. (2010). Record of invasive *Quadrastichus erythrinae* Kim, gall wasp of *Erythrina variegata* L. from eastern India with notes on gall morphologies. *Insect Environment* 17(1): 9-11.

Das, P. and Deka, B.C. (2002). Efficacy of neem-based pesticides against *Meloidogyne graminicola* on rice as seed treatment. *Indian Journal of Nematology* 32: 203-205.

Daszak, P. and Cunningham, A.A. (2003). Anthropogenic change, biodiversity loss and a new agenda for emerging diseases. *J. Parasitol.* 89: S37–S41.

Datta, S.C. (2005). Possible use of amaranth as catch crop for root-knot nematodes intercropped with mulberry. *J Environ Sociobiol* 2: 61-65.

Datta, S.C. and Datta, R. (2007). Intercropping amaranth with mulberry for managing root-knot nematodes and improving sericulture. *Sericol* 47: 297-306.

David, B.V. and Ananthakrishnan, T.N. (2004).The swarming caterpillar or armyworm, *Spodoptera mauritia.* In *General and Applied Entomology (Second Edition),* Tata McGraw-Hill Publishing Coy. Ltd, New Delhi, 1184 pp.

David, B.V. and Regu, K. (1995). *Aleurodicus dispersus* Russell (Aleyrodidae: Homoptera), a whitefly pest new to India. *Pestology* 19 (3): 5-7.

David, H. and Easwaramoorthy, S. (1986). In H. David, S. Easwaramoothy and R. Jayanthi (eds.) *Sugarcane Entomology in India,* pp. 383–422. Sugarcane Breeding Institute, Coimbatore.

Davidson, J.M., Werres, S., Garbelotto, M., Hansen, E.M., Rizzo, D.M. (2003). *Sudden Oak Death and Associated Diseases Caused by* Phytophthora ramorum. Plant Health Prog. doi: 10.1094/PHP-2003-0707-01-DG. Available: http://www.plantmanagementnetwork.org/.

De Waele, D. and Elsen, A. (2007). Challenges in tropical plant nematology. *Annu. Rev. Phytopathol.* 45: 457–485.

DEFRA (2010). *Science Search.* http://randd.defra.gov.uk/. Accessed 26 April 2010.

Deguine, J.P., Ferron, P. and Russell, D. (2008). Sustainable pest management for cotton production: A review. *Agron Sustain Dev* 28: 113-137.

Deka, B.C. and Das, P. (2003). Efficacy of different groups of chemicals in managing *Meloidogyne graminicola* on rice as seed treatment. *Indian Journal of Nematology* 33: 65-66.

Demirozer, O., Tyler-Julian, K., Funderburk, J., Reitz, S. and Leppla, N. (2012). Integrated pest management program for thrips and Tospoviruses in fruiting vegetables. *Pest Management Science* 68 (12): 1537–15458.

Deng, G.R., Yang, H.H. and Jin, M.X. (1987). Augmentation of Coccinellid beetles for controlling sugarcane woolly aphid. *Chin. J. Biol. Control* 3: 166–168.

Deshmukh P.S., Chougale, A.K., Shahasane, S.S., Desai, S.S. and Gaikwad S.G. (2012). Studies on biology of Hadda beetle, *Epilachna vigintioctopunctata* (Coleptera, Coccinillidae): a serious pest of wild better gourd, *Momordica dioica. Trends in Life Sciences* 1(3): 46-48.

Deshpande, V. (2009) Mealy bug species, the Bt cotton killer, is exotic. *Indian Express,* 17 July 2009. http://www.indianexpress.com/news/mealybug-species-the-bt-cotton-killer-is exotic-experts/490540/0.

Devi, L.S. (1997). Evaluation of green manures against the pigeon pea cyst nematode *Heterodera cajani. National Academy Science Letters* 20(1/2): 1-2.

Devi, S. and Gupta, P. (1995). Effect of four green manures against *Heterodera cajani* on pigeon pea sown with or without *Rhizobium* seed treatment. *Indian Journal of Mycology and Plant Pathology* 25(3): 254-256.

Devi, S. and Gupta, P. (1996). Larval emergence from egg sacs of *Heterodera cajani* in extracts of cakes in various media and their effect on cowpea. *Indian Journal of Nematology* 25: 190-193.

Dey, P.K. (1948). *Plant Pathology*. Administrative Report of Agricultural Department, UP 1945-46, pp. 43-46.

Dhaliwal, G.S. (2008). Pest management in global context: From green revolution to gene revolution. Key Note Address. *2nd Congress on Insect Science*, Indian Society for the Advancement of Insect Science, Ludhiana, Punjab.

Dhaliwal, G.S. and Koul, O. (2010). *Quest for Pest Management: From Green Revolution to Gene Revolution*. Kalyani Publishers, New Delhi.

Dhaliwal, G.S., Arora, R. and Dhawan, A.K. (2004). Crop losses due to insect pests in Indian agriculture: An update. *Indian J. Ecol.* 31(1): 1-7.

Dhaliwal, G.S., Dhawan, A.K. and Singh, R. (2007). Biodiversity and ecological agriculture: Issues and perspectives. *Indian J. Ecol.* 34(2): 100-109.

Dhawan, A.K. and Saini, S. (2009). Study on biology of mealy bug, Phenacoccus solenopsis Tinsley on cotton in Punjab. Symposium Abstracts. *Proceedings of the National Symposium on IPM Strategies to Combat Emerging Pests in the Current Scenario of Climate Change*, pp. 35-36. CAU, Pasighet, Arunachal Pradesh.

Dhawan, A.K., Singh, K., Saini, S., Mohindru, B., Kaur, A., Singh, G. and Singh, S. (2007). Incidence and damage potential of mealy bug, *Phenacoccus solenopsis* Tinsley on cotton in Punjab. *Indian Journal of Ecology* 34: 110-116.

Dhillon, M.K., Gujar G.T. and Kalia, V. (2011). Impact of Bt cotton on insect biodiversity in cotton ecosystem in India. *Pak. Entomol.* 33(2): 161-165.

Dhillon, M.K., Singh, R., Naresh, J.S. and Sharma, H.C. (2005). The melon fruit fly, *Bactrocera cucurbitae*: A review of its biology and management. *Journal of Insect Science* 5: 40.

Dhobi, C.B., Bharpoda, T.M. and Borad, P.K. (2014). First record of *Paracoccus marginatus* Williams and Granara de Willink and *Rastrococcus iceryoides* (Green) on Bt cotton (*Gossypium hirsutum* L.) in Gujarat. *Insect Environment* 20(1): 6-7.

Dias, W.P., Freitas, V.M., Ribeiro, N.R., Moita, A.W. and Carneiro, R.M.D.G. (2010). Reação de genótipos de milho a *Meloidogyne mayaguensis* e *M. ethiopica*. *Nemat. Bras.* 34(2): 99-105.

Dickson, D.W. (1998). Peanut. In K.R. Barker, G.A. Peterson, and G.L. Windham (eds.) *Plant and Nematode Interaction*, pp. 523–566. American Society of Agronomy, Madison, WI.

Dillon M.E., Wang G. and Huey R.B. (2010). Global metabolic impacts of recent climate warming. *Nature* 467: 704-707.

Dina, S.O. (1988). Timing the application of deltamethin and cypermethrin for the control of insect pests of cowpea *Vignu unguiculata* (L.) Walp. *Tropical Pest Management* 34: 65-67.

Ding, S.W. and Voinnet, O. (2007). Antiviral immunity directed by small RNAs. Cell 130(3): 413–426.

Dosdall, L.M. (1994). Evidence for successful overwintering of diamondback moth, *Plutella xylostella* (L.) (Lepidoptera: Plutellidae), in Alberta. *Canadian Entomologist* 126: 183-185.

Downes, S., Mahon, R. and Olsen, K. (2007). Monitoring and adaptive resistance management in Australia for Bt-cotton: Current status and future challenges. *J. Invertebr. Pathol.* 95(3): 208-213.

Durairaj, C., Ganapathy, N. and Senguttuvan, T. (2000). New pest of pigeon pea. *The Hindu*, January 20, 2000.

DuToit, L.J. (2005). Iris yellow spot virus of onion. In: *Proceedings of the Great Lakes Fruit, Vegetable, and Farm Market Expo*, Michigan State Horticultural Society/Michigan Vegetable Council/Michigan State University, Grand Rapids, Michigan, USA. Michigan, USA.

DuToit, L.J. and Pelter, G.Q. (2005). Susceptibility of storage onion cultivars to iris yellow spot in the Columbia Basin of Washington, 2004. *Biological and Cultural Tests*, 20 (V006).

Duveiller, E., Hodson, D. and Tiedmann, A. (2010). Wheat blast caused by *Magnaporthe grisea*: A reality and new challenge for wheat research (Abstr.). In: *8th. Int. Wheat Conf.*, pp. 247-248. Vavilov Res. Inst. Plant Ind., St. Petersburg, Russia.

Dwivedi, B.P. and Shukla D.N. (2002). Biocontrol of *Fusarium* wilt of guava (*Psidium guajava* L.) using *Trichoderma* and *Gliocladium* species. *Karnataka J. Agric. Sci.* 15: 399-400.

Dwivedi, S.K. (1990). Efficacy of some antibiotics on *Fusarium oxysporum* f. sp. *Psidii* causing wilt disease of guava. *Hindusthan Antibiotics Bull.* 32: 89-90.

Dwivedi, S.K. (1991). Population dynamics of microfungi including pathogenic forms in the beds of completely healthy partially wilted and completely wilted guava trees grown on a line. *J. Trop. Pl. Dis.* 9: 95-109.

Dwivedi, S.K. (1992). Effect of culture filterate of soil microbes of pathogens inciting wilt disease of guava (*Psidium guajava* L.) under *in vitro* conditions. *Nat. Acad. Sci. Lett.* 15: 33-35.

Dwivedi, S.K. (1993). Fungitoxicity of *Foeniculum vulgare* seed oil used against a guava wilt pathogen. *Natl. Acad. Sci. Lett.* 16: 207-207.

Dwivedi, S.L., Nigam, S.N., Reddy, D.V.R., Reddy, A.S. and Ranga Rao, G.V. (1995). Progress in breeding groundnut varieties resistant to peanut bud necrosis virus and its vector. In A.A.M. Buiel, J.E. Parlevliet and J.M. Lenne (eds.) *Recent Studies on Peanut Bud Necrosis Disease*, pp. 35-40. *Proceedings of a Meeting, ICRISAT Asia Center*, Patancheru 502 324, Andhra Pradesh, India, 80 pp.

Edward, J.C. (1960a). Penetration and establishment of *Fusarium oxysporum* f. *psidii* in the guava root. *Indian Phytopath.* 13: 168-170.

Edward, J.C. (1960b). Variation in the guava wilt pathogen, *Fusarium oxysporum* f. *psidii*. *Indian Phytopath.* 13: 30-36.

Edward, J.C. (1960c). Wilt disease of guava. *Allahabad Farmer* 34: 289-293.

Edward, J.C. (1961). Root stock trials for guava wilt control. *Allahabad Farmer* 35: 5-9.

Edward, J.C. and Gaurishanker. (1964). Rootstock trial for guava (*Psidium guajava* L.). *Allahabad Farmer* 38: 249-250.

Edward, J.C. and Srivastava, R.N. (1957). Studies on guava wilt. *Allahabad Farmer* 31: 144-146.

Edwards, R. (1996). Tomorrow's bitter harvest – the genetic diversity of our agriculture is rapidly vanishing, leaving our crops prone to pest and plague. *New Sci.* 17: 14–15.

El Borai, F.E. and Ducan, L.W. (2005). Nematodes parasites of subtropical and tropical fruits tree crops. p.467-492. In: M. Luc, R.A. Sikora and J. Bridge (eds.) *Plant Parastic Nematodes in Subtropical and Tropical Agriculture.* CAB International, Wallingford.

Elbert, A., Overbeck, H., Iwaya, K. and Tsuboi, S. (1990). Imidacloprid, a novel systemic nitromethylene analogue insecticide for crop protection. *Brighton Crop Protection Conference, Pests and Diseases - 1990. Vol. 1*, pp. 21-28. British Crop Protection Council, Thornton Heath, UK.

Elyas, Z. and Sharma, S.B. (1997). Mechanism of resistance to *Heterodera cajani* in *Cajanus platycarpus* accessions. *International Journal of Nematology* 7(2): 119-121.

EPPO. (2014). EPPO Reporting Service, No. 2014/015. Paris, France: European and Mediterranean Plant Protection Organization.

Estrella, N., Sparks, T.H. and Menzel, A. (2007). Trends and temperature response in the phenology of crops in Germany. *Global Change Biology* 13: 1737–1747.

Evans, C.K., Bag, S, Frank, E., Reeve, J.R., Ransom, C., Drost, D. and Pappu, H.R. (2009). Natural infection of Iris yellow spot virus in twoscale saltbush (*Atriplex micrantha*) growing in Utah. *Plant Disease* 93(4): 430.

Fademi, O.A. (1984). Influence of rate and time of carbofuran application to control root-knot nematodes in upland rice. *International Rice Research Newsletter* 9: 22-23.

Faizal, M.H., Prathapan, K.D., Anith, K.N., Mary, C.A., Lekha, M. and Rini, C.R. (2006). Erythrina gall wasp *Quadrastichus erythrinae*, yet another invasive pest new to India. *Current Science* 90(8): 1061-1062.

FAO (1995). *International Standards for Phytosanitary Measures: Principles of Plant Quarantine as Related to International Trade, Food and Agriculture Organisation of the United Nations.* World Agricultural Information Centre (WAICENT). Publication No. 1, (http://www.ippc.int /servlet/BinaryDownloaderServlet/ISPM_01_English.pdf?filename Z1027428004562_ ispm1_final_en.pdf).

Fatah, T., Rafii, M.Y., Rahim, H.A., Meon, S., Azhar, M. and Latif, M.A. (2014). Cloning and analysis of QTL linked to blast disease resistance in Malaysian rice variety Pongsu Seribu 2. *International Journal Agriculture Biol*ogy 16: 395–400.

Ferreira, S.A., Pitz, K.Y., Manshardt, R., Zee, F., Fitch, M. and Gonsalves, D. (2002). Virus coat protein transgenic papaya provides practical control of papaya ringspot virus in Hawaii. *Plant Disease* 86 (2): 101.

Fitter, A.H. and Fitter, R.S.R. (2002). Rapid changes in flowering time in British plants. *Nature* 296: 1689-1691.

Fleming R.A. and Tatchell G.M. (1995). Shifts in the flight periods of British aphids: a response to climate warming? In: R. Harrington and N. Stork (eds.) *Insects in a Changing Environment,* pp. 505- 508. Academic Press.

Frais, T. and Garcia, E.R. (1981). Effectiveness of some microorganisms antagonistic to *Phytophthora palmivora* (Butl) in controlling black pod rot of cocoa. *Rev Mexicana Fitopatol* 1(3): 16–20.

Fry, W. and Goodwin, S.B. (1997). Resurgence of the Irish Potato famine fungus. *Bioscience* 47: 363–371.

Fuchs, T.W., Stewart, J.W., Minzenmayer, R. and Rose, M. (1991). First record of *Phenacoccus Solenopsis* Tinsley in cultivated cotton in the United States. *Southwestern Entomologist* 16: 215-221.

Fuentes, A., Ramos, P.L., Fiallo, E., Callard, D., Sánchez, Y., Peral, R., Rodríguez, R. and Pujol, M. (2006). Intron-hairpin RNA derived from replication associated protein C1 gene confers immunity to Tomato Yellow Leaf Curl Virus infection in transgenic tomato plants. *Transgenic Research* 15(3): 291-304.

Fuller, G.B. (2005). *Asian Productivity Organization* pp. 31–21 .http://www.apo-tokyo.org/ publications/files/pjrep-ag-ge-stm-04-b.pdf#page=32.

Funasaki, G.Y., Lai P.Y. and Nakahara, L. (1990). Status of natural enemies of *Heteropsylla cubana* Crawford (Homoptera: Psyllidae) in Hawaii. In B. Napompeth and K.G. MacDicken (eds.) *Leucaena psyllid: Problems and Management,* pp. 153-158. Winrock International/IDRC/NFTA/F-FRED, Bangkok, Thailand.

Ganguly, A.K., Pankaj, S. and Sirohi, A. (1996). Effect of soil solarization of rice nursery beds to suppress plant parasitic nematodes. *International Rice Research Notes* 21: 80-81.

Ganguly, S., Singh, M. and Ganguly, A.K. (2010). Record of potato cyst nematode, *Globodera rostochiensis* and *G. pallida* in Shimla, Himachal Pradesh, India. *Indian J Nematol* 40: 96-102.

Garg, R.N., Gaur, H.S. and Singh, G.C. (1995). Effect of tillage practices and water regimes on the soil physical properties and plant parasitic nematodes in rice. *Annals of Plant Protection Science* 3: 121-126.

Gaur, H.S. (2003). *Root-Knot Disease of Rice and Wheat: Problem and Management.* Technical Bull. (TB – ICN: 1/2003), IARI, New Delhi, 23 pp.

Gaur, H.S. and Singh, I. (1977). Pigeon-pea cyst nematode, *Heterodera cajani*, associated with the moong crop in the Punjab State. *Journal of Research, Punjab Agricultural University* 14(4): 509.

Gaur, H.S., Singh, J., Sharma, S.N. and Chandel, S.T. (1996). Distribution and community analysis of plant parasitic nematodes in rice growing areas of Haryana, India. *Annals of Plant Protection Science* 4: 115-121.

Gawande, S.J., Gurav, V.S., Ingle, A.A. and Gopal, J. (2014). First report of Iris yellow spot virus infecting *Allium tuberosum* in India. *Plant Disease* 98(8): 1161.

Geetha, B. and Swamiappan, M. (2001). Record of a parasitoid and a parasite mite on spiralling whitefly, *Aleurodicus dispersus* Russell in Tamil Nadu. *Insect Environment* 6: 179-180.

Gent, D.H., duToit, L.J., Fichtner, S.F., Mohan, S.K., Pappu, H.R. and Schwartz, H.F. (2006). Iris yellow spot virus: an emerging threat to onion bulb and seed production. *Plant Disease* 90(12): 1468-1480.

Gent, D.H., Schwartz, H.F. and Khosla, R. (2004). Distribution and incidence of Iris yellow spot virus in Colorado and its relation to onion plant population and yield. *Plant Disease* 88(5): 446-452.

Gergon, E.B. and Prot, J.C. (1993). Susceptibility of wild rice species to nematode *Meloidogyne graminicola. International Rice Research Notes* 18: 16.

Gerling, D. and Kravchenko, V. (1995). Pest management of *Bemisia* out of doors. In: D. Gerling and T. Mayer (eds.) *Bemisia: Taxonomy, Biology, Damage, Control and Management,* pp. 667-680. Intercept, UK.

German, T.L., Ullman, D.E. and Moyer, J.W. (1992). *Tospoviruses*: Diagnosis, molecular biology, phylogeny and vector relashionships. *Ann. Rev. Phytopathol.* 30: 315-348.

Gertsson, C. A. (2005). New species and new province records of scale insects from Sweden (Hemiptera: Coccoidea) up to the year 2004. *Entomologisk Tidskrift. Stockholm* 126: 35-42. [In Swedish with English Summary].

Gewin, V. (2003). Bioterrorism: agricultural shock. *Nature* 421: 106–108.

Ghosal, A., Chatterjee, M.L. and Manna, D. (2012). Studies on some insecticides with novel mode of action for the management of tomato fruit borer (*Helicoverpa armigera* Hub.). *Journal of Crop and Weed* 8(2): 126-129.

Ghosh, A.K. (1988). Homoptera: Aphidoidea, 4. Subfamily Phloemyzinae, Anoeciinae and Hormaphidinae. In *The Fauna of India and Adjacent Countries.* Zoological Survey of India, Kolkata, 429 pp.

Gibb, A.R., Bunn, B., Pinese, B., Tenakani, D., Kawi, A.P., Ramankutty, P. and Suckling, D.M. (2006). *Identification of the Sex Pheromone of the Red Banded Mango Caterpillar* Deanolis sublimbalis *Snellen.* Hortresearch, Aukland, New Zealand, 20 pp.

Gibb, A.R., Pinese, B., Tenakanai, D., Kawi, A.P., Bunn, B., Ramankutty, P. and Suckling, D.M. (2007). (Z)-11-Hexadecenal and (3Z, 6Z, 9Z)-Tricosatriene: Sex pheromone components of the red banded mango caterpillar *Deanolis sublimbalis*. *J. Chem. Ecol.* 33: 79-589.

Gilbertson, R.L., Rojas, M.R., Kon, T. and Jaquez, J. (2007). Introduction of Tomato yellow leaf curl virus into the Dominican Republic: the development of a successful integrated pest management strategy. In: H. Czosnek (ed.). *The Tomato Yellow Leaf Curl Virus Disease: Management, Molecular Biology, Breeding for Resistance,* pp. 283-307. Springer, Dordrecht, The Netherlands.

Gill, K.S., Dhaliwal, H.S., Randhawa, A.S., Sharma, S.K. and Karam, C. (1986b). PBW 138 variety of bread wheat (*Triticum aestivum* L.). *Journal of Research of the Punjab Agricultural University* 23: 162.

Gill, K.S., Dhaliwal, H.S., Randhawa, A.S., Sharma, S.K., Aujla, S.S. and Karam, C. (1986a). PBW 120 variety of bread wheat (*Triticum aestivum*). *Journal of Research of the Punjab Agricultural University* 23: 161.

Giraddi, R.S., Amaranath, K., Chandra Shekar, Kedanuri, R. and Patil, R.S. (2000). Bioefficacy of new molecules of insecticides against gram pod borer (*Helicovera armigera*) in pigeon pea (*Cajanus cajun*). *Insect Environment* 6: 1–24.

Gogoi, N.K., Phookan, A.K. and Narzary, B.D. (2002). Management of collar rot of Elephant's foot yam. *Indian Phytopathol* 55(2): 238–240.

Gokte, N., Maheshwari, M.L. and Mathur, V.K. (1993). Nematicidal activity of new essential oils against root-knot and cyst nematode species. *Indian Journal of Nematology* 21(2): 123-127.

Golez, H.G. (1991). Bionomics and control of the mango seed borer, *Noorda albizonalis* Hampson (Pyralidae, Lepidoptera). *Acta Horticulturae* 291: 418-424.

Gomes V.M., Souza R.M., Silva, M.M. and Dolinski, C. (2008). Caracterizac, a˜o do estado nutricional de goiabeiras em decli´nio parasitadas por *Meloidogyne mayaguensis*. *Nematol Bras* 32: 154–160.

Gomes, V.M., Souza, R.M., Midorikawa, G., Miller, R. and Almeida, A.M. (2012). Guava decline: Evidence of nationwide incidence in Brazil. *Nematropica* 42: 153-162.

Gomes, V.M., Souza, R.M., Mussi-Dias, V., da Silveira, S.F. and Dolinski, C. (2011). Guava decline: A complex disease involving *Meloidogyne mayaguensis* and *Fusarium solani*. *J. Phytopathol.* 159: 45-50.

Gomes. V.M., Souza. R.M., Corrêa. F.M. and Dolinski. C. (2010). Management of *Meloidogyne mayaguensis* in commercial guava orchards with chemical fertilization and organic amendments. *Nematologia Brasileira* 34: 23-30.

Gonsalves, D. (1998). Control of papaya ringspot virus in papaya: a case study. *Annual Review of Phytopathology* 36: 415–437.

Gopal, M. and Gupta, A. (2002). An opportunistic bacterial pathogen, *Pseudomonas alcaligenes,* may limit the perpetuation of *Oryctes* virus, a biocontrol agent of *Oryctes rhinoceros* L. *Biocontrol Science and Technology* 12(4): 507-512.

Gopal, K., Reddy, M.K., Reddy, D.V.R. and Muniyappa, V. (2010). Transmission of peanut yellow spot virus (PYSV) by Thrips, *Scirtothrips dorsalis* Hood in groundnut. *Archives of Phytopathology and Plant Protection* 43(4/6): 421-429.

Goto, K. and Fukatsu, R. (1956). Studies on the white tip of rice plant. III. Analysis of varietal resistance and its nature. *Bulletin of the National Institue of Agricultural Sciences, Tokyo. Series C. Plant Pathology and Entomology* 6: 123-149.

Goulart, A., Paiva, F. and Mesquita, A (1992). Perdas en trigo (*Triticumaestivum*) causadas por *Pyricularia oryzae*. *Fitopatologia Brasileira* 17: 115-117.

Goulart, A., Paiva, F. and Mesquita, A. (1990). Ocorrência da brusone (*Pyricularia oryzae*) do trigo (*Triticum aestivum*) em Matto Grosso do Sul. *Fitopatologia Brasileira* 15: 112-114.

Goulart, A.C.P., Paiva, F.A. and Andrade, P.J.M. (1995). Relação entre a incidência da brusone em espigas de trigo e a presença de *Pyricularia grisea* nas sementes colhidas. *Fitopatol. Bras.* 20: 184-189.

Goulart, A.C.P., Sousa, P.G. and Urashima, A.S. (2007). Damages in wheat caused by infection of *Pyricularia grisea*. *Summa Phytopathol.* 33: 358–363.

Govindaiah, Dandin, S.B., Philip, T. and Datia, R.K. (1990). Effect of marigold (*Tagetes patula*) intercropping against *Meloidogyne incognita* infecting mulberry. *Indian Journal of Nematology* 20: 96-99.

Govindaiah, Dandin, S.B. and Sharma, D.D. (1991). Pathogenicity and avoidable leaf yield loss due to *Meloidogyne incognita* in mulberry (*Morus alba* L.). *Indian Journal of Nematology* 21: 52-57.

Greer, C.A. and Webster, R.K. (2001). Occurrence, distribution, epidemiology, cultivar reaction, and management of rice blast disease in California. *Plant Disease* 85: 1096-1102.

Gujar, G.T., Kalia, V. and Dhillon, M.K. (2010). Bt-cotton sustainability in India – Environmental concern. In: *Souvenir of the National Symposium on Perspectives and Challenges of Integrated Pest management for Sustainable Agriculture*, pp. 64-70. Indian Society of Pest Management and Economic Zoology and Dr. Y.S. Parmar University of Horticulture and Forestry, Nauni, Solan, Himachal Pradesh, India.

Gupta, G.P., Mahapatro, G.K. and Puri, S.N. (2009). Contemporary scenario of pests and pesticides in the global climate change: Realities, apprehensions and possibilities. In: V.V. Ramamurthy, G.P. Gupta and S.N. Puri (eds) *Proc. Natn. Symp. IPM Strategies to Combat Emerging Pests in the Current Scenario of Climate Change.* pp. 399-412. CAU, Pasighat, Arunachal Pradesh.

Gupta, L., Mishra, H.O., Kalra, A. and Khanuja, S.P.S. (2004). Root-rot and wilt: A new disease of ashwagandha (*Withania somnifera*) caused by *Fusarium solani*. *J. Medicinal and Aromatic plants Sci.* 26: 285-287.

Gupta, M.K. and Goswami, P.K. (1995). Incidence of sugarcane woolly aphid and its effect on yield attributes and juice quality. *Indian Sugar* 44: 883–885.

Gupta, V.K. and Misra A.K. (2009). Efficacy of bioagents against *Fusarium* wilt of guava. *J. Mycol. Plant Pathol.* 39: 101-106.

Gupta, V.K., Misra A.K., Pandey B.K., Ram R.A., Mishra S.P. and Chauhan U.K. (2009a). Eco-friendly management of *Fusarium* sp. causing wilt disease of guava (*Psidium guajava* L.) by bio-dynamic antagonists. *J. Eco-Friendly Agric.* 4: 77-79.

Gupta, V.K., Misra, A.K., Gaur, R.K., Jain, P.K., Gaur, D. and Sharma S. (2010). Current status of wilt disease of guava (*Psidum guajava* L.). *Biotechnology* 9(2): 176-195.

Halder, J., Rai, A.B. and Kodandaram, M.H. (2013). Compatibility of neem oil and different entomopathogens for the management of major vegetable sucking pests. National Academy Science Letters 36(1): 19-25.

Halder, J., Rai, A.B., Dey, D. and Kodandaram, M.H. (2014). Is Apanteles paludicole Cameron, a limiting biotic factor for minor pest status of Sphenarches caffer (Zeller)? Journal of Biological Control 28(2): 119-121.

Halder, J., Rai, A.B., Kodandaram, M.H. and Dey, D. (2011). *Pediobius foveolatus* (Crowford): A promising bioagent against *Epilachna vigitionctopunctata*. *National Conference on Horti-Business Linking Farmers with Market*, p. 133. Dehradun, Uttarakhand.

Halder, J. and Srinivasan, S. (2007). Biochemical basis of resistance to spotted pod borer, Maruca vitrata (Geyer) in urd bean, Annals of plant Protection Sciences 15(2): 287-290.

Halder, J. and Srinivasan, S. (2011). Varietal screening and role of morphological factors on distribution and abundance on spotted pod borer, Maruca vitrata (Geyer) on cowpea. Annals of Plant Protection Sciences 19(1): 71-74.

Halder, J., Srinivasan, S. and Muralikrishna, T. (2006). Role of various biophysical factors on distribution and abundance of spotted pod borer, *Maruca vitrata* (Geyer) in mung bean. *Annals of Plant Protection Sciences* 14(1): 49-51.

Hall, R. (1995). Challenges and prospects of integrated pest management. In: R. Reuveni (ed.) *Novel Approaches to Integrated Pest Management,* pp.1-19. Lewis Publishers, Boca Raton, Florida, USA.

Hamilton, J.G., Dermody, M., Aldea, A.R., Zangerl, A., Rogers, M.R., Berenbaum. and Delucia, E. (2005). Anthropogenic Changes in tropospheric composition increase susceptibility of soybean to insect herbivory. *Environ Entomol* 34: 479-485.

Hammon, R. (2004). Managing thrips in western Colorado onions. In: Proceedings of the 2004 *National Allium Research Conference,* p. 107. Grand Junction, Colorado, USA. Colorado State University, Fort Collins, USA.

Handoo, Z.A. (1998). *Plant-Parasitic Nematodes.* http://www.ars.usda.gov/Services/docs. htm.

Harathi, P.N. (2008). *Screening of Field Bean Genotypes against Pod Borer Complex and Studies on Mechanisms of Insect Pest Resistance.* M. Sc. (Agri.) thesis, Department of Entomology, Sri Venkateswara Agricultural College, Tirupati, Acharya N.G. Ranga Agricultural University, Rajendranagar, Hyderabad.

Harathi, P.N., Hari Prasad, K.V. and Srinivasan, S. (2015). Biology of plume moth, *Sphenarches caffer* zeller on natural and artificial diet. *International Journal of Tropical Agriculture* 33(2): 1013-1016.

Hardwick, N.V., Armstrong, A.C., Ellis, S.A., Gladders, P., Yarham, D.J., Cannon, R.J.C., Baker, R.H.A., Polley, R.W. and Turner, J.A. (1996). The impact of climate change on crop pests and diseases. In R.J. Froud-Williams, R. Harrington, T.J. Hocking, H.G. Smith and T.H. Thomas (eds.) *Implications of "Global Environment Change" for crops in Europe, Aspects of Applied Biology* 45: 261-268.

Harrington, R., Clark, S.J., Welham, S.J., Verrier, P.J., Denholm, C.H., Hullé, M., Maurice, D., Rounsevell, M.D. and Cocu, N. (2007). Environmental change and the phenology of European aphids. *Global Change Biology* 13: 1550–1564.

Harvell, C.D., Mitchell, C.E., Ward, J.R., Altizer, S., Dobson, A.P., Ostfeld, R.S. and Samuel, M.D. (2002). Climate warming and disease risks for terrestrial and marine biota. *Science* 296: 2158–2162.

Hasan, A. (1984). Synergism between Heterodera cajani and Fusarium udum attacking Cajanus cajan. *Nematologia Mediterranea* 12(1): 159-162.

Haseeb, A., Kumar, V., Shukla, P.K. and Ahmad, A. (2006). Effect of different bioinoculants, organic amendments and pesticides on the management of *Meloidogyne incognita – Fusarium solani* disease complex on tomato cv. K-25. *Indian Journal of Nematology* 36: 65-69.

Hassall, C., Thompson D.J., French, G.C. and Harvey, I.F. (2007). Historical changes in the phenology of British Odonata are related to climate. *Global Change Biology* 3: 923-1099.

Hayat, M. (2009). Description of a new species of *Aenasius* Walker (Hymenoptera: Enbcyrtidae), parasitoid of the mealybug, *Phenacoccus solenopsis* Tinsley (Homoptera: Pseudococcidae) in India. *Biosystematica* 3: 21-26.

Heu, R.A., Nagata, N.M., Fukada, M.T. and Yonahara, B.Y. (2002). *Papaya Ringspot Virus Established on Maui*. Department of Agriculture, Hawaii.

Hobbs, P.R., Sayre, K. and Gupta, R. (2008). The role of conservation agriculture in sustainable agriculture. *Phil. Trans. Research Society* 363: 543–555.

Hodar J.A., Castro J. and Zamora R. (2003). Pine processionary caterpillar *Thaumetopoea pityocampa* as a new threat for relict Mediterranean Scots pine forests under climatic warming. *Biological Conservation* 110: 123-129.

Holtzmann OV. (1968). A foliar disease of tuberose caused by *Aphelenchoides besseyi*. *Plant Disease Reporter* 52: 56.

Huger, A.M. (1966). A virus disease of the Indian rhinoceros beetle *Oryctes rhinoceros* (L.) caused by a new type of insect virus, *Rhabdionvirus oryctes* gen. n., sp. n. *J. Invertebr. Path.* 8: 38-51.

Humphreys, D.A., Williamson, V.M., Salazar, L., Flores-Chaves, L. and Gómez-Alpizar, L. (2012). Presence of *Meloidogyne enterolobii* Yang & Eisenback (=*M. mayaguensis*) in guava and acerola from Costa Rica. *Nematology* 14: 199-207.

Hunter, D., Brunt, J. and Delp, C. (2001). AusAID/SPC Taro genetic resources: conservation and utilization. In *A bibliography of taro leaf blight*, pp. 2–15. Secretariat of the Pacific Community, Noumea.

ICRISAT (International Crops Research Institute for the Semi-Arid Tropics) (1992). *Medium Term Plan 1994-98*. 3 Vols. ICRISAT Patancheru, Andhra Pradesh, India.

Igarashi, S. (1990). Update on wheat blast (*Pyricularia oryzae*) in Brazil. In: D Saunders (ed.) *Proc. Int. Conf. Wheat Non-Traditional Warm Areas*, pp. 480-483. Mexico, D.F.

Igarashi, S., Utiamada, C.M., Igarashi, L.C., Kazuma, A.H. and Lopes, R.S. (1986). *Pyricularia* em trigo. 1. Ocorrência de *Pyricularia* sp. no estado do Paraná. *Fitopatol. Bras.* 11: 351-352.

Inagaki, H. (1985). The plant parasitic nematodes important in Japan and the related researches. *JARQ* 18: 194-201.

Ioannou, N. (1987). Cultural management of tomato yellow leaf curl disease in Cyprus. *Plant Pathology* 36(3): 367-373.

IPCC. (2007). Climate change 2007: Impacts, adaptation and vulnerability. In: M.L. Parry, O.F. Canziani, J.P. Palutikof, P.J. van der Linden and C.E. Hanson (eds.) *Contribution of Working Group II to the Fourth Assessment Report of the Intergovernmental Panel on Climate Change*, Cambridge University Press, Cambridge, UK, 976 pp.

Islam, K., Islam, M.S. and Ferdousi, Z. (2011). Control of *Epilachna vigintioctopunctata* Fab. (Coleoptera: Coccinellidae) using some indigenous plant extracts. *Journal of Life and Earth Sciences* 6: 75-80.

Jackson, G. (1999). *Taro Leaf Blight*. Pest Advisory Leaflet No. 3, Plant Protection Service, Secretariat of the Pacific Community, Suva, Fiji Islands, 2 pp.

Jackson, G.V.H. (1996). Strategies for taro leaf blight research in the region. In: *Proceedings of the Taro Leaf Blight Seminar*, pp. 95–100. Alafua, Western Samoa, 22–26 November, 1993. South Pacific Commission, New Caledonia, Noumea.

Jackson, G.V.H., Gollifer, D.E and Newhook, F.J. (1980). Studies on the taro leaf blight fungus *Phytophthora colocasiae* in Solomon Islands: control by fungicides and spacing. *Ann Appl Biol* 96: 1–10.

Jacob, J.P. and Ramesh Kumar, A. (2009). Incidence of galls induced by *Leptocybe invasa* on seedlings of *Eucalyptus camaldulensis* and *E. tereticornis* from different seed sources in southern India. *International Journal of Ecology and Environmental Sciences* 35(2-3): 187-198.

Jacob, T.K. and Devasahayam, S. (2010). Incidence of Erythrina gall wasp (*Quadrastichus erythrinae* Kim), an invasive insect pest on *Erythrina* spp., in major black pepper (*Piper nigrum* L.) areas of Kerala and Karnataka, India. *Journal of Plantation Crops* 38(2): 161-164.

Jadhav, V.T. and Sharma, K.K. (2009). Integrated management of diseases in pomegranate (Lead Paper). *Int. Symp. on Pomegranate and other Minor including Mediterranean Fruits*, pp. 48-52. Univ. of Agri. Sci., Dharwad, Karnataka, India.

Jain, R.K., Pappu, H.R., Pappu, S.S., Reddy, M.K. and Vani, A. (1998). Watermelon bud necrosis *Tospovirus* is a distinct virus species belonging to serogroup IV. *Arch. Virol.* 143: 1637-1644.

Jain, S.S. (1956). A preliminary note on the inactivation of *Fusarium oxysporum* f. *psidii* in guava plants by chemotherapeutic treatment. *Indian J. Hort.* 13: 102-104.

Jamadar, M.M., Jawadagi, R.S., Sataraddi, A.R., Patil, D.R. and Patil, P.V. (2009). Status of pomegranate diseases and their management in North Karnataka in India (Abstr.). *IInd Int. Symp. on Pomegranate and other Minor including Mediterranean Fruits*, p. 134. Univ. of Agri. Sci., Dharwad, Karnataka, India.

James, C. (2009). *Global Status of Commercialized Biotech/GM Crops: 2009.* ISAAA Brief No. 41, International Service for the Acquisition of Agro-Biotech Applications, Ithaca, New York, USA.

Jamwal, V.V.S. Ahmad, H. Sharma, D. Srivastava, K. and Kumar, V. (2013). Comparative biological and morphometric attributes of *Epilachna vigintioctopunctata* Fabr. on brinjal and bitter Gourd. *Vegetos* 26 (2): 426-437.

Janse, J. (1996). Potato Brown Rot in Western Europe – History, Present Occurrence and Some Remarks on Possible Origin, Epidemiology and Control Strategies. *Bulletin OEPP/EPPO* 26: 679–695.

Jatala, P. and Bridge, J. (1993) Nematode parasites of root and tuber crops. In: M. Luc, R.A. Sikora and J. Bridge (eds.) *Plant Parasitic Nematodes in Subtropical and Tropical Agriculture*, pp. 137–180. CAB International, Wallingford.

Javaregowda. (2006). Incidence of snail, *Achatina fulica* (Bowdich) in betel vine and its management. *Pest Management in Horticultural Ecosystems* 12(1): 41-43.

Jayashankar, M., Sridhar, V. and Verghese, A. (2013). Management of the giant African snail, *Achatina fulica* (Bowdich) (Stylommatophora: Achatinidae) in India. *Pest Management in Horticultural Ecosystems* 19(1): 1-9.

Jena, R.N. and Rao, Y.S. (1976). Nature of root knot nematode (*Meloidogyne graminicola*) resistance in rice (*Oryza sativa* L.) I. Isolation of resistant varieties. *Proceedings of the Indian Academy of Sciences* 83: 177- 184.

Jhooty, J.S., Chand, J.N. and Krishnamurthy, V. (1984). *Report of Committee constituted by ICAR on guava decline in Punjab and Haryana.* Submitted to ICAR, New Delhi.

Johnson, N. (2000). *Studies on Nematode Pests of Cut Flowers with Special Reference to Carnation, Cystera, Gladiolus and Asiatic Lily.* Ph. D. thesis, Tamil Nadu Agri. Univ., Coimbatore, 160 pp.

Jones, F.G.W. (1961). The potato root eelworm *Heterodera rostochiensis* Woll. in India. *Current Science* 30: 187.

Jones, R.A.C. (2004). Using epidemiological information to develop effective integrated virus disease management strategies. *Virus Res.* 100: 5-30.

Jotwani, M.G. and Butani, D.K. (1978). Crop pests and their control: Pearl millet. *Pesticides* 12(2): 20-30.

Joubert, M.H. and Frean, R.T. (1993). An investigative evaluation of fungicides against guava wilt. *Inlightingsbull. Inst. Vir Tropiese Subtropiese Gewasse* 246: 3-3.

Kalita, M. and Phukan, P.N. (1990). Reactions of some rice cultivars to *Meloidogyne graminicola. Indian Journal of Nematology* 20: 215-216.

Kamala Jayanthi P.D. and Abraham Verghese. (2010). Sapodilla seed borer, *Trymalitis margarias* Meyrick: An invasive or indigenous species? *Pest Management in Horticultural Ecosystems* 16(2): 141-147.

Kannan, N., Shanmugasundaram, S. and Lakshmanan, M. (1980). Isolation of a bacterial pathogen from the coconut pest *Oryctes rhinoceros* L. *Entomon* 5(4): 285-289.

Kannan, R. and James, D.A. (2009). Effects of climate change on global diversity: a review of key literature. *Tropical Ecol* 50: 31-39.

Kanwar R.S. and Bajaj, H.K. (2010). Cereal cyst nematode infestation in wheat. In MR Khan, MS Jairajpuri (eds.) *Nematode Infestations, Part I: Food Crops*, pp. 192-217. The National Academy of Sciences, Allahabad, India.

Kapoor, V.C. (1993). *Indian Fruit Flies (Insecta: Diptera:Tephritidae)*. Oxford and IBH Publishing Co. Pvt. Ltd., New Delhi, 228 pp.

Kapoor, V.C. and Agarwal, M.L. (1980). Identified collection of fruit flies (Diptera: Tephritidae) in national Pusa collection, New Delhi. *Bulletin of Entomology* 17: 1-15.

Kataki, P.K., Hobbs, P. and Adhikary, B. (2001). The rice-wheat cropping system of South Asia: Trends, constraints and productivity- A prologue. In: P.K. Kataki (ed.) *The Rice-Wheat Cropping System of South Asia: Trends, Constraints, Productivity and Policy. Journal of Crop Production* 3 (2): 1-26.

Kato, K., Onuki, M., Fuji, S. and Hanada, K. (1998). The first occurrence of *Tomato yellow leaf curl virus* in tomato (*Lycopersicon esculentum* Mill.) in Japan. *Ann. Phytopathol. Soc. Japan.* 64: 552-559.

Kauffman, W.C., Meyerdirk, D.E., Miller, D., Schauff, M., Hernandez, H.G. and Villanueva Jimenez J.A. (2001). Papaya mealybug biological control in Puerto Rico and Dominican Republic. *ESA Annual Meeting*, San Diego, CA

Kaushal, K.K., Sharma, G.L. and Paruthi, I.J. (2001). Nematode diseases of wheat and barley and their management. *National Congress on Centenary of Nematology in India-Appraisal and Future Plans*, pp. 23-24. Division of Nematology, Indian Agri. Res. Inst., New Delhi, India.

Kerlin, K. (2002). As global warming rises, so do tree-killing infestations of bark beetles. *E/The Environmental Magazine*, 4 April. Available: http://www.enn.com/news/enn-stories/ 2002/04/ 04042002/s_46495.asp (accessed 8 July 2003).

Khan, A., Shaukat, S.S. and Siddiqui, I.A. (2005). A survey of nematodes of pomegranate orchards in Balochistan province, Pakistan. *Nematol Medit* 33: 25-28.

Khan, M. (2003). Salt mixtures for mirid management. *The Australian Cotton Grower* 24(3): 10.

Khan, M.R. (2001). White-tip nematode, *Aphelenchoides besseyi* in rice-tuberose cropping system. *In* S.K. Mukhopadhyay, D.C. Ghosh and G.C. De (eds.) *Proceedings of National Seminar on Frontiers of Crop Management.* Visva-Bharati, Sriniketan, India.

Khan, M.R. (2004). Observation on foliar nematode, *Aphelenchoides besseyi* in tuberose (*Polianthes tuberosa* L.) in West Bengal. *Annals of Plant Protection Sciences* 12(1): 106-109.

Khan, M.R. (2006a). Managing foliar nematode-a threat for tuberose cultivation. *Indian Horticulture* 51: 17.

Khan, M.R. (2006b). Current options for managing nematode pest of crops in India, pp: 16-50. In: N. Mohilal and R. K. Gambhir (eds.) *Plant Nematology in India*, p.178. Parasitology Lab., Department of Life Sciences, Manipur University, India.

Khan, M.R. and Ghosh, S. (2007). Evaluation of tuberose cultivars against foliar nematode, *Aphelenchoides besseyi*, In: *National Symposium on Nematology in 21st Century: Emerging paradigms*, p. 77. Nematological Society of India, held at Assam, India.

Khan, M.R., Pathak, B. and Ghosh Dostidar, K. (2008). Management approaches of tuberose foliar nematode, *Aphelenchoides besseyi* in West Bengal, India. *Indian Journal of Nematology* 38: 25-29.

Khan, M.R., Shit, S., Pal, A.K. and Biswas B. (2005). Integrated approach for management of foliar nematode, *Aphelenchoides besseyi* infecting tuberose in West Bengal, India. *International Journal of Nematology* 15: 30-34.

Khan, M.R. Shit, S., Pal, A.K., Biswas, B. (2006). Management of foliar nematode, *Aphelenchoides besseyi* infecting tuberose in West Bengal, India. *Indian J. Nematol.* 36 (1): 44-47

Khoo, K.C., Ooi, P.A.C. and Ho, C.T. (1991). *Crop pests and their Management in Malaysia.* Tropical Press Sdn. Bhd, Kuala Lumpur, Malaysia, 242 pp.

Kimber, I. (2008). UK moths: Oak Processionary *Thaumetopoea processionea.* http://ukmoths.org.uk/show.php?id=4228.

Kinh, D.N., Huong, N.M. and Ut, N.U. (1982). Root-knot disease of rice in the Mekong Delta, Vietnam. *Int. Rice Res. Newsl.* 7(4): 15.

Kiritani, K. (2006). Predicting impact of global warming on Population dynamics and distribution of arthropods in Japan. *Popul Ecol* 48: 5-12.

Kiritani, K. (2007). The impact of global warming and land-use change on the pest status of rice and fruit bugs (Heteroptera) in Japan. *Global Change* 13: 1586-1595.

Kirk, W.D.J. (1997). Distribution, abundance and population dynamics. In: T. Lewis (ed.) *Thrips as Crop Pests,* pp. 217-257. CAB International, Wallingford, UK.

Kishore, P. (2005). Changing scenario of the pest status in sorghum and pearl millet. *J Entomol Res* 29(3): 183-188.

Kishore, P. and Srivastava, K.P. (1976). Occurrence of cotton grey weevil as a serious pest of sorghum. *Entomologists' Newsletter* 6(3): 30-31.

Kobayashi, Y. and Sugiyama, T. (1977). Dissemination and reproduction of white tip nematode, *Aphelenchoides besseyi*, in mechanical transplanting rice culture. *Japanese Journal of Nematology* 7: 74-77.

Koenning, S.R. and Barker, K.R. (1992). Relative damage functions and reproductive potentials of *Meloidogyne arenaria* and *M. hapla* on peanut. *J. Nematol.* 24: 187-192.

Koenning, S.R., Overstreet, C., Noling, J.W., Donald, P.A., Becker, J.O. and Fortnum, B.A. (1999) Survey of crop losses in response to phytoparasitic nematodes in the United States for 1994. *J Nematol* 31(4S): 587–618.

Kohli, M.M., Mehta, Y.R., Guzman, L., Viedma, L.D. and Cubilla, L.E. (2011). *Pyricularia* blast - a threat to wheat cultivation. *Czech Journal of Genetics and Plant Breeding* 47(Special Issue): S130–134.

Kore, B. (2006). Erythrina gall wasp, *Quadrastichus erythrinae. Current Science* 91(1): 8.

Koshy, P.K. and Swarup, G. (1971a). Investigations on the life history of the pigeon-pea cyst nematode, *Heterodera cajani. Indian Journal of Nematology* 1: 44-51.

Koshy, P.K. and Swarup, G. (1971b). On the number of generations of *Heterodera cajani*, the pigeon-pea cyst nematode in a year. *Indian Journal of Nematology* 1: 88-90.

Koshy, P.K. and Swarup, G. (1971c). Factors affecting emergence of larvae from cysts of *Heterodera cajani* Koshy, 1967. *Indian Journal of Nematology* 1: 209-219.

Koshy, P.K. and Swarup, G. (1973). Susceptibility of plants to pigeon-pea cyst nematode, *Heterodera cajani. Indian Journal of Nematology* 2(1972): 1-6.

Kranthi, K.R. (2012). *Bt cotton Question and Answers.* Indian Society for Cotton Improvement (ISCI), Mumbai, 70 pp.

Kranthi, K.R., Kranthi, S., Gopalakrishnan, N., Asokan, R. and Mayee, C.D. (2009) Bt resistance-Its management and prospects in the current context of climate change. In: V.V. Ramamurthy, G.P. Gupta and S.N. Puri (eds.) *Proc. Natn. Symp. IPM Strategies to Combat Emerging Pests in the Current Scenario of Climate Change,* pp. 237-261. CAU, Pasighat, Arunachal Pradesh.

Kreye, C., Bouman, B.A.M., Reversat, G., Fernandez, L., Vera Cruz, C., Elazegui, F., Faronilo, J.E. and Llorca L. (2009). Biotic and abiotic causes of yield failure in tropical aerobic rice. *Field Crops Research* 112: 97-106.

Krishanappa, K., Reddy, B.R, Ravichandra, N.G. and Ravindra, H. (2001). Incidence of Root-knot nematode on rice in Karnataka (Abstr.). *National Congress on centenary of Nematology in India Appraissal and future plans,* Indian Agricultural Research Institute, New Delhi.

Krishna Kumar, N.K. (1998). *Bio-Ecology and the Management of the Serpentine Leaf Miner,* Liriomyza trifolii *on Vegetable Crops.* Final report of ICAR AD-hoc scheme, IIHR, Bangalore.

Krishna Kumar, N.K., Krishna Moorthy, P.N., Easther, A.P. and Rama, N. (1997). Seasonality and management of the gall midge, *Aspondylia* sp. on brinjal and sweet peppers. *Proc. First Natl. Symp. Hortil. Crops,* pp. 98-101. AAPMHE, Bangalore.

Krishna Moorthy, P.N. and Krishna Kumar, N.K. (2002). Advances in the use of botanicals in the IPM of major vegetable crops. *Proc. Intn. Conf. Vegetables,* PNASF, Bangalore.

Krishna Moorthy, P.N., Tewari, G.C. and Srinivasan, K. (1988). Action thresholds for the management of pod borer, *Adisura atkinsoni* Moore on field bean, *Lablab purpureus. Insect Sci. Applic.* 9: 489-492.

Krishna Prasad, K.S. (1986) Potato nematodes. In G. Swarup and D.R. Dasgupta (eds.) *Plant Parasitic Nematodes in India.* Indian Agri Res. Inst., New Delhi.

Krishnaiah, K. (1993). Changing scenario in rice insect pest problems. In: H.C. Sharma and M. Veerabadra Rao (eds.) *Pests and pest management in India– The changing scenario,* pp.11-18. Plant Protection Association of India, Hyderabad.

Krishnamoorthy, A. and Mani, M. (1996). Suppression of brinjal mealy bug *Coccidohystrix insolita* with *Cryptolaemus montrouzieri. Insect Envt.* 2: 50.

Krupashanker, S. (1998). *Studies on Bud Necrosis Virus Disease of Watermelon.* M.Sc. (Ag) thesis, University of Agricultural Sciences, Bangalore, 110 pp.

Kumar, B. (2008). *Studies on Root-Knot and Wilt Complex in* Coleus forskohlii *(Wild.) Briq. Caused by* Meloidogyne incognita *(Kofoid and White) Chitwood and* Fusarium chlamydosporum *(Frag. and Cif.) Booth.* M. Sc. (Agri) thesis in Plant Pathology, University of Agricultural Sciences, Dharwad.

Kumar, K., Kumar, A. and Nath, V. (2011). Emerging pests and diseases of litchi (*Litchi chinensis* Sonn.). *Pest Management in Horticultural Ecosystems* 17(1): 11-13.

Kumar, N.G., Nirmala, P. and Jaayappa, A.H. (2010). Effect of various methods of application of insecticides on the incidence of serpentine leaf miner, *Liriomyza trifolii*

(Burgess) and other pests in soybean. *Karnataka Journal of Agricultural Sciences.* 23(1):130-132.

Kumar, N.K.K. and Rawal, R.D. (1999). Onion thrips, *Thrips tabaci*, a vector of onion tospovirus. *Insect Environment* 5(2): 52.

Kumar, P.K.V., Prakasan, C.B. and Vijayalakshmi, C.K. (1990). Coffee berry borer, *Hypothenemus hampei* (Coleoptera : Scolytidae): First record from India. *J. Coffee Res* 20(2): 161-164.

Kumar, R. Nitharwal, M., Chauhan, R., Pal, V. and Kranthi, K.R. (2011). Evaluation of ecofriedly control methods for the management of mealy bug, *Phenacoccus solenopsis* Tinsley in cotton. *Journal of Entomology* 9: 32-40.

Kumar, S. and Singh, D. (2014). Integrated management of Karnal bunt of wheat. *J Industrial Pollution* 30(2): 247-250.

Kumarasinghe, N.C. and Basnayake, B.R.S.B. (2009). Influence of monsoonal weather on sudden establishment of the sugarcane woolly aphid in Sri Lanka. *Sugar Tech* 11(3): 267-273.

Kumashiro, B.R., Lai, P.Y., Funasaki, G.Y. and Teramoto, K.K. (1983). Efficacy of *Nephaspis amnicola* and *Encarsia? haitiensis* in controlling *Aleurodicus dispersus* in Hawaii. *Proceedings of the Hawaiian Entomological Society* 24: 261-269.

Kumutha, K. (2001). *Symbiotic Influence of AM Fungi and Rhizobacteria on Biochemical and Nutritional Changes in Mulberry* (Morus alba). Ph. D. thesis, Tamil Nadu Agri. Univ., Coimbatore.

Kunik, T., Salomon, R., Zamir, D., Navot, N., Zeidan, M., Michelson, I., Gafni, Y. and Czosnek, H. (1994). Transgenic tomato plants expressing the tomato yellow leaf curl virus capsid protein are resistant to the virus. *Bio/Technology* 12(5): 500-504.

Kushwaha, K.S., Allah Noor and Rathore, R.R.S. (1980). Insect pest management in arid agriculture. *Annals of Arid Zone* 19: 503-509.

Lateef, S.S. (1977). A new hymenopteran pest *Tanaostigmodes* sp. recorded on pigeonpea at ICRISAT, Hyderabad, India. *Trop. Grain Leg. Bull.* 7: 6-7.

Laurila, J., Ahola, V., Lehtinen, A., Joutsjoki, T., Hannukkala, A., Anne Rahkonen, A. and Pirhonen, M. (2008). Characterization of *Dickeya* strains isolated from potato and river water samples in Finland. *European Journal of Plant Pathology* 122: 213-225.

Lee, M.D., Chen, C.H., Tsay, T.T. and Lin, Y.Y. (1998). Survey and control of guava nematode diseases. *Plant Protection Bulletin (Taipei)* 40: 265-276.

Lee, Y.B., Park, J.S. and Han, S.C. (1972). Studies on the chemical control of white-tip nematode, Aphelenchoides besseyi Christie, before transplanting. *Korean Journal of Plant Protection* 11(1): 37-40.

Leu, L.S. and Kao C.W. (1979). Artificial inoculation of guava with *Myxosporium psidii. Pl. Dis. Reptr.* 63: 1077-1079.

Leu, L.S., Kao, C.W., Wang, C.C., Liang, W.J. and Hsieh, S.P.Y. (1979). Myxosporium wilt of guava and its control. *Pl. Dis. Reptr.* 63: 1075-1077.

Lewis, T. (1997). *Thrips as Crop Pests.* CAB Int., New York.

Liao, C.T. and Lin, C.S. (2000). Occurrence of the legume pod borer *Maruca testulalis* (Geyer) (Lepidoptera: pyralidae) on cowpea (*Vigna unguiculata* Walp) and its insecticidal application trial. *Plant protection Bulletin* 42: 213 – 222.

Lima, I.M., Souza, R.M., Silva, C.P. and Carneiro, R.M.D.G. (2005). *Meloidogyne* spp. from preserved areas of Atlantic Forest in the State of Rio de Janeiro, Brazil. *Nemat. Bras.* 29: 31-38.

Linford, M.B. (1932). Transmission of the *Pineapple yellow spot virus* by *Thrips tabaci*. *Phytopathol.* 22: 301-349.

Lingappa, S. (2003). Combating the woolly menace. *Deccan Herald*, 4 March 2003.

Lingaraju, S. and Mallesh, S.B. (2010). Emerging nematode diseases in two horticultural crops: Their management through novel approaches. *National Conf. on Innovations in Nematological Research for Agricultural Sustainability – Challenges and A Roadmap Ahead*, p. 51. Tamil Nadu Agri. Univ., Coimbatore.

Logani, R., Pandey R.C., Dwivedi B.K. and Singh S.P. (2002). An ecofriendly management of wilt disease of guava by wiltnema (a botanical fungicide). *Bioved* 13: 47-50.

Lopez, A.W. (1931). *Ann. Rep. Res. Bur. Philipp. Sugar Assoc.* Entomology Department Report, 273 pp.

López, R., CaRmona, D., VinCini, A.M., Rubbianesi, G. and CaLDiz, D. (2010). Population dynamics and damage caused by the Leaf miner *Liriomyza huidobrensis* Blanchard (Diptera: Agromyzidae), on seven potato processing varieties grown in temperate environment. *Neotropical Entomology* 39(1): 108-114.

Lord, J.S., Lazzeri, L., Atkinson, H.J. and Urwin, P.E. (2011). Biofumigation for control of pale potato cyst nematodes: activity of Brassica leaf extracts and green manures on *Globodera pallida* in vitro and in soil. *Journal of Agricultural and Food Chemistry* 59: 7882-7890.

Maciel, J.L.N., Ceresini, P.C., Castroagudin, V.L., Zala, M., Kema, G.H.J. and McDonald, B.A. (2014). Population structure and pathotype diversity of the wheat blast pathogen *Magnaporthe oryzae* 25 years after its emergence in Brazil. *Phytopathology* 104: 95–107.

Maciel, J.L.N., Paludo, E.A., Só e Silva, M., Scheeren, P.L. and Caierão, E. (2008). Reação à brusone de genótipos de trigo do programa de melhoramento da Embrapa Trigo no estádio de planta adulta. Embrapa Trigo, Passo Fundo, RS, Brazil.

Magina, F.L. (2005). *A Review of Coffee Pest Management.* www.aaec.vt.edu/ ipmcrspuganda.

Mahto, T.P. and Yadav, R.P. (2009). Effect of manuring with oilcake based vermicompost on incidence of red spider mite of Okra crop. *Annals of Plant Protection Sciences* 17: 40-45.

Makkouk, K.M. and Laterrot, H. (1983). Epidemiology and control of tomato yellow leaf curl virus. In: R.T. Plumb and J.M. Thresh (ed.) *Plant Virus Epidemiology. The Spread and Control of Insect-Borne Viruses,* pp. 315-321. Blackwell Scientific Publications, Oxford, United Kingdom.

Malaker, P.K., Verma, N.C.D., Tiwari, T.P., Collis, W.J., Duveiller, E., Singh, P.K., Joshi, A.K., Singh, R.P., Braun, H.J., Peterson, G.L., Pedley, K.F., Ferma, M.L. and Valent, B. (2016). First report of wheat blast caused by *Magnoporthe oryzae* pathotype triticum in Bangladesh. *Plant Dis*ease 100(11): 2330.

Malik, M.S., Sangwan, N.K., Dhindsa, K.S. and Bhatti, D.S. (1987). Nematicidal activity of extracts of *Xanthium strumarium*. *Pesticides* 21(10): 19-20.

Mallesh, S.B. 2008. *Plant Growth Promoting Rhizobacteria, Their Characterization and Mechanisms in the Suppression of Soil Borne Pathogens of Coleus and Ashwagandha.* Ph. D. thesis, University of Agricultural Sciences, Dharwad, 206 pp.

Malumphy, C. (1997). Laurel scale, *Aonidia lauri* (Bouché)(Homoptera, Coccoidea, Diaspididae), a pest of bay laurel, new to Britain. *Entomologists Gazette* 48: 195-198.

Mamatha, T. and Rai V.R. (2004). Evaluation of fungicides and plant extracts against *Fusarium solani* leaf blight in *Terminalia catappa*. *J. Mycol. Plant Pathol.* 34: 306-307.

Mathur, R.S. (1956). Guava diseases in India. *Indian J. Hort.* 13: 26-29.

Mani, M. (2010). Origin, introduction, distribution and management of the invasive spiralling whitefly *Aleurodicus dispersus* Russell in India. *Karnataka Journal of Agricultural Sciences* 23(1): 59-75.

Mani, M. and Krishnamoorthy, A. (2002). Classical Biological Control of the Spiralling whitefly, *Aleurodicus dispersus* Russell—An Appraisal. *International Journal of Tropical Insect Science* 22(4): 263-273.

Mani, M., Dinesh, M.S. and Krishnamoorthy, A. (2000b). Presence of *Encarsia* spp. on spiralling whitefly *Aleurodicus dispersus* (Russell) in peninsular India. *Insect Environment* 6: 100

Mani, M., Krishnamoorthy, A. and Dinesh, M.S. (2000a). Biological control studies on the spiraling whitefly *Aleurodicus dispersus* Russel. *Entomological Congress: Perspectives for New Millenium*, Trivandrum, India.

Manoj Kumar, R., Rao, M.S. and Ramachandran, N. (2010). Biomanagement of disease complex caused by *Meloidogyne incognita* and *Phytophthora parasitica* Dastur in protected cultivation of gerbera using formulation of *Trichoderma harzianum* and *Pseudomonas fluorescens*. *National Conf. on Innovations in Nematological Research for Agricultural Sustainability – Challenges and A Roadmap Ahead*, pp. 97-98. Tamil Nadu Agri. Univ., Coimbatore.

Manser, P.D. (1968). *Meloidogyne graminicola* a cause of root knot of rice. *FAO Plant Prot. Bull.* 16: 11.

Mansour, A. and Al-Musa, A. (1992). Tomato yellow leaf curl virus: Host range and virus-vector relationships. *Plant Pathol.* 41: 122-125.

Marangoni, M.S., Nunes, M.P., Fonseca, N. Jr. and Mehta Y.R. (2013). *Pyricularia* blast on white oats: a new threat to wheat cultivation. *Trop. plant pathol.* 38(3): 198-202.

Maranhão, S.R.V.L., Moura, R.M. and Pedrosa, E.M.R. (2001). Reação de indivíduos segregantes de araçazeiro a *Meloidogyne incognita* raça 1, *M. javanica* e *M. mayaguensis*. *Nemat. Bras.* 27: 173-178.

Marshall, D., Work, T.T. and Cavey, J.F. (2003). Invasion pathways of Karnal bunt of wheat into the United States. *Plant Disease* 87: 999-1003.

Maryana, N., Anastasia, D. and Prima, R. (2006). *Asphondylia* sp. (Diptera: Cecidomyiidae) infesting chilli pods in Bogor, West Java, Indonesia. *6th Int. Cong. of Dipterology*, p. 157. Fukuoka, Japan.

Mathur, R.S. and Jain S.S. (1960). Selecting guavas for wilt resistance. *Proc. Natl. Acad. Sci. B.* 30: 33-36.

Mathur, R.S., Jain S.S. and Swarup J. (1964). Chemical treatment for guava wilt. *Proc. Natl. Acad. Sci.* 34: 33-36.

Mathur, V.K. (1986). Quarantine: an effective and possible method to control plant parasitic nematodes. In: G. Swarup and D.R. Dasgupta, eds. *Plant-Parasitic Nematodes of India, Problems and Progress*, pp. 490-497. Indian Agricultural Research Institute, New Delhi, India.

Mathur, V.K., Arya, H.C., Mathur, R.L. and Handa, D.K. (1974). The occurrence of biotypes of cereal cyst nematode (*Heterodera avenae*) in the light soils of Rajasthan and Haryana, India. *Nematologica* 20:19-26

Maurice, N.G. and Ramteke, P.W. (2012). Development of *Epilachna vigintioctopunctata* (Fabricius) (Coleoptera: Coccinellidae) on essential and alternative foods. *International Journal of Advance Pharmaceutical and Biological Sciences* 2(3): 199-204.

Mazyad, H.M., Nakhla, M.K., El-Amrety, A.A. and Dos, S.A. (1986). Further studies on the epidemiology of tomato yellow leaf curl virus in Egypt. Egypt *Acta Horticulturae* 190: 121-130.

McGee, D.C. (1997). *Plant Pathogens and the Worldwide Movement of Seeds.* APS Press.

McGregor, A.J. and Moxon, J.E. (1985). Potential for biological control of tent building ants associated with *Phytophthora palmivora* (Butl) pod rot of cocoa in Papua New Guinea. *Ann Appl Biol* 107: 271–277.

Meagher, J.W. (1970). Seasonal fluctuations in numbers of the cereal cyst nematode (*Heterodera avenae*) and of *Pratylenchus minyus* and *Tylenchorhynchus brevidens* in soil. *Nematologica* 16:333-347.

Meagher, J.W. (1974a). *The Biology of the Cereal Cyst Nematode* (Heterodera avenae *Woll.*) *and its Interaction with a Fungus in a Soil-Borne Disease of Wheat.* Ph.D. Thesis, University of Melbourne.

Meagher, J.W. (1974b). Cryptobiosis of the cereal cyst nematode (*Heterodera avenae*) and effects of temperature and relative humidity on survival of eggs in storage. *Nematologica* 20: 323-336.

Mehta, Y.R. and Baier, A. (1998). Variação patogênica entre isolados de *Magnaporthe grisea* atacando triticale e trigo no estado do Paraná. *Summa Phytopathologica* 24: 119–125.

Mehta, Y.R., Nunes, M.P. and Oliveira, J.C. (2006). Occurrência de brusone em aveia no Estado do Paraná. In: *Resultados Experimentais. XXVI Reunião da Comissão Brasileira de Pesquisa de Aveia*, pp. 55–57. FAPA, Guarapuava, Paraná.

Meyerdirk, D.E. and Kauffman, W.C. (2001). *Status on the Development of a Biological Control Program for* Paracoccus marginatus *Williams, Papaya Mealybug.* Internal USDA, APHIS, PPQ Report.

Michelson, I., Zamir, D. and Czosnek, H. (1994). Accumulation and translocation of *Tomato yellow leaf curl virus* (TYLCV) in a *Lycopersicon esculentum* breeding line containing the *L. chilense* TYLCV tolerance gene Ty-1. *Phytopathology* 84: 928-933.

Milan, A.R. (2007). Breeding of *Psidium* species for root knot nematode resistance in Malaysia. *Acta Hort.* 735: 61-69.

Milan, A.R. (2010). Collection and evaluation of guava (*Psidium guajava* L.) for nematode resistance in Malaysia. *Acta Hort.* 849: 357-362.

Mishra, S. (2006). *Molecular Studies of Virulence Factor Genes (PG, KHS, CHS, PDA, PELA and FMK/MAPK) of* Fusarium oxysporum *in Guava Wilt.* M.Sc. Thesis, Molecular Diagnostic Laboratory, CDRI, Lucknow, U.P., India.

Misra, A.K. (2005). Important diseases of guava in India with special reference to wilt. *Acta Hortic.* 735: 507-523.

Misra, A.K. and Gupta V.K. (2009). *Trichoderma*: Biology, biodiversity and biotechnology. *J. Eco-Friendly Agric.* 4: 99-117.

Misra, A.K. and Pandey B.K. (1994). *Studies on Guava Wilt.* CIHNP, Lucknow, 28 pp.

Misra, A.K. and Pandey, B.K. (2000b). Progressive natural wilting of guava plants during different months. *Indian Phytopath.* 53: 423-427

Misra, A.K. and Prasad B. (2003). Relative efficacy of different bio-agents for the control of guava wilt. *J. Mycol. Plant Pathol.* 33: 494-494.

Misra, A.K. and Prasad D. (2004). *Aspergillus niger* strain AN 17 potent bioagent to control wilt disease and its easy multiplication. *Proceedings of the Symposium on Recent Advances in Fungal Bioagents and Their Social Benefits*, pp. 12-12. NBRI, Lucknow.

Misra, A.K. and Shukla, S.K. (2002). Assessment of loss due to guava wilt around Lucknow. *National Seminar on Production and Post-Harvest Technology of Guava,* pp. 34-35. Dept. of Hort., CSAUA&T, Kanpur.

Misra, A.K., Pandey B.K., Prasad B. and Shukla S.K. (2003a). Pathogenic diversity in the cause of wilt disease of guava. *Indian Phytopath.* 56: 314-314.

Misra, A.K., Prakash O, and Sen B. (2000). Biological control of guava wilt by *Aspergillus niger* strain AN17 (Pusa Mrida). *Proceedings of the National Seminar on Hi-tech Horticulture,* pp. 149-149. Bangalore.

Misra, A.K., Prasad D., Prasad B. and Shukla S.K. (2004). Effective management of wilt disease of guava. *Proceedings of the National Symposium on Crop Surveillance: Disease Forecasting and Management,* pp. 92-93. Div. of Plant Pathology, IARI, New Delhi.

Misra, A.K., Rajan S., Prasad B., Shukla S.K. and Prasad D. (2003b). Resistant source of wilt disease of guava. *Seminar on Plant Diseases of National Importance with Special Reference to Guava Wilt and Mango Malformation* (Souvenir and Abstrct), pp: 51-52. Zonal Conference (East Zone), Indian Society of Mycology and Plant Pathology, CISH and IISR, Lucknow.

Mittal, V. and Ujagir, R. (2005). Evaluation of naturalyte spinosad against pod borer complex in early pigeon pea. *Indian Journal of Plant Protection* 33(2): 211-215.

Mohandas, C. (2001). Nematode diseases of tuber crops and their management. *Natl. Cong. on Centenary of Nematol. in India – Appraisal & Future Plans,* pp. 35-36. Indian Agri. Res. Inst., New Delhi.

Mohindru, B., Jindal, V. and Dhawan, A.K. (2009). Record of parasitoid on mealy bug *Phenacoccus solenopsis* in Tomato. *Indian J. Ecol.* 36(1): 101-102.

Mojtahedi, H., Santo, G.S. and Wilson, J.H. (1993). Managing *Meloidogyne chitwoodi* on potato with rapeseed as green manure. *Plant Dis.* 77(1): 42–46.

Mojumdar, V., Mishra, S.D., Haque, M.M. and Goswami, B.K. (1989). Nematicidal efficacy of some wild plants against pigeon pea cyst nematode, *Heterodera cajani. International Nematology Network Newsletter* 6(2): 21-24.

Molero, T., Molina, J. and Casassa-Padron, A.M. (2010). Advances in the genetic study of resistance of cultivars of *Psidium* spp. to *Meloidogyne* spp. in a tropical dry forest. *Acta Hort.* 849: 309-318.

Montoya, E.C. (1999). Caracterización de la infestación del café por la broca y efecto deldaño en la calidad de la bebida. *Cenicafé (Colombia)* 50: 245-258.

Morales, F.J. and Anderson, P.K. (2001). The emergence and dissemination of whitefly-transmitted geminiviruses in Latin America– Brief review. *Arch. Virol.* 146: 415–441.

Morilla, G., Janssen, D., García-Andrés, S., Moriones, E., Cuadrado, I. M. and Bejarano, E. R. (2005). Pepper (*Capsicum annuum*) is a dead-end host for *Tomato yellow leaf curl virus. Phytopathology* 95: 1089-1097.

Morimoto, R.N., Imura, O. and Kiura, T. (1998). Potential effects of global warming on the occurence of Japanese pest insects. *Appl Entomol Zool* 38: 147-155.

Morin, S., Ghanim, M., Zeidan, M., Czosnek, H., Verbeek, M. and van den Heuvel, J.F.J.M. (1999). A GroEL homologue from endosymbiotic bacteria of the whitefly *Bemisia tabaci* is implicated in the circulative transmission of *Tomato yellow leaf curl virus. Virology* 256: 75–84.

Moriones, E. and Navas-Castillo, J. (2000). *Tomato yellow leaf curl virus,* an emerging virus complex causing epidemics worldwide. *Virus Res.* 71: 123-134.

Moyer, J.W., German, T., Sherwood, J.L. and Ullman, D. (1999). An update of *Tomato spotted wilt virus* and related *Tospoviruses* (On-line).Available: http://www.apsnet.org./online/featurestospovirus (05 Nov. 2013).

Mugniery, D. and Balandras, C. (1984). Examen des possibilit6s d'6radication du n~matode ft kystes, *Globodera pallida* Stone. *Agronomie* 4: 773- 778.

Mukherjee, P.K. and Raghu, K. (1997). Effect of temperature on antagonistic and biocontrol potential of *Trichoderma* sp. on *Sclerotium rolfsii*. *Mycopathologia* 139(3): 151–155.

Muniappan, R., Meyerdirk, D.E., Sengebau, F.M., Berringer, D.D. and Reddy, G.V.P. (2006). Classical biological control of the papaya mealy bug, *Paracoccus marginatus* (Hemiptera: Pseudococcidae) in the republic of Palau. *Florida Entomologist* 89(2): 212-217.

Muniappan, R., Shepard, B.M., Watson, G.W., Carner, G.R., Sartiami, D., Rauf A. and Hammig, M.D. (2008). First report of the papaya mealybug, *Paracoccus marginatus* (Hemiptera: Pseudococcidae), in Indonesia and India. *J Agric Urban Ent.* 25 (1): 37–40.

Muthulakshmi M., Devrajan K. and Jonathan E. I. (2010a). Biocontrol of root knot nematode, *Meloidogyne incognita* (Kofoid and White) Chitwood in mulberry (*Morus alba* L.). *Journal of Biopesticides* 3(2): 479 – 482.

Muthulakshmi M., Jonathan E. I. and Devrajan K. (2010b). Interaction of *Meloidogyne Incognita* and *Macrophomina Phaseouna* on Mulberry (*Morus alba* L.), *National Conference on Innovations in Nematological Research for Agricultural Sustainability - Challenges and A Roadmap Ahead*, p. 108. Nematological Society of India & Tamil Nadu Agricultural University, Coimbatore.

Nadagouda, S., Patil, B.V., Venkateshalu. and Sreenivas, A.G. (2010). Studies on development of resistance in serpentine leaf miner, *Liriomyza trifolii* (Burgess) (Agromyziidae; Diptera) to insecticides. *Karnataka Journal of Agricultural Sciences* 23(1): 56-58.

Nagaraju, D.K. (2000). *Biology, Ecology and Management of the Capsicum Gall Midge,* Asphondylia capparis *Rubsaaman (Diptera: Cecidomyiidae) and other Insects Associated with Gall on Bell Pepper.* Ph. D. Thesis. Univ. Agric. Sci., Bangalore (India).

Nagesh, M. and Parvatha Reddy, P. (2000). Crop loss estimation in carnation and gerebera due to the root-knot nematode, *Meloidogyne incognita* (Kofoid & White) Chitwood. *Pest Management in Horticultural Ecosystems* 6: 158-159.

Nagesh, M. and Parvatha Reddy, P. (2005). Management of carnation and gerbera to control the root-knot nematode, *Meloidogyne incognita,* in commercial polyhouses. *Nematologia Mediterranea* 33: 157-162.

Nagesh, M., Hussaini, S.S., Ramanujam, B. and Chidanandaswamy, B.S. (2006). Management of *Meloidogyne incognita* and *Fusarium oxysporum* f. sp. *lycopersici* wilt complex using antagonistic fungi in tomato. *Nematologia Mediterranea* 34: 63-68.

Nagrare, V.S., Kranthi, S., Biradar, V.K., Zade, N.N., Sangode, V., Kakde, G., Shukla, R.M., Shivare, D., Khadi, B.M. and Kranthi, K.R. (2009). Widespread infestation of the exotic mealybug species *Phenacoccus solenopsis* (Tinsley) (Hemiptera: Pseudococcidae) on cotton in India. *Bulletin of Entomological Research* 99: 537-541.

Naik, D. (2004). *Biotechnological Approaches for the Management of Wilt Disease Complex in Capsicum* (Capsicum annum *L.*) and *Eggplant* (Solanum melongena) *with Special Emphasis on Biological Control.* Ph. D. thesis, Kuvempu Univ., Shimoga.

Naik, D., Rao, M.S., Shylaja, M., Reddy, M.S. and Rahiman, B.A. (2003). Management of wilt disease complex in capsicum using *Bacillus pumilus* and *Paecilomyces lilacinus. Proc. of Int. Plant Growth Promoting Rhizobacteria Workshop*, pp. 276-281. Calicut, India.

Nair, C.P.R. (2000). Studies on coconut eriophyid mite *Aceria guerreronis* K. in India. *Proceedings of the International Workshop on Coconut Eriophyid Mite* (Aceria gurerreronis), pp. 9-12. Coconut Research Institute, Sri Lanka.

Nair, M.R.G.K. (1970). *Insects and Mites of Crops in India.* New Jack Printing Works Private Limited, Bombay, 32 pp.

Nair, M.R.G.K. (1995). *Insects and Mites of Crops in India.* ICAR, New Delhi, pp. 55, 163, 165.

Nasira, K., Shaheen, N. and Shahina, F. (2011). Root-knot nematode *Meloidogyne incognita wartellei* on pomegranate in Swat, KPK. *Pakistan J Nematol* 29(1): 117-118.

Naveen, V., Naik. M.I., Manjunatha, M., Pradeep, S., Shivanna, B.K. and Sridhar, S. (2009). Biology of legume pod borer, *Maruca testulalis* Geyer on cowpea. *Karnataka J. Agric. Sci.* 22 (3-Spl. Issue): 668-669.

Neeraj Joon, M.S. and Bhatia, S.K. (2005). Guava wilt. An investigative report of its occurrence causes and cures. *Haryana J. Hort. Sci.* 34: 201-207.

Nelson, S. (2013). *Bacterial Wilt of Edible Ginger in Hawai'i.* Plant Disease PD-99, College of Tropical Agriculture and Human Resources, University of Hawai'i at Mānoa, Honolulu, Hawai'i.

Netscher, C. and Erlan. (1993). A root-knot nematode, *Meloidogyne graminicola*, parasitic on rice in Indonesia. *Afro-Asian J. Nematol.* 3: 90–95.

Nicotina, M. and Cioffi, E. (2000). Diffusion of *Polyphagotarsonemus latus* (Acarina: Tarsonemidae) in the Campania and Lazio regions (Central-South Iatly): A real threat for many agricultural and floral crops. *Informative Fitopatologico* 11:59-64.

Nirmala, R. (2003). *Evaluation of Entomopathogenic Fungi against Sugarcane Woolly Aphid,* Ceratovacuna lanigera *Zehntner (Homoptera: Aphididae) and Other Aphid Species.* Report of the 'Hands on Training' submitted to the Project Directorate of Biological Control, Bangalore.

Nisha, M.S and Sheela, M.S. (2006). Bio-management of *Meloidogyne incognita* on Coleus, *Solenostemon rotundifolius* by integrating solarization, *Paecilomyces lilacinus, Bacillus macerans* and neem cake. *Indian J. Nematol.* 36: 136-138.

Nishizawa, T. (1953). Studies on the varietal resistance of rice plants to the rice nematode disease "senchu shingar byo" (VI). *Bulletin of the Kyushu Agricultural Experimental Station* 1: 339-349.

Noris, E., Accotto, G.P., Tavazza, R., Brunetti, A., Crespi, S. and Tavazza, M. (1996). Resistance to tomato yellow leaf curl geminivirus in *Nicotiana benthamiana* plants transformed with a truncated viral C1 gene. *Virology (New York)* 224(1): 130-138.

Noris, E., Lucioli, A., Tavazza, R., Caciagli, P., Accotto, G.P. and Tavazza, M. (2004). Tomato yellow leaf curl Sardinia virus can overcome transgene-mediated RNA silencing of two essential viral genes. *Journal of General Virology* 85(6): 1745-1749.

Norman, K., Basri, M.W. and Ramle, M. (2005). Environmental factors affecting the population density of *Oryctes rhinoceros* in a zero-burn oil palm replant. *Journal of Oil Palm Research* 17: 53-63.

Nour El-Deen, A.H., Al-Barty Amal, F., Darwesh Hadeer, Y. and Al-Ghamdi, A.S. (2016). Eco-Friendly Management of Root-Knot Nematode, *Meloidogyne incognita* Infecting Pomegranate at Taif Governorate, KSA. *Research Journal of Pharmaceutical, Biological and Chemical Sciences* 7(1): 1070-1076.

Odigie, E.E. and Ikotun, T. (1982). *In-vitro* and *in-vivo* inhibition of growth of *Phytophthora palmivora* (Butl.) by antagonistic microorganisms. *Fitopatol Bras* 7(2): 157–169.

OEPP/EPPO. (1996). Data sheets on quarantine organisms No. 23, *Tilletia indica.* OEPP/EPPO Bulletin.

Oerke, E.C. (2006). Crop losses to pests. *J. Agric. Sci.* 144: 31-43.

Oerke, E.C., Dehne, H.W., Schonbeck F. and Weber A. (1994). *Crop Production and Crop Protection: Estimated Losses in Major Food and Cash Crops.* Elsevier, Amsterdam, 808 pp.

Oros, G. and Ujvary I. (1999). Botanical fungicides: Natural and semi-synthetic ceveratrum alkaloids. *Pesticide Sci.* 55: 253-264.

Orsenigo, M. (1954). Suscettibilita di varieta Italiane di riso alla malattia della 'white-tip'. *Annali della Facolta di Agraria Universita Cattolica del S. Cuore, Milano* 51: 1-7.

Otim-Nape, G.W., Bua, A., Baguma, Y. and Thresh, J.M. (1997). Epidemic of severe cassava mosaic disease in Uganda and efforts to control it. *Afr. J. Root Tuber Crops* 2: 42–43.

Ou, S.H. (1985). *Rice Diseases.* CAB International, Wallingford, UK, 380 pp.

Padgham, J.L., Abawi, G.S., Duxbury, J.M. and Mazid, M.A. (2004a). Impact of wheat on *Meloidogyne graminicola* populations in the rice–wheat system of Bangladesh. *Nematropica* 34: 183-190.

Padgham, J.L., Duxbury, J.M., Mazid, A.M., Abawi, G.S. Hossain, M. (2004b). Yield loss caused by *Meloidogyne graminicola* on lowland rainfed rice in Bangladesh. *J. Nematol.* 36: 42–48.

Pandey, P.K. and Pandey, K.K. (2001). A note on source of resistance on watermelon bud necrosis caused by Tospovirus. *Veg. Sci.* 28: 199-200.

Pandey, R. and Kalra, A. (2003). Root-knot disease of Ashwagandha, *Withania somnifera* and its ecofriendly and cost effective management. *J. Mycol. Pl. Pathol.* 33: 240-245.

Pandey, R. and Kalra, A. (2005). Chemical activators: A novel and sustainable management approach for *Meloidogyne incognita* (Kofoid and White) Chitwood in *Chamomile recutita* L. *Arch. Phytopathol. Pl. Prot.* 38: 107-111.

Pandey, R., Mishra, A.K., Tiwari, S., Singh, H.N. and Kalra, A. (2011). Enhanced tolerance of *Mentha arvensis* against *Meloidogyne incognita* (Kofoid and White) Chitwood through mutualistic endophytes and PGPRs. *J. Plant Interactions* Aug: 1-7.

Pankaj, Ahlawat, J.S. and Saha, M. (2006). Predominant nematode pests of rice nursery in North-Western India. In: *2nd International Rice Congress*, ICAR New Delhi.

Pappu, H.R., Jones, R.A.C. and Jain, R.K. (2009). Global status of tospovirus epidemics in diverse cropping systems: successes achieved and challenges ahead. [Plant virus epidemiology: controlling epidemics of emerging and established plant viruses - the way forward. Papers presented at the *Xth International Plant Virus Epidemiology Symposium*, Andhra Pradesh, India, 14-18 October 2007. *Virus Research* 141(2): 219-236.

Paret, M.L., Cabos, R., Kratky, B.A. and Alvarez, A.M. (2010). Effect of plant essential oils on *Ralstonia solanacearum* race 4 a nd bacterial wilt of edible ginger. *Plant Disease* 94: 521–527.

Park, J.S. and Lee, J.O. (1976). Varietal resistance of rice to the white-tipped nematode *Aphelenchoides besseyi. International Rice Research Newsletter* 1(1): 15.

Parrella, M.P., Yost, J.T., Heinz, K.M. and Ferrentino, G.W. (1989). Mass rearing of *Diglyphus begini* (Hymenoptera: Eulophidae) for biological control of *Liriomyza trifolii* (Diptera: Agromyzidae). *Journal of Economic Entomology* 82: 420-425.

Parsons, M.S. and Davey, P. (2007). The rise of the Scarce bordered straw *Helicoverpa armigera* (Hübner) (Lep.: Noctuidae). *Entomologist's Record* 119: 185-192.

Parvatha Reddy, P., Singh, D.B. and Nagalingam, B. (1981). Fighting citrus nematodes in Andhra Pradesh. *Indian Horticulture* 26(3): 24-25.

Patel, B.A., Patel, D.J., Sharma, S.B. and Patel, H.V. (1996). Nematode problems of groundnut and their management in Gujarat, India. *International Arachis Newsletter* 16: 38-39.

Patel, D., Patel, B.A., Patel, S.K., Patel, R.L. and Patel, R.G. (1999). Root-knot nematode, *Meloidogyne indica* on kagzi lime in North Gujarat. *Indian J Nematol.* 29(2): 185-205.

Patel, D.J., Patel, B.A., Chavda, J.C. and Patel H.V. (1988). Record of *Meloidogyne javanica* on groundnut in Gujarat, India. *International Arachis Newsletter* 3: 16–17.

Patel, H.R. (1983). *Studies on Root-Knot Nematodes of Groundnut.* Ph. D. thesis, Gujarat Agri. Univ., Sardarkrushinagar, Gujarat, India, 125 pp.

Patel, H.R., Patel, B.A. and Patel, H.V. (1993). Pathotypes of *Meloidogyne javanica* in India. *Nematol Medit* 21: 207-208.

Patel, H.R., Vaishnav, M.U. and Dhruj, I.U. (1985). Interaction of *Meloidogyne arenaria* and *Fusarium solani* on groundnut. *Indian J Nematol* 15: 98-99.

Patel, H.R., Vaishnav, M.U. and Dhruj, I.U. (1986). Efficacy of aldicarb sulfone and carbofuran flowable seed treatment on plant growth and against *Meloidogyne arenaria* on groundnut. *Pesticides* 20: 29-31.

Patel, M.G., Jhala, R.C., Vaghela, N.M. and Chauhan, N.R. (2010). Bio-efficacy of buprofezin against mealy bug, *Phenacoccus solenopsis* Tinsley (Hemiptera : Pseudococcidae) an invasive pest of cotton. *Karnataka Journal of Agricultural Sciences* 23(1): 14-18.

Patel, N.N. and Vala D.G. (2004). Evaluation of phytoextracts against the growth of *Fusarium solani. Plant Dis. Res.* 19: 204-204.

Pathak, B. and Khan, M. R. (2009). Yield loss potential of foliar nematode, *Aphelenchoides besseyi* in tuberose. *Indian J. Nematol.* 39: 38-40.

Pathak, K.N. and Kumar, B. (1996). Effect of culture filtrates of *Gliocladium virens* and *Trichoderma harzianum* on the penetration of rice roots by *Meloidogyne graminicola. Indian Journal of Nematology* 25: 223- 224.

Pathak, K.N. and Kumar, B. (2003). Effect of culture filtrates of *Gliocladium virens* and *Trichoderma harzianum* on the penetration of rice roots by *Meloidogyne graminicola. Indian Journal of Nematology* 33: 149- 151.

Patil, R.K., Rachappa, V., Thipannavar, P.S., Ramegowda, G.K., Lingappa, S. and Krishna Naik, L. (2002). Kabbina bele marakavada beeli unne henu (in Kannada). *Krishi Munnade* 15: 14–16.

Pawan Kumar and Poonam Dhawan. (2013). Biological and serological diagnosis of Iris yellow spot virus in onion from northern India. *Annals of Agri Bio Research* 18(1): 48-54.

Pawar S.A., Mhase N.L., Kadam D.B. and Chandele A.G. (2013). Management of Root-Knot Nematode, *Meloidogyne incognita,* Race-II Infesting Pomegranate by using Bioinoculants. *Indian Journal of Nematology* 43(1): 92-94.

Pereira, F.O.M., Souza, R.M., Souza, P.M., Dolinski, C. and Santos, G.K. (2009). Estimativa do Impacto Econômico e Social Direto de Meloidogyne mayaguensis na Cultura da Goiaba no Brasil. *Nemat. Bras.* 33: 176-181.

Pillai, K.S. (1976). Nematicidal control of root-knot nematode on *Coleus parviflorus. J. Root Crops* 2: 60-63.

Pimental, D. (ed.) (1997). *Techniques for Reducing Pesticide Use.* Wiley, New York.

Pimentel, D. (2007). Area-wide pest management: Environmental, economic and food issues. *In:* M.J.B. Vreysen, A.S. Robinson and J. Hendrichs (eds.) *Area-wide Control of Insect Pests: From Research to Field Implementation,* pp. 35-47. Springer, Dordrecht, The Netherlands.

Pimentel, D. (2009) Pesticides and pest control. In: R. Peshin and A.K. Dhawan (eds.) *Integrated Pest Management: Innovation - Development.* Vol. 1, pp. 83-88. Springer, Dordrecht, The Netherland.

Pimentel, D., Acquay, H., Biltonen, M., Rice, P., Silva, M., Nelson, J., Lipner, V., Giodano, S., Horowitz, A. and D'Amore, M. (1993). Assessment of environmental and economic impacts of pesticide use. In D. Pimentel and H. Lehman (eds.) *The Pesticide Question: Environment, Economics and Ethics*, pp. 47–84. Chapman & Hall, New York.

Pittman, H.A. (1927). Spotted wilt of tomatoes. Preliminary note concerning the transmission of the spotted wilt of tomatoes by an insect vector (*Thrips tabaci* Lind). *J. Sci. Indian Res.* 1: 74-77.

Plowright, R.A. and Bridge, J. (1990). Effect of *Meloidogyne graminicola* (Nematoda) on the establishment, growth and yield of rice cv. IR 36. *Nematologica* 36: 81-89.

Polston, J.E. and Anderson, P.K. (1997). The emergence of whitefly-transmitted geminiviruses in tomato in the western hemisphere. *Plant Dis.* 81: 1358-1369.

Polston, J.E. and Lapidot, M. (2007). Management of Tomato yellow leaf curl virus: US and Israel perspectives. In: H. Czosnek (ed.) *Tomato Yellow Leaf Curl Virus Disease,* pp. 251-262. Springer, Dordrecht, The Netherlands.

Polston, J.E., Bois, D., Serra, C.A. and Concepcion, S. (1994). First report of a tomato yellow leaf curl-like *Geminivirus* in the Western Hemisphere. *Plant Dis.* 78: 831.

Polston, J.E., Cohen, L., Sherwood, T.A., Joseph, R.B. and Lapidot, M. (2006). Capsicum species: Symptomless hosts and reservoirs of *Tomato yellow leaf curl virus. Phytopathology* 96: 447-452.

Popova M.B., Andrusenko, V.V. and Korsakova, L.A. (1980). Degree of resistance of rice variet,ies to *Aphelenchoides besseyi* Christie, 1942. *Byulleten' Vsesoyuznogo Instituta Gel'mintologii im. K.I. Skryabina (Fitogel'mintologiya)* 26: 43-49.

Porter, J. (1995). The effects of climate change on the agricultural environments for crop insect pests with particular reference to the European corn borer and grain maize. In: R. Harrington and N.E. Stork (eds.) *Insect in A Changing Environment*, pp. 93-123. Academic Press.

Prabhu, A.S., Araujo, L.G., Faustina, C. and Berni, R.F. (2003). Estimativa de danos causados pela brusone na produtividade de arroz de terras altas. *Pesqui. Agropecu. Bras.* 38: 1045-1051.

Prabhu, A.S., Filippi, M.C. and Castro, N. (1992). Pathogenic variation among isolates of *Pyricularia grisea* infecting rice, wheat, and grasses in Brazil. *Tropical Pest Management* 38: 367–371.

Pradhan, S. (1964). Assessment of losses caused by insect pests of crops and estimation of insect population. In: N.C. Pant (ed.) *Entomology in India*, pp. 17-58. Entomological Society of India, New Delhi.

Prasad, D., Shukla, S.K., Prasad, B. and Misra, A.K. (2003). Effect of intercrop and different doses of NPK on the incidence of guava wilt. *Proceedings of the Seminar on Plant Diseases of National Importance with Special Reference to Guava Wilt and Mango Malformation*, pp. 53-53. CISH and IISR, Lucknow.

Prasad, J.S. and Somasekhar, N. (2009). *Nematode Pests of Rice: Diagnosis and Management.* Technical Bulletin No. 38, Directorate of Rice Research (ICAR), Rajendranagar, Hyderabad, A.P. India. 29 pp.

Prasad, J.S., Panwar, M.S. and Rao, Y.S. (1986). Screening of some rice cultivars against the root-knot nematode, *Meloidogyne graminicola*. *Indian Journal of Nematology* 16: 112-113.

Prasad, J.S., Panwar, M.S. and Rao, Y.S. (1987). Nematode problems of rice in India. *Tropical Pest Management* 33: 127-136.

Prasad, N., Mehta, P.R. and Lal, S.B. (1952). Fusarium wilt of guava (*Psidum guajava* L.) in Uttar Pradesh, India. *Nature* 169: 753.

Prot, J.C. and Matias, D.M. 1995. Effects of water regime on the distribution of *Meloidogyne graminicola* and other root-parasitic nematodes in a rice field toposequence and pathogenicity of *M. graminicola* on rice cultivar UPL R15. *Nematologica* 41: 219–28.

Purcifull, D.E. and Hiebert, E. (1979). serological distinction of watermelon mosaic virus isolates (PDF). *Phytopathology* 69(2): 112–116.

Puri, S.N. and Mote, U.N. (2004). Emerging pets problems in India and critical issues in their management: An overview. In: B.Subrahamaniyam, V.V. Ramamurthy and V.S. Singh (eds.) *Frontier Areas of Entomological Research,* pp. 13-24. *Proceedings on the National Symposium on Frontier Areas of Entomological Research*, Division of Entomology, IARI, New Delhi, India.

Puri, S.N. and Ramamurthy, V.V. (2009). Insects and integrated pest management in the context of climate change-An overview. In: V.V. Ramamurthy, G.P. Gupta and S.N. Puri (eds.) *Proc. Natn. Symp. IPM Strategies to Combat Emerging Pests in the Current Scenario of Climate Change,* pp.1-7. CAU, Pasighat, Arunachal Pradesh.

Puri, S.N., Murthy, K.S. and Sharma, O.P. (2000). Integrated Pest management in vegetables: Issues and strategies. *In.* G. Kalloo and K. Singh (eds.) *Emerging Scenario in Vegetable Research and Development,* pp. 293-303. Research Periodicals and book publishing house, Texas, USA.

Rabindra, R.J., Mohanraj, P., Poorani, J., Jalali, S.K., Joshi, S.S. and Ramani, S. (2002). *Ceratovacuna lanigera.* Zehntner (Homoptera: Aphididae), a serious pest of sugarcane in Maharashtra and attempts at its management by biological means. *J. Biol. Control* 16: 171–172.

Rahman, M.L. and Das, P. (1994). Seed soaking with chemicals for reducing infestation of Meloidogyne graminicola in rice. *Journal of the Agricultural Science Society of North East India* 7: 107-108.

Rai, A.B., Loganathan, M., Halder, J., Venkataravanappa, V. and Naik, P.S. (2014). *Eco-friendly Approaches for Sustainable Management of Vegetable Pests.* IIVR Technical Bulletin-53, Indian Institute of Vegetable Research, Varanasi, Uttar Pradesh, 104 pp.

Rai, A.B., Satpathy, S., Gracy, R.G., Swamy, T.M.S. and Rai, M. (2007). Yellow mite (*Polyphagotarsonemus latus* banks) menace in chilli crop. *Vegetable Science* 34(1): 1-13.

Rai, P.K. and Singh, K.P. (1995). Efficacy of certain oilcake amendments on *Heterodera cajani, Fusarium udum* and associated wilt of pigeonpea. *International Journal of Tropical Plant Diseases* 13(2): 213-219.

Rai, P.K. and Singh, K.P. (1996). Efficacy of certain oilcake amendments on *Heterodera cajani,* Fusarium udum and associated wilt of pigeonpea. *International Journal of Tropical Plant Diseases* 14(1): 51-58.

Rajasekhara Rao, K. (2013). Corm-borer *Aposonyx chalybaeus* (Hope) – a serious pest of taro in eastern Himalayan region of India. In: R.S. Misra and M. Neduchezhiyan (eds) *Aroids: Opportunities and Challenges,* pp. 229–232. Allied Publishers Pvt Ltd, New Delhi.

Rajasekharam, T., Patil, M.S. and Kamanna, B.C. (2011). Management of bud necrosis disease on watermelon. *Karnataka J. Agric. Sci.* 24(4): 577-580.

Rajendran, T.P. (2009). Integrated pest management-Policy directions in the context of climate change. In: V.V. Ramamurthy, G.P. Gupta and S.N. Puri (eds.) *Proc. Natn. Symp. IPM Strategies to Combat Emerging Pests in the Current Scenario of Climate Change*, pp. 8-13. January 28-30, 2009, Pasighat, Arunachal Pradesh.

Raji, A., Ladeinde, O. and Dixon, A. (2008). Screening land races for additional sources of field resistance to cassava mosaic disease and green mite for integration into the cassava improvement program. *J Integr Plant Biol* 50: 311–318.

Raju, S.V.S., Bar, U.K., Uma Shankar and Kumar, S. (2007). Scenario of Infestation and management of eggplant shoot and fruit borer, *Leucinodes orbonalis* Guen., in India. *Resistant Pest Management Newsletter* 16(2): 14-16.

Ram, P., Saini, R.K. and Vijaya. (2009). Preliminary studies on field parasitization and biology of solenopsis mealybug parasitoid, *Aenasius bambawalei* Hayat (Encyrtidae: Hymenoptera). *J. Cotton Res. Dev.* 23(2): 313-315.

Ramakrishnan, S., Varadharajan, G. and Sutharsan, P.D. (1984). TKM 9 is resistant to rice-root-knot nematode. *International Rice Research Notes* 9: 20.

Ramamurthy, V.V., Gupta, G.P. and Puri, S.N. (eds.) (2009). *Proc. Natn. Symp. IPM Strategies to Combat Emerging Pests in the Current Scenario of Climate Change*. CAU, Pasighat, Arunachal Pradesh.

Ramana, K.V. and Mohandas. (1988). Occurrence of potato cyst nematode *Globodera pallida* in Kerala. *Indian J Nematol* 18: 141.

Ramani, S. (2000). Fortuitous introduction of an aphelinid parasitoid of the spiralling whitefly, *Aleurodicus dispersus* Russell (Homoptera: Aleyrodidae) into the Lakshadweep islands, with notes on host plants and other natural enemies. *Journal of Biological Control* 14: 55-60.

Ramaprasad Shresti, A.Y. (2005). *Studies on Collar Rot Complex of* Coleus forskohlii *(Wild.) Briq.* M. Sc. (Agri.) thesis, University of Agricultural Sciences, Dharwad, Karnataka, India.

Ramesh Babu, S., Kalyan, R.K., Ameta, G.S. and Meghwal, M.L. (2015). Analysis of outbreak of tobacco caterpillar, *Spodoptera litura* (Fabricius) on soybean. *Journal of Agrometeorology* 17(1): 61-66.

Ramraju, K., Natrajan, K., Babu, P.C.S., Planiswmy, S., Rabindra, R.J., Fernando, L.C.P., Moraes de, G.J. and Wickramananda, I.R. (2002). Studies on coconut eriophyid mite, *Aceria guerreronis* Keifer in Tamil Nadu, India. In L.C.P. Fernando and G.J. Moraes de (eds.) *Proceedings of the International Workshop on Coconut Mite* Aceria guerreronis, Coconut Research Institute, Sri Lanka.

Rangarajan, A.V. and Mahadevan, N.R. (1974). Incidence of gall midge *Asphondylia capsici* Barnes. on chilli (*Capsicum annuum* L.) in Tamil Nadu. *Indian J. Ent.* 36(1): 66-67.

Rao, J., Anandprakash, Sujatha, A., Rao, J. and Prakash, A. (2001). Incidance of the coconut mite *Aceria guerreronis* Keifer (Acari: Eriophyidae) on coconut in Orissa. *J. App. Zoological Res.* 12: 2-3.

Rao, M.S., Naik, D., Shylaja, M. and Parvatha Reddy, P. (2002). Prospects for the management of nematode disease complex in capsicum using biological control agents. *Proc. of Int. Conf. on Vegetables*, pp. 347-351. Bangalore.

Rao, Y.S. (1985). Research on rice nematodes. In: S.Y. Padmanabhan (ed.) *Rice in India*, pp. 591-615. ICAR Monograph.

Rao, Y.S. and Biswas, H. (1973). Evaluation of yield losses in rice due to the root-knot nematode, *Meloidogyne incognita*. *Indian Journal of Nematology* 3: 4.

Rao, Y.S. and Israel, P. (1973). Life history and bionomics of *Meloidogyne graminicola*, the rice root-knot nematode. *Indian Phytopathol.* 26: 333–340.

Rao, Y.S., Prasad, J.S. and Panwar, M.S. (1986). Nematode problems in rice: Crop losses, symptomatology and management. In G. Swarup, and D.R. Dasgupta, (eds.) *Plant Parasitic Nematodes of India – Problems and Progress*, pp. 279-299. Indian Agricultural Research Institute, New Delhi.

Rao, Y.S., Prasad, J.S., Yadava, C.P. and Padalia, C.R. (1984). Influence of rotation crops in rice soils on dynamics of plant parasitic nematode populations. *Biology Agriculture Horticulture* 2: 69-78.

Raut, S. K. and Barker, G.M. (2002). *Achatina fulica* Bowdich and other Achatinidae as pests in tropical agriculture. In: G.M. Barker (ed.) *Mollusc as Crop Pests,* pp. 55-114. CABI Publishing, Wallingford.

Ravi, K.S., Kitkaru, A.S., and Winter, S. (2006). Iris yellow spot virus in onions: a new tospovirus record from India. *Plant Pathology* 55(2): 288.

Ravikumara, N.M.I., Manjunatha, M. and Pradeep, S. (2007). Evaluation of attractant waste material and bait for the management of giant African snail, *Achatina fulica* Bowdich. *Karnataka Journal of Agricultural Sciences* 20(2): 288-290.

Ravindra, H., Sehgal M., Pradeep, S., Shruti, S. A., Imran Khan and Narasimhamurthy, H.B. (2013). Evaluation of rice cultivars against root knot nematode *Meloidogyne graminicola* (Golden and Brichfield) in the field. *National Symposium on Nematodes: A Friend and Foe to Agricultural and Horticultural crops*, Dr. Y. S. Paramar University of Horticulture and Forestry, Nauni, Solan, Himachala Pradesh.

Real, L.A. (1996). Sustainability and the ecology of infectious disease. *Bioscience* 46: 88.

Reddy, D.V.R., Ratna, A.S., Sudarshana, M.R., Poul, F. and Kiran Kumar, L. (1992). Serological relationship and purification of bud necrosis virus, a tospovirus occurring in peanut. *Annals of Applied Biology* 120: 279-286.

Reddy, L.S. and Naik, S.L. (2000). Spread of Eriophyid mites in to the coconut gardens of Chittoor district, Andhra Pradesh. *Indian Coconut J.* 31: 8-9.

Reina, J., Morilla, G., Bejarano, E.R., Rodriguez M.D. and Janssen, D. (1999). First report of *Capsicum annuum* plants infected by tomato yellow leaf curl virus. *Plant Dis.* 83: 1176-1176.

Renitha Govind and Ushakumari, R. (2010) Efficacy of different bioagents in controlling the root-knot nematode (*Meloidogyne graminicola* Golden and Birchfield) of rice under dry condition. *National Conf. on Innovations in Nematological Research for Agricultural Sustainability – Challenges and A Roadmap Ahead*, Tamil Nadu Agri. Univ., Coimbatore.

Reversat, G. and Soriano, I. (2002). The potential role of bananas in spreading rice root-knot nematode, *Meloidogyne graminicola. International Rice Research Newsletter* 27: 23-24.

Rocha, J.R.A.S.C., Pimentel, A.J.B., Ribeiro, G. and Souza, M.A. (2014). Efficiency of fungicides in wheat blast control. *Summa Phytopathol.* 40: 347-352.

Rodriguez, F.M.E. and Landa, J.B. (1977). Chemical soil disinfection against parasitic nematode in guava nurseries. *Centrol Agricola Cieneias Agricolos* 4: 57-77.

Rodríguez-Kábana, R., Robertson, D.G., Weaver, C.F. and Wells, L. (1991b). Rotations of bahiagrass and castorbean with peanut for the management of *Meloidogyne arenaria. Suppl. J. Nematol.* 23: 658-661.

Rodríguez-Kábana, R., Robertson, D.G., Wells, L., Weaver, C.F. and King, P.S. (1991a). Cotton as a rotation crop for management of *Meloidogyne arenaria* and *Sclerotium rolfsii* in peanut. *Suppl. J. Nematol.* 23: 652-657.

Rogers, C.A., Wayne, P.M., Macklin, E.A., Muilenberg, M.L., Wagner, C.J. Epstein, P.R. and Bazzaz, F.A. (2006). Interaction of the onset of spring and elevated atmospheric CO_2 on ragweed (*Ambrosia artemisiifolia* L.) pollen production. *Environmental Health Perspectives* 114: 865-869.

Rojas, M.R. and Gilbertson, R.L. (2008). Emerging plant viruses: A diversity of mechanisms and opportunities, Chapter 3. In: M.J. Roossinck (ed.) *Plant Virus Evolution*, pp. 27-51. Springer-Verlag, Berlin, Heidelberg, Germany.

Rosenzweig, C. and Hillel, D. (1998). *Climate Change and the Global Harvest.* Oxford University Press, New York.

Rosenzweig, C., Iglesius, A., Yang, X.B., Epstein, P.R. and Chivian, E. (2000). Climate Change and U.S. Agriculture: The Impacts of Warming and Extreme Weather Events on Productivity, Plant Diseases, and Pests. NASA Publications. Paper 24. http://digitalcommons.unl.edu/nasapub/24.

Roy, A.K. (1973). Reaction of some rice cultivars to the attack of *Meloidogyne graminicola. Indian Journal of Nematology* 3: 72-73.

Roy, A.K. (1976). Effect of decaffeinated tea waste and water hyacinth compost on the control of *Meloidogyne graminicola* on rice. *Indian Journal of Nematology* 6(1): 73-77.

Roy, D.B., Rothery, P., Moss, D., Pollard, E. and Thomas, J.A. (2001). Butterfly numbers and weather: predicting historical trends in abundance and the future effects of climate change. *Journal of Applied Ecology* 25: 819-828.

Roy, S., Singh, B.P. and Bhattacharyya, S.K. (1991). Biocontrol of late blight of potato (Abstr.). *Phytophthora Newsletter* 17: 18.

Rubinstein, G., Morin, S. and Czosnek, H. (1999). Long-term effect of imidacloprid on mortality of the whitefly *Bemisia tabaci* caged with treated eggplant and tomato, and on transmission of tomato yellow leaf curl geminivirus (TYLCV) to tomato. *Journal of Economical Entomology* 92: 658-662.

Rugman-Jones, P.F., Hoddle, M.S. and Stouthamer, R. (2010). Nuclear mitochondrial barcoding exposes the global pest western flower thrips (Thysanoptera: Thripidae) as two sympatric cryptic species in its native California. *Journal of Economic Entomology* 103: 877–886.

Sadakathulla, S. and Ramachandran, T.K.A. (1990). A novel method to control rhinoceros beetle, *Oryctes rhinoceros* L. in coconut. *Indian Coconut Journal* 21: 7-8, 10-12.

Sahoo, S.K. and Jha, S. (2009). Bioecology of mango fruit borer, *Autocharis* (=*Noorda*) *albizonalis* Hampson (Pyralidae, Lepidoptera)-a recent threat to mango growers in West Bengal, India. In S.A. Oosthuyse (ed.) *Proc. VIII Int. Mango Symposium*, ISHS. *Acta Horticulture* 820: 1345-1425.

Sajeev, T.V. (2011). Public Engagement tools in invasive species management. *APFISN Workshop on Forest Health Technology and Phytosanitary Standards,* Beijing, China.

Sakhuja, P.K. and Sethi, C.L. (1985). Frequency of occurrence of various plant-parasitic nematodes and root-rot fungi on groundnut in Punjab. *Indian Journal of Nematology* 15: 191–194.

Sangeetha, G. and Kaimal. (2015). The coconut mite, *Aceria Guerreronis* Keifer (Acari: Eriophydae) – A plague in Kerala. *International Journal of Science and Research* 4(3): 693-696.

Sarode, S.V., Kolhe, A.V. and Sable, V.R. (2009). IPM strategies for cotton in relation to climate change. In: V.V. Ramamurthy, G.P. Gupta and S.N. Puri (eds.) *Proc Natn Symp. on IPM Strategies to Combat Emerging Pests in the Current Scenario of Climate Change,* pp.181-205. CAU, Pasighat, Arunachal Pradesh.

Sasser, J.N. and Freckman, D.W. (1987). A world perspective on Nematology: The role of the Society. In J.A. Veech and D.W. Dickson (eds.) *Vistas on Nematology*, pp. 7-14. Society of Nematologists, Hyattsville, Maryland.

Sathiamma, B., Nair, C.P.R. and Koshy, P.K. (1998). Outbreak of nut infesting eriophyid mite *Aceria guerreronis* (Keifer) in coconut plantations in India. *Indian Coconut Journal* 29: 1-3.

Satpathy, S., Gotyal, B.S. and Krishnan, S. (2016). First report of cotton mealybug, *Phenacoccus solenopsis* Tinsley on cultivated jute (*Corchorus olitorius* L.) in India. *Entomologia Generalis* 36(1): 55-61.

Satpathy, S., Kumar, A., Singh, A.K. and Pandey, P.K. (2005). Chlorfenapyr: A new molecule for diamondback moth (*Plutella xylostella* L.) management in cabbage. *Annals of Plant Protection Sciences* 13: 88-90.

Satyanarayana, T., Mitchell, S.E., Brown, S., Kresowich, S., Reddy, D.V.R., Naidu, R.A., Jarret, R. and Demski, J.W. (1995). Nucleotide sequence of the nucleocapsid gene of peanut bud necrosis virus (Abstr.). In A.A.M. Buiel, J.E. Parlevliet, and J.M. Lenne (eds.) *Recent Studies on Peanut Bud Necrosis Disease*, p. 9. *Proceedings of a Meeting, ICRISAT Asia Center*, Patancheru 502 324, Andhra Pradesh, India, 80 pp.

Scherer, A. (2009). *Ocorrência e hospedabilidade de Meloidogyne mayaguensis em goiabeiras e em plantas de cobertura de solo no Paraná (Tese de Doutorado em Agronomia)*. Universidade Estadual de Londrina, Londrina, 75p.

Seal, S.E., VandenBosch, F. and Jeger, M.J. (2006). Factors influencing Begomovirus evolution and their increasing global significance: Implications for sustainable control. *Crit. Rev. Plant Sci.* 25: 23-46.

Seenivasan, N. (2010). Biointensive management of *Meloidogyne incognita* and *Macrophomina phaseolina* disease complex in medicinal coleus. *Indian J. Pl. Prot.* 38: 186-192.

Seenivasan, N. and Devrajan, K. (2008). Integrated Approach for the management of root-knot nematode, *Meloidogyne incognita* in medicinal coleus. *Indian J. Nematol.* 38(2): 154-158.

Segeren van den Oever, H., Sanchit. and Bekker, M.L. (1985). Observations on *Meloidogyne oryzae* Mass, Sanders Dede 1978 in irrigated rice in Surinam. *Surinaamse Landbouw* 32: 51-59.

Sehgal, M., Somasekhara, Y., Ravichandra, N.G., Ravindra, H., Jain, R.K. and Sardana, H.R. (2012). An outbreak of rice root-knot nematode, *Meloidogyne graminicola* in Shivamogga, Karnataka, India. *Indian J. Nematol.* 42: 102.

Seni, A. and Sahoo, A.K. 2013. First Report of papaya mealybug *Paracoccus marginatus* on papaya from West Bengal. *Indian Journal of Plant Protection* 41(4): 353-355.

Senthamarai, M., Poornima, K. and Subramanian, S. (2006a). Bio-management of root-knot nematode, *Meloidogyne incognita* on *Coleus forskohlii* Briq. *Indian J. Nematol.* 36: 206-208.

Senthamarai, M., Poornima, K. and Subramanian, S. (2006b). Assessment of avoidable yield loss on *Coleus forskohlii* due to *Meloidogyne incognita. Indian J. Nematol.* 36: 296-297.

Senthamari, K., Poornima, K., Subramanian, S. and Sudheer, J. (2008). Nematode-fungal disease complex involving *Meloidogyne incognita* and *Macrophomina phaseolina* on medicinal coleus, *Coleus forskohlii* Briq. *Ind. J. Nematol.* 38: 30-33.

Senthilkumar, P., Ramakrishnan, S., Jonathan, E.I. and Prabhu, S. (2010). Comparison of resistant and susceptible rice cultivaas infested with root-knot nematode *Meloidogyne graminicola. National Conference Innovations in Nematological Research for Agricultural Sustainability: Challenges and A Roadmap Ahead*, Nematological Society of India & Tamil Nadu Agricultural University, Coimbatore, p. 96.

Sertkaya, E., Telli, S. and Yigit, A. (2006). Pest status and parasitoids of gall midge *Asphondylia capsici* Barnes. (Diptera: Cecidomyiidae) in Antakya province. *Turkeys Entomol. Derg.* 30(3): 223-234.

Shankara Murthy, M. and Nagaraj, S.K. (2014). Outbreak of maize leaf roller, *Cnaphalocrocis trapezalis* Guenee (= *Marasmia trapezalis*) (Crambidae: Lepidoptera) on sorghum in Karnataka, south India. *Insect Environment* 19(4): 250-252.

Shanmuga Priya, M. (2015). Biomanagement of rice root knot nematode, *Meloidogyne graminicola* Golden and Brichfield in aerobic rice. *International Journal in Management and Social Science* 3(4): 591-598.

Sharma, H.C. (2005). *Heliothis/Helicoverpa Management: Emerging Trends and Strategies for Future Research.* Oxford & IBH, and Science Publishers, New Delhi, India, 469 pp.

Sharma, H.C. (2014). Climate change effects on insects: Implications for crop protection and food security. *Journal of Crop Improvement* 28: 229-259.

Sharma, H.C., Dhillon, M. and Pampapathy, G. (2006). Multiple-resistance to sorghum shoot fly, spotted stem borer and sugarcane aphid in sorghum. *International Journal of Tropical Insect Science* 26(4): 239 – 245.

Sharma, H.C., Srivastava, C.P., Durairaj, C. and Gowda, C.L.L. (2010). Pest management in grain legumes and climate change. In S.S. Yadav, D.L. McNeil, R. Redden and S.A. Patil (eds.) *Climate Change and Management of Cool Season Grain Legume Crops*, pp. 115-140. Business Media, Springer Science, Dordrecht, The Netherlands.

Sharma, K.K., Yadav, H.S. and Chandra, A. (1999). Reaction of field bean varieties to pod borer omplex. *JNKVV Research Journal* 33(1-2): 78-79.

Sharma, P. and Pandey, R. (2009). Biological control of root-knot nematode; *Meloidogyne incognita* in the medicinal plant; *Withania somnifera* and the effect of biocontrol agents on plant growth. *African J. Agri. Res.* 4: 564-567.

Sharma, S.B. (1995). Resistance to *Rotylenchulus reniformis, Heterodera cajani,* and *Meloidogyne javanica* in accessions of *Cajanus platycarpus. Plant Disease* 79(10): 1033-1035.

Sharma, S.B. (ed.) (1998). *The Cyst Nematodes.* Dordrecht, Netherlands: Kluwer Academic Publishers.

Sharma, S.B., Ali, S.S., Patel, D.J., Patel, H.V., Patel, B.A. and Patel, S.K. (1993). Distribution and importance of plant parasitic nematodes associated with pigeonpea in Gujarat State, India. *Afro-Asian Journal of Nematology* 3(1): 55-59.

Sharma, S.B., Ashok Kumar, P. and McDonald, D. (1991). A greenhouse technique to screen pigeonpea for resistance to *Heterodera cajani. Annals of Applied Biology* 118(2): 351-356.

Sharma, S.B., Rego, T.J., Mohiuddin, M. and Rao, V.N. (1996). Regulation of population densities of *Heterodera cajani* and other plant-parasitic nematodes by crop rotations on vertisols in semi-arid tropical production systems in India. *Journal of Nematology* 28(2): 244-251.

Sharma, S.B., Remanandan, P. and Jain, K.C. (1993). Resistance to cyst nematode (*Heterodera cajani*) in pigeonpea cultivars and in wild relatives of Cajanus. *Annals of Applied Biology* 123(1): 75-81.

Shinde, V.K.R., Verma, J.P. and Sharma, S.K. (1970). Laboratory evaluation of some insecticides against adults of *Myllocerus undecimpustulatus maculosus* Desbrochers. *Indian Journal of Entomology* 32(4): 283-285.

Shivalingaswamy, T.M.S. Satpathy, S. and Banerjee, M.K. (2002). Estimation of crop losses due to insect pests in vegetables, In: B. Sarat Babu, K.S. Vara Prasad, K. Anita, R.D. Rao, S.K. Chakraborty and P.S. Chandukar (eds.) *Resource Management in Plant Protection* Vol. I: 24-31. Plant Protection Association of India, Hyderabad.

Shivalingaswamy, T.M.S., Satpathy, S., Rai, A.B. and Rai, M. (2006). *Insect Pests of Vegetable Crops: Identification and Management.* Technical Bulletin No. 30, IIVR, Varanasi, 15 pp.

Shukla, R.P. and Prasad, V.G. (1885). Field evaluation of some new insecticides against mango leaf weevil, *Rhynchaenus mangiferae* Marshall (Coleoptera: Curculionidae). *Indian Journal of Entomology* 47: 300-303.

Shukla, S.K., Prasad B., Prasad D. and Misra A.K. (2003). A cheap substrate for the mass multiplication of bio-agents of guava wilt. *Proceedings of the Seminar on Plant Diseases of National Importance with Special Reference to Guava Wilt and Mango Malformation*, pp. 54-55. CISH and IISR, Lucknow.

Shylaja, M. (2005). *Studies on the Development of Bio-Intensive Disease Management Strategies in Tuberose (Polyanthes tuberosa L.) and Carnations* (Dianthus caryophyllus L.) *Using Biological Control Agents.* Ph. D. thesis, Kuvempu Univ., Shimoga, 132 pp.

Siddiqui M.A. and Alam M.M. (2001). Integrated management of the root-knot and reniform nematodes with cropping sequences and ploughing. *Arch. Phytopathol. Plant Protect.* 33(5): 415–430.

Siddiqui, Z.A. and Mahmood, I. (1993). Occurrence of races of *Heterodera cajani* in Uttar Pradesh, India. *Nematologia Mediterranea* 21(2): 185-186.

Siddiqui, Z.A. and Mahmood, I. (1995a). Biological control of Heterodera cajani and *Fusarium udum* by *Bacillus subtilis, Bradyrhizobium japonicum* and *Glomus fasciculatum* on pigeonpea. *Fundamental and Applied Nematology* 18(6): 559-566.

Siddiqui, Z.A. and Mahmood, I. (1995b). Some observations on the management of the wilt disease complex of pigeonpea by treatment with a vesicular arbuscular fungus and biocontrol agents for nematodes. *Bioresource Technology* 54(3): 227-230.

Siddiqui, Z.A. and Mahmood, I. (1996). Biological control of *Heterodera cajani* and *Fusarium udum* on pigeonpea by *Glomus mosseae, Trichoderma harzianum,* and *Verticillium chlamydosporium. Israel Journal of Plant Sciences* 44(1): 49-56.

Siddiqui, Z.A., Mahmood, I. and Hayat, S. (1998). Biocontrol of *Heterodera cajani* and Fusarium udum on pigeonpea using *Glomus mosseae, Paecilomyces lilacinus* and *Pseudomonas fluorescens. Thai Journal of Agicultural Science* 31(3): 310-321.

Sidhu, A.K. and Mehta, H.S. (2008). Role of butterflies in the natural ecosystem with special reference to high altitude (Pangi Valley, Himachal Pradesh). In *Proceedings of International Conference on Climate Change, Biodiversity and Food Security in the South Asian Region*, pp. 36. Punjab State Council for Science and Technology, Chandigarh and United Nations Educational, Scientific and Cultural Organization, New Delhi.

Simpson, C.E. and Starr, J.L. (2001). Registration of 'COAN' peanut. *Crop Science* 41: 918.

Singh K.P. and Singh V.K. (1992). *Terminelia arjuna* leaf powder reduces population density of *Heterodera cajani. Int. Pigeonpea Newsl.* 16: 17-18.

Singh, B. and Lal, S.B. (1953). Wilt of guava. *Agric. Husb.* 111(10, 11 & 12): 78-79.

Singh, B.B., Pannu, P.P.S. and Singh, N. (1999). Appraisal of Karnal bunt of wheat in Punjab. *Plant Disease Research.* 14(2): 201-203.

Singh, J. (2004). Mite pest scenario in India. In: B. Subrahamaniyam, V.V. Ramamurthy and V.S. Singh (eds.) *Frontier Areas of Entomological Research,* pp. 85-92. *Proceedings on the National Symposium on Frontier Areas of Entomological Research*, November, 5-7, 2003, Division of Entomology, IARI, NewDelhi-110012, India.

Singh, J. and Raghuraman, M. (2010). Paradigm of human resource development in agricultural acarology (*Abstr.*). In: *Proceedings of International Symposium-cum-Workshop in Acarology,* pp. 11–13. BCKV, Kalyani, West Bengal.

Singh, J. and Singh, R.N. (2004). *Major Harmful Mites of Vegetables and Other Crops and their Control.* Bulletin, Dhanuka Group, New Delhi, 5 pp.

Singh, K.P., Jaiswal, R.K. and Kumar, N. (2007). *Catenaria anguillulae* Sorokin: a natural biocontrol agent of *Meloidogyne graminicola* causing root knot disease of rice (*Oryza sativa* L.). *World Journal of Microbiology and Biotechnology* 23:291-294.

Singh, N.N. (1997). Economics and use of insecticides against pod fly and pod borer complex on pigeon pea. *Indian Journal of Entomology* 59: 228-231.

Singh, P. and Khanna, A. (2015). incidence of phytoparasitic nematodes in vegetable crops grown under protected cultivation in Himachal Pradesh. *International Journal of Science, Environment and Technology* 4(6): 1640–1646.

Singh, P.J., Dhaliwal, H.S. and Gill, K.S. (1989). Chemical control of Karnal bunt (*Neovossia indica*) of wheat (*Triticum aestivum*) by single spray of fungicides at heading. *Indian Journal of Agricultural Science*, 59(2): 131-133.

Singh, R., Singh, P.P. and Singh, V. (2006). Integrated management of collar rot of *Amorphophallus paeoniifolius* Blume caused by *Sclerotium rolfsii* Saccardo. *Veg Sci* 33(1): 45–49.

Singh, S.H., Ved Ratan, Gaur, G.S. and Gangwar, D. (2003). Evaluation of antagonists against guava wilt pathogen. *Farm Sci. J.* 12: 58-59.

Singh, S.J. and Krishna Reddy, M. (1995). *Thrips flavus* Shrank (Thysanoptera: Thripidae), a new insect ector of a *Tospovirus* infecting watermelon in India. *Pest Management Hortic. Ecosyst.* 1: 115-118.

Singh, S.K. and Singh, K.P. (2009). A first record of root-knot nematode (*Meloidogyne graminicola*) on wheat in Uttar Pradesh, India. *Journal of Mycology and Plant Pathology* 39: 340-342.

Singh, S.K., Singh, A.K. and Singh, H.M. (2007). Relative resistance of okra germplasm to shoot and fruit borer, *Earias vittella* Fabr. under field conditions. *Journal of Applied Zoological Researches* 18(2): 121-123.

Singh, S.P. and Ballal, C.R. (1991). *Status of Coffee Berry Borer* [Hypothenemus hampei *(Ferr.)*] *(Coleoptera: Scolytidae).* Technical Document No. 36, All India Coordinated Research Project on Biological Control of Crop Pests and Weeds, Bangalore. 6 pp.

Singh, V.K. and Singh K.P. (1995). Development and pathogenicity of *Heterodera cajani* on a susceptible and moderately susceptible cultivars of pigeonpea. *International Journal of Tropical Plant Diseases* 13(2): 237-244.

Siqueira, K.M.S., Freitas, V.M., Almeida, M.R.A., Santos, M.F.A., Cares, J.E., Tigano, M.S. and Carneiro, R.M.D.G. (2009). Detecção de *Meloidogyne mayaguensis* em goiabeira e mamoeiro no estado de Goiás, usando marcadores moleculares. *Trop. Plant Pathol.* 34: 256-260.

Skule, B. and Vilhelmsen, F. (1997). *Thaumetopoea processionea* L. found in Denmark. http://www.lepidoptera.dk/process.htm.

Smilanick, J.L., Dupler, M., Goates, B.J., Hoffman, J.A., Clark, D. and Dobson, D. (1986). Germination of teliospores of Karnal, dwarf, and common bunt fungi after ingestion by animals. *Plant Disease* 70(3): 242-244.

Smilanick, J.L., Prescott, J.M., Hoffmann, J.A., Secrest, L.R. and Weise, K. (1989). Environmental effects on survival and growth of secondary sporidia and teliospores of Tilletia indica. *Crop Protection* 8(2): 86-90.

Smith, N.A., Singh, S.P., Wang MingBo, Stoutjesdijk, P.A., Green, A.G. and Waterhouse, P.M. (2000). Total silencing by intron-spliced hairpin RNAs. *Nature (London)* 407(6802): 319-320.

Smitha M.S. and Mathew, M.P. (2010b). Evaluation of banana cultivars against root mealy bugs, *Geococcus* spp. *Pest Management in Horticultural Ecosystems* 16(2): 176-178.

Smitha, M.S. and Mathew, M.P. (2010a). Management of root mealybugs, *Geococcus* spp. in banana cv. Nendran. *Pest Management in Horticultural Ecosystems* 16(2): 108-119.

Smitha, M.S., Mathew, M.P., Thomas, J., Ushakumari, R. and Nair, S. (2005). Root mealy bug, *Geococcus citrinus* Kuwana – A threat to banana cultivation in Kerala. *Insect Environment* 11: 112-113.

Somasekhar, N. and Prasad, J.S. (2009). Root-knot nematode *Meloidogyne graminicola* –An emerging threat to rice cultivation. *DRR Newsletter* 7: 3-4.

Somasekhara, Y.M. (2006). Spacious distribution of wilt (*Ceratocystis fimbriata* Ell & Halst) of pomegranate (*Punica granatum* L.) in India. *Research in Crops* 7: 844-853.

Somasekhara, Y.M. and Wali, S.Y. (1999). New record of *Ceratocystis fimbriata* causing wilt of pomegranate. *Plant Dis.* 83 (4): 400.

Somasekhara, Y.M., Wali, S.Y. and Shaik, M.K. (2009). Studies and the management of pomegranate *(Punica granatum)* wilt *(Ceratocystis fimbriata,* Latin American group*).* Paper Presented In: *2nd Inter. Symp. Pomegranate and Minor including Mediterranean Fruits,* pp. 132-133. Univ. Agric. Sci., Dharwad.

Sonyal, S. (2010). *Studies on Pomegranate Wilt Complex.* M. Sc. (Agri.) thesis, Dept. of Plant Path., Univ. of Agri. Sci., Dharwad, India.

Soriano, I.R. and Reversat, G. (2003). Management of *Meloidogyne graminicola* and yield of upland rice in South-Luzon, Philippines. *Nematology* 5(6): 879-884.

Soriano, I.R., Espiritu, M.J., Schmit, V., Brar D. and Reversat G. (1998). Resistance to rice root-knot nematode *Meloidogyne graminicola* in *Oryza longistaminata* and *Oryza glaberrima. Philippine Journal of Crop Science* 23: 89.

Soriano, I.R.S., Prot, J.C. and Matias, D.M. (2000). Expression of tolerance for *Meloidogyne graminicola* in rice cultivars as affected by soil type and flooding. *J. Nematol.* 32:309–317.

Sparks T.H. and Yates, T.J. (1997) The effect of spring temperature on the appearance dates of British butterflies 1883-1993. *Ecography* 20: 368-374.

Sreejith, A. (2011). Socioeconomic and livelihood impact of invasive species on marginal homesteads: The case of *Aceria guerreronis* on coconut palms in India. *Conference on International Research on Food Security, Natural resource management and Rural development,* University of Bonn, Tropentag.

Sreekanth, M., Sreeramalu, M., Raot, R.D.V.J.P., Balu, B.S. and Babu, T.R. (2006). Virus vector relashionships of *Peanut bud necrosis virus* and *Thrips palmi* in green gram. *Indian J. Plant Prot.* 34: 66-71.

Srinivasa, M.V., Viraktamath, C.A. and Reddy, C. (1999) A new parasitoid of the spiralling whitefly *Aleurodicus dispersus* Russell (Hemiptera: Aleyrodidae) in South India. *Pest Management in Horticultural Ecosystems* 5: 59-61.

Srinivasan, K. and Krishna Moorthy, P.N. (1991). Indian mustard as a trap crop for management of major lepidopterous pests on cabbage. *Trop. Pest Mangmt.* 37: 26-32.

Srinivasan, K. and Krishna Moorthy, P.N. (1993). Evaluation of neem products and other standard chemicals for the management of major pest complex on cabbage: Comparison between standard spray regime and IPM involving mustard as a trap crop. In R.P. Singh, M.S. Chari, A.K. Raheja and W. Kraus (eds.) *Neem and Environment, Vol. I,* pp. 447-458. Oxford and IBH Publishing Co. (Pvt.) Ltd., New Delhi.

Srinivasan, K., Virakthmath, C.A., Gupta, M. and Tewari, G.C. (1995). Geographical distribution, host range and parasitoids of serpentine leaf miner, *Liriomyza trifolii* (Burgess) in South India. *Pest management in Horticultural Ecosystem* 1(2): 93-100.

Srinivasan, R., Sundaraj, S., Pappu, H.R., Diffie, S., Riley, D.G. and Gitaitis, R.D. (2012). Transmission of iris yellow spot virus by *Frankliniella fusca* and *Thrips tabaci* (Thysanoptera: Thripidae). *J Econ. Ent.* 105(1): 40-47.

Sriram, S. and Misra, R.S. (2007). Biological control of taro (*Colocasia esculenta* (L) Schott) leaf blight caused by *Phytophthora colocasiae* (Racib) and storage losses with rhizobacteria. *J Biol Control* 22(2): 181–188.

Sriram, S., Misra, R.S., Sahu, A.K. and Maheswari S.K. (2003). Rhizobacteria: Potential biological control agent against taro leaf blight pathogen, *Phytophthora colocasiae* (Racib). *J Root Crops* 29(1): 50–53.

Srivastava, A.K., Ahmad, R., Kumar, S. and Mohandos, S. (2001). Role of VA -Mycorrhiza in the management of wilt disease of guava in the alfivols of Chotanagpur. *Indian Phytopath.* 54: 78-81.

Staviskv. I.. Funderburk. I.. Broadbeck, B.V.. Olson. S.M. and Andersen. P.C. (2002). Population dynamics of *Frankliniella* spp. and tomato spotted wilt incidence as influenced by cultural management tactics in tomato. *Journal of Economic Entomology* 95: 1216–1221.

Stigter, H. and Romeijn, G. (1992). *Thaumetopoea processionea* na ruim een eeuw weer plaatselijk massaal in Nederland (Lepidoptera: Thaumetopoeidae). *Entomologische Berichten, Amsterdam* 52: 66–69.

Stirling, G.R. (1991). *Biological Control of Plant Parasitic Nematodes.* CABI International, Wallingford.

Stonehouse, J. (2001). An overview of fruit fly research knowledge and needs in the Indian Ocean region. In *Proceedings on Second National Symposium on Integrated Pest Management (IPM) in Horticultural Crops: New Molecules, Biopesticides and Environment,* pp. 21-23. Univ. of Agri. Sci., Bangalore.

Straw, N.A. (1995) Climate change and the impact of the green spruce aphid, *Elatobium abietinum* (Walker), in the UK. *Scottish Forestry* 49: 134-145.

Suhag, L.S. (1976) Observations in guava decline in Haryana and its control. *Pesticides* 10: 42-44.

Suhag, L.S. and Khera A.P. 1986. Studies on the variation in nutritional level of wilted, regenerated and healthy trees of guava cultivar Banarasi-Surkha. *Indian Phytopath.* 39: 90-92.

Sujatha, A. and Zaheruddeen, S.M. (2002) Biology of pyralid fruit borer, *Deanolis albizonalis* Hampson: A new pest of mango. *Journal of Applied Zoological Research* 13(1): 1-5.

Sujatha, M. and Lakshminarayana, M. (2007). Resistance to *Spodoptera litura* (Fabr.) in *Helianthus* species and backcross derived inbred lines from crosses involving diploid species. *Euphytica* 155: 205–213.

Sujithra, M., Srinivasan, S. and Muralikrishna, T. (2010). Bionomics and seasonal occurrence of plume moth, *Sphenarches caffer* of field bean. *Annals of Plant Protection Sciences* 18(1): 241-243.

Sundara Babu, P.C., Balasubramanian, M. and Jayaraj, S. (1983). *Studies on the Pathogenicity of* Metarhizium anisopliae *(Metschnikoff) Sorokin* var. major *Tulloch in* Oryctes rhinoceros *(L.).* Research Publication No. 1, Tamil Nadu Agricultural University, Coimbatore, 32 pp.

Suresh, L.M., Malathi, V.G. and Shivanna, M.S. (2013). Serological diagnosis and host range studies of important viral diseases of a few cucurbitaceous crops in India. *Arch. Phytopathol.* 46: 1919-1930.

Swain, B.N. and Prasad, J.S. (1988). Chlorophyll content in rice as influenced by the root-knot nematode, *Meloidogyne graminicola* infection. *Current Science* 57: 895-896.

Swanson, J.K., Yao, J., Tans-Kersten, J.K. and Allen, C. 2005. Behavior of *Ralstonia solanacearum* race 3 biovar 2 during latent and active infection of geranium. *Phytopathology* 95: 136–140.

Swarup, G., Sethii, C.L. and Kaushal, K.K. (1979). On the biotypes of *Heterodera avenae*, the causal organism of molya disease of wheat and barley in India. *Indian J. Nematol.* 9: 164-168.

Tabashnik, B.E., Gassman, A.J., Crowder, D.W. and Carriere, Y. (2008). Insect resistance to Bt crops: Evidence versus theory. *Nature Biotechnol.* 26(2): 199-202.

Tanwar, R.K., Jeyakumar, P. and Monga, D. (2007). *Mealybugs and Their Management.* National Centre for Integrated Pest Management, New Delhi.

Tanwar, R.K., Jeyakumar P. and Vennila, S. (2010). *Papaya Mealybug and its Management Strategies.* NCIPM Technical Bulletin 22, 26 pp.

Taylor, T.A. (1967). The bionomics of *Maruca testulalis.* (Geyer) (Lepidoptera : Pyralidae), a major pest of cowpeas in Nigeria. *Journal of West African Science Association* 12: 111-129.

Tewari, G.C. and Krishna Murthy, P.N. (1986). *Eurytoma* sp. A new parasite on *Asphondylia* sp. infesting eggplant. *Entomon* 12: 227-228.

Tewari, G.C., Krishna Murthy, P.N. and Sardana, H.R. (1987). Nature of damage and chemical control of gall midge, *Asphondylia* sp. infesting eggplant. *Indian J. Agric. Sci.* 57: 745-748.

Tilialo, R., Greenough, D. and Trujillo, E.E. (1996). The relationship between balanced nutrition and disease susceptibility in Polynesian taro. In: *Proceedings of a workshop on mineral nutrient disorders of root crops in the Pacific*, pp. 105–109. ACIAR Proceedings No. 65, Nuku'alofa, Kingdom of Tonga, 17–20 April 1995.

Tobin, J., Haydock, P., Hare, M., Woods, S. and Crump, D. (2008). Effect of the fungus Pochonia chlamydosporia and fosthiazate on the multiplication rate of potato cyst nematodes (*Globodera pallida* and *G. rostochiensis*) in potato crops grown under UK field conditions. *Biological Control* 46: 194-201.

Todd, E.H. and Atkins, J.G. (1959). White tip disease of rice. III. Yield tests and varietal resistance. *Phytopathology* 49: 189-191.

Tomar, R.K.S., Yadav, H.S. and Agarwal, R.K. (1997). Parasitoids of chilli gall midge, *Asphondylia capsici* and their role in chilli ecosystem. *Indian J. Ent.* 59(2): 173-178.

Tripathi, G.M. (1995). Record of parasite and predator complex of sugarcane woolly aphid, *Ceratovacuna lanigera* Zehntner in Nagaland. *Indian Sugar* 44: 883–885.

Tripathi, S., Suzuki, J.N.Y., Ferreira, S.A. and Gonsalves, D. (2008). Papaya ringspot virus-P: Characteristics, pathogenicity, sequence variability and control. *Molecular Plant Pathology* 9(3): 269–280.

Trnka M., Muska F., Semeradova D., Dubrovsky M., Kocmankova E. and Zalud Z. (2007). European corn borer life stage model: Regional estimates of pest development and spatial distribution under present and future climate. *Ecological Modelling* 207: 61–84.

Trudgill, D.L. (1987). Effects of rates of a nematicide and of fertilizer on the growth and yield of cultivars of potato which differ in their tolerance of damage by potato cyst nematodes (*Globodera rostochiensis* and *G. pallida*). *Plant and Soil* 104: 235-243.

Trudgill, D.L., Phillips M.S. and Alphey T.J.W. (1987). Integrated control of potato cyst nematode. *Outlook on Agriculture* 16: 167-172.

Udikeri, S.S., Kranthi, K.R., Patil, S.B., Modagi, S.A. and Vandal, N.B. (2010). Bionomics of mirid bug *Creontiades biseratense* (Distant) and oviposition pattern in Bt cotton. *Karnataka Journal of Agricultural Sciences* 23(1): 153-156.

Urashima, A.S., Galbieri, R. and Stabili, A. (2005). DNA fingerprinting and sexual characterization revealed two distinct populations of *Magnaporthe grisea* in wheat blast from Brazil. *Czech J. Genet. Plant Breed.* 41: 238-245.

Urashima, A.S., Hashimoto, Y., Don, L.D., Kusaba, M., Tosa, Y., Nakayashiki, H. and Mayama, S. (1999). Molecular analysis of the wheat blast population in Brazil with a homolog of retrotransposon MGR583. *Ann. Phytopathol. Soc. Japan.* 65: 429-436.

Urashima, A.S., Igarashi, S. and Kato, H. (1993). Host range, mating type, and fertility of *Pyricularia grisea* from wheat in Brazil. *Plant Dis.* 77: 1211-1216.

Urashima, A.S., Leite, S.F. and Galbieri, R. (2007). Eficiência da disseminação aérea em *Pyricularia grisea*. *Summa Phytopathol.* 33: 275- 279.

USAID (2000). *Making Progress in Africa – 2000*, USAID (http://www. dec.org/pdf_docs/ PDABT805.pdf).

Valent, B. and Chumley, F.G. (1991). Molecular genetic analysis of the rice blast fungus, *Magnaporthe grisea*. *Annu. Rev. Phytopathol.* 29: 443- 467.

Vandenbussche, D. (1997). The oak processionary caterpillar (*Thaumetopoea processionea* Linnaeus), presentation of an interregional approach. *Parasitica (Belgium)* 53: 103-109.

Vandermast, D.B. and Van Lear, D.H. (2002). Riparian vegetation in the southern Appalachian mountains (USA) following chestnut blight. *For. Ecol. Manage.* 155: 97–106.

Varma, A. and Malathi, V.G. (2003). Emerging geminivirus problems: A serious threat to crop production. *Ann. Appl. Biol.* 142: 145-164.

Vazquez, Y.V. (2012). *The Effect of Brassicaceous Green Manures on the Potato Cyst Nematode,* Globodera rostochiensis. A thesis submitted to the University of Ghent for the award of Doctor of Applied Biological Sciences. Ghent University, Ghent.

Veeresh, G.K. (1990). The status of leucaena psyllid, *Heteropsylla cubana* Crawford in India. In B. Napompeth and K.G. MacDicken (eds.) *Leucaena psyllid: Problems and Management,* pp 14-16. Winrock International/IDRC/NFTA/F-FRED, Bangkok, Thailand.

Vennila, S. (2008). Pest management for cotton ecosystems or ecosystem management for cotton production? *Curr Sci* 94(11): 1351-1352.

Vennila, S., Deshmukh, A.J., Pinjarkar, D., Agarwal, M., Ramamurthy, W., Joshi, S., Kranthi, K.R. and Bambawale, O.M. (2010). Biology of the Mealybug, *Phenacoccus solenopsis* on Cotton in the Laboratory. *J. Insect Sci.* 10: 115.

Vennila, S., Ramamurthy, V.V., Deshmukh, A., Pinjarkar, D.B., Agarwal, M. Pagar, P.C., Prasad, Y.G., Prabhakar, M., Kranthi, K.R. and Bambawale, O.M. (2010). *A Treatise on Mealy bugs of Central Indian Cotton Production System.* Technical Bulletin No. 24, NCIPM, Pusa Campus, New Delhi, 39 pp.

Verghese, A., Jayanthi, P.D.K. and Sudha Devi, K. (2000). Developing an IPM based export protocol and forewarning model for mango fruit fly, *Bactrocera dorsalis* Hendel. *National Seminar on Hitech Hort.,* Bangalore.

Verma, A.N., Sharma, P.U. and Ramzan, M. (1969). Relative toxicity of different insecticides to the adults of cotton grey weevil, *Myllocerus undecimpustulatus maculosus* Desbrochers (Curculionidae: Coleoptera). *Punjab Agricultural University, Journal of Research* 6: 170-173.

Vidavski, F., Czosnek, H., Gazit, S., Levy, D. and Lapidot, M. (2008). Pyramiding of genes conferring resistance to *Tomato yellow leaf curl virus* from different wild tomato species. *Plant Breed.* 127: 625-631.

Viedma L. (2005). Wheat blast occurrence in Paraguay (Abstr.). *Phytopathology* 95: 152.

Vijaya Kumari, N. and Sujathamma, P. (2014). Integrated management of root knot nematode *Meloidogyne incognita* Chitwood in mulberry (*Morus* spp). *Journal of Pharmacy and Biological Sciences* 9(1): 30-33.

Villota, F.J. and Agudelo, F.V. (1997. Evaluation of guava material (*Psidium guajava* L.) for the damage behaviour of *Meloidogyne incognita. Fitopatologia Colombiana* 21: 31- 38.

Virakthmath, C.A., Tewari, G.C., Gupta, M. and Srinivasan, K. (1993). American serpentine leaf miner is a new threat to crops. *Indian Farming* 43(2): 10-12.

Vishakantaiah, M. and Jagadeesh Babu, C.S. (1980). Bionomics of the tur web worm, *Maruca testulalis* (Lepidoptera : Pyralidae). *The Mysore Journal of Agricultural Sciences* 14: 529-532.

Vishwakarma, R., Prasad, P.H., Ghatak, S.S. and Mondal, S. (2011). Bio-efficacy of plant extracts and entomopathogenic fungi against Epilachna beetle, *Henosepilachna vigintioctopunctata* (Fabricius) infesting bottle gourd. *Journal of Insect Science* 24(1): 65-70.

Waage, J., Woodhall, J.W., Bishop, S.J., Smith, J.J. Jones D.J. and Spence, N.J. (2006). T15: Patterns of new plant disease spread: a plant pathogen database analysis. *Foresight: Infectious Diseases: Preparing for the Future.* Office of Science and Innovation. http://www.foresight.gov.uk/Previous_Projects/Detection_and_Identification_of_Infectious_Diseases/Reports_and_Publications/Final_Reports/T/t15.pdf.

Wallace, J.B., Webster, J.R., Eggert, S., Meyer, J.L. and Siler, E. (2001). Large woody debris in a headwater stream: long-term legacies of forest disturbance. *Intl. Rev. Hydrobiol.* 86:501–513.

Wang, Y., Ye, G., Luan, I.X., Xiao, J., Chen, Y. and Chen, D. (2009). Boll size affects the insecticidal protein content in *Bacillus thuringiensis* (*Bt*) cotton. *Field Crop Res* 110: 106-110.

Warham, E.J. (1986). Karnal bunt disease of wheat: a literature review. *Tropical Pest Management* 32(3): 229-242.

Warham, E.J. (1988). Screening for Karnal bunt (*Tilletia indica*) resistance in wheat, triticale, rye and barley. *Canadian Journal of Plant Pathology* 10(1): 57-60.

Warham, E.J., Prescott, J.M. and Griffiths, E. (1989). Effectiveness of chemical seed treatments in controlling Karnal bunt disease of wheat. *Plant Disease* 73(7): 585-588.

Watson, G.W. and Malumphy. C.P. (2004) *Icerya purchasi* Maskell, cottony cushion scale (Hemiptera: Margarodidae), causing damage to ornamental plants growing outdoors in London. *British Journal of Entomology and Natural History* 17: 105-109.

Watts, W.D.J. (2012). *Means to Determine the Efficacy of Several Biofumigant Swards at Suppressing the Potato Cyst Nematodes,* Globodera rostochiensis *(Wollenweber) and Globodera pallida (Stone) respectively.* A thesis submitted to Harper Adams University College for the award of Bachelor of Science with Honours, Newport.

Wayne, P., Foster, S., Connolly, J., Bazzaz, F. and Epstein, P. (2002). Production of allergenic pollen by ragweed (*Ambrosia artemisiifolia* L.) is increased in CO_2-enriched atmospheres. *Annals of Allergy, Asthma and Immunology* 88: 279-282.

Webb, S. (1965). Isolation and identification of watermelon on mosaic viruses 1 and 2. *Phytopathology* 55: 895–900.

Webb, S., Hiebert, E. and Kucharek, T.A. (2003). *Identity and Distribution of Viruses Infecting Cucurbits in Florida.* 93 (S1). American Phytopathological Society.

Whitehead, A.G. (1968). Taxonomy of *Meloidogyne* (Nematodea: Heteroderidae) with descriptions of four new species. *Transactions of Zoological Society of London* 31: 263-401.

Whitehead, A.G. (1986). Chemical and integrated control of cyst nematodes. In F. Lamberti and C. E. Taylor (eds.) *Cyst Nematodes,* pp. 413-432. Plenum Press, New York.

Whittaker, J.B. (1999). Impacts and responses at population level of herbivorous insects to elevated CO_2. *Eur J Ent* 96: 149-156.

Wood, B.J. (1968). *Pest of Oil Palms in Malaysia and their Control.* Incorporated Society of Planters, Kuala Lumpur, Malaysia.

World Food Summit (1996). *Monitoring Progress since the World Food Summit.* FAO, Rome.

Wright, A.J. and Woodhall, J. (2006). T5.11: Review: Case Study of *Phytophthora Ramorum*, an Emerging Plant Pathogen in Cultivated and Wild Environments. *Foresight Infectious Diseases: Preparing for the Future.* Office of Science and Innovation. http://www.foresight.gov.uk/Previous_Projects/Detection_and_Identification_of_Infectious_Diseases/Reports_and_Publications/Final_Reports/T/t5_11.pdf.

Xu, J., Liu, P., Meng, Q. and Long, H. (2004). Characterization of *Meloidogyne* species from China using isozyme, phenotypes and amplified mitochondrial DNA restriction fragment length polymorphism. *Eur. J. Plant Pathol.* 110: 309-315.

Yadav, B.S. (1986). Nematode problems of pulse crops. In: G. Swarup and D.R. Dasgupta (eds.) *Plant-Parasitic Nematodes of India, Problems and Progress,* pp. 328-335. Indian Agricultural Research Institute, New Delhi, India.

Yadav, J.S., Ogwok, E., Wagaba, H., Patil, B.L., Bagewadi, B. et al (2011). RNAi-mediated resistance to cassava brown streak Uganda virus in transgenic cassava. *Mol Plant Pathol* 12: 677–687.

Yamada, W., Shiomi, T. and Yamamoto, H. (1953). Studies on the white tip disease of rice plants. III. Hibernation of the causal nematode, varietal resistance of rice plants to the disease, and effect of rice culture upon the disease occurrence. *Special Bulletin. Okayama Prefecture Agricultural Experiment Station* (48): 27-36.

Yamamura, K. and Kiritani K. (1998). A simple method to estimate the potential increase in the number of generations under global warming in temperate zones. *App Ent Zool* 33: 289-298.

Yang, Y., Sherwood, TA, Patte, CP, Hiebert, E. and Polston, J.E. (2004). Use of Tomato yellow leaf curl virus (TYLCV) Rep gene sequences to engineer TYLCV resistance in tomato. *Phytopathology* 94(5): 490-496.

Yeh, S.D., Lin, Y.C., Cheng, Y.H., Jih, C.L., Chen, M.J. and Chen C.C. (1992). Identification of tomato spotted wilt-like virus on watermelon in Taiwan. *Plant. Dis.* 76: 835-840.

Yenjerappa, S.T. (2009). *Epidemiology and Management of Bacterial Blight of Pomegranate caused by* Xanthomonas axonopodis *pv.* punicae *(Hingorani and Singh) Vauterin et al.* Ph.D. thesis, University of Agricultural Sciences, Dharwad.

Yoshida, K., Saunders, D.G., Mitsuoka, C., Natsume, S., Kosugi, S., Saitoh, H., Inoue, Y., Chuma, I., Tosa, Y., Cano, L.M., Kamoun, S. and Terauchi, R. (2016). Host specialization of the blast fungus *Magnaporthe oryzae* is associated with dynamic gain and loss of genes linked to transposable elements. *BMC Genomics* 18: 370.

Yoshii, H. and Yamamoto, S. (1951). On some methods for the control of the rice nematode disease. *Science Bulletin of the Faculty of Agriculture, Kyushu University* 12: 123-131.

Young, E.C. (1975). *A Study of Rhinoceros Beetle Damage in Coconut Palms.* Technical Paper, South Pacific Commission, No. 170, 63 pp.

Zaka, S.M., Saeed, S., Bukhari, S.A. and Baksh, E. (2006). Mealybug, *Phenacoccus solenopsis* (Homoptera: Pseudococcidae): A novel pest of cotton in Pakistan. *Proceedings of 34th SAARC Countries Science Conference*, University of Veterinary and Animal Sciences, Lahore, Pakistan, 32 pp.

Zaki, F.A. and Bhatti, D.S. (1986). Control of pigeon-pea cyst nematode, *Heterodera cajani* Koshy, 1967 by chemical seed treatment. *Indian Journal of Nematology* 16(1): 106-108.

Zamir, D., Ekstein-Michelson, I., Zakay, Y., Navot, N., Zeidan, M., Sarfatti, M., Eshed, Y., Harel, E., Pleban, T., Oss, H.V., Kedar, N., Rabinowitch, H.D. and Czosnek, H. (1994). Mapping and introgression of a *Tomato yellow leaf curl virus* tolerance gene, TY-1. *Theor. Appl. Genet.* 88: 141-146.

Zhou, X., Harrington, R., Woiwod, I.P., Perry, J.N. and Clark, S.J. (1996). Impact of climate change on aphid flight phenology. In: R.J. Froud-Williams, R. Harrington, T.J. Hocking, H.G. Smith and T.H. Thomas, (eds.) *Implications of Global Environment Change" for Crops in Europe. Aspects of Applied Biology* 45: 299-305.

Zrachya, A., Kumar, P.P., Usha Ramakrishnan, Levy, Y., Loyter, A., Arazi, T., Lapidot M. and Gafni, Y. (2007). Production of siRNA targeted against TYLCV coat protein transcripts leads to silencing of its expression and resistance to the virus. *Transgenic Research* 16(3): 385-398.

Zuckerman, B.M., Dicklow, M.B. and Acosta, N. (1993). A strain of *Bacillus thuringiensis* for the control of plant parasitic nematodes. *Biocontrol Science and Technology* 3: 41-46.

Acronyms

AMF Arbuscular mycorrhizal fungi
AMV Alfalfa mosaic virus
APHIS Animal and Plant Health Inspection Services
BDMV Bean dwarf mosaic virus
BG-I Bolgard I (transgenic *Bt* cotton)
BG-II Bolgard II (transgenic *Bt* cotton)
BGMV Bean golden mosaic virus
BIPM Biointensive integrated pest management
BPH Brown plant hopper (on rice)
Bt *Bacillus thurigiensis*
BYDV Barley yellow dwarf virus
CFU Colony forming units
CGIAR Consultative Group on International Agricultural Research
CICR Central Institute for Cotton Research, Nagpur
CIMMYT International Maize and Wheat Improvement Center, Mexico
CMD Cassava mosaic disease
CMV Cassava mosaic virus
CTCRI Central Tuber Crops Research Institute, Thiruvananthapuram
DAT Days after transplanting
DBM Diamondback moth
EGW Eucalyptus gall wasp
EIL Economic injury level
EIDs Emerging infectious diseases
EPA Environmental Protection Agency
EPPO European and Mediterranean Plant Protection Organisation
ETL Economic threshold level
FORS Fish oil resin soap
FYM Farm yard manure
GPA Green peach aphid
GPS Global positioning system
IAS Invasive alien species
ICAR Indian Council of Agricultural Research, New Delhi
ICMV Indian cassava mosaic virus
IDM Integrated disease management
IGP Indo-Gangatic plains
INM Integrated nematode management
INSV Impatiens necrotic spot virus
IPPC International Plant Protection Convention
IPCV Indian Peanut Clump Virus
IPM Integrated pest management
IYSD Iris yellow spot disease
IYSV Iris yellow spot virus
LDCs Least-developed countries
MAT Male annihilation technique
MSL Mean sea level
NBAIR National Bureau of Agricultural Insect Resources, Bangalore
NPV Nuclear polyhedrosis virus
NSKE Neem seed kernel extract
NSPE Neem seed powder extract
PBND Peanut bud necrosis disease
PBNV Peanut bud necrosis virus
PCN Potato cyst nematode
PDI Percent disease index
PDR Pathogen-derived resistance
PepMV Pepino mosaic virus
PGPR Plant growth promoting rhizobacteria
PLRV Potato leaf roll virus
PRSV Papaya ring spot virus
PYSV Peanut yellow spot virus
QTL Quantitative trait loci
RCTs Resource conservation technologies
RKI Root-knot index
RWCS Rice- wheat cropping system
SAR Systemic acquired resistance
SLCMV Sri Lankan cassava mosaic virus
ToTV Tomato torrado virus
TCSV *Tomato chlorotic spot virus*
TSWV Tomato spotted wilt virus
TYLCD Tomato yellow leaf curl disease
TYLCV Tomato yellow leaf curl virus
WBNV Watermelon bud necrosis virus
WFT Western flower thrips
WSMV Watermelon silver mottle tospovirus
WTO World Trade Organizatio

Subject Index

A

Aceria guerreronis 10, 93, 100, 261
Achatina (Lissachatina) fulica 75
Acrosternum hilare 34
Aleurodicus dispersus 10, 53, 54, 55, 67, 70
Allim cepa 145
Amla 56, 64, 65, 67
Amorphophallus paeoniifolius 152, 168
Anacardium occidentale 98
Aplosonyx chalybaeus 90, 99, 100, 152, 172
Anar butterfly 64
  Damage 64
  Management 64
Aphelenchoides besseyi 8, 9, 174, 183, 195, 240, 242, 259
Apoderous blandus 65
Arachis hypogaea 112, 191, 198, 199, 200
Asphondylia capparis 11, 14, 69, 82, 89
Aswagandha 255
Aswagandha root-knot nematode 255
  Management 255
  Symptoms 255
Aswagandha disease complex 258
  Management 258
  Symptoms 258

B

Bactrocera caryeae 53, 61, 67
Bactrocera spp. 73, 261
Banana 60
Banana root mealy bug 60
  Damage 60
  Host range 60
  Key facts 60
  Management 60
Barley 6, 186, 187, 188, 195, 262, 271
Bell Pepper 234
Bell pepper root-knot nematodes 234
  Management 235
  Symptoms 234
Root-knot and bacterial wilt complex 236
  Interaction studies 236
  Management 236
Bemisia argentifolii 76
Bemisia tabaci 13, 18, 68, 69 74, 76, 138, 139, 160, 261
Brassica oleracea var. *botrytis* 84
Brinjal/Egg plant 81
Brinjal gall midge 82
  Damage 82
  Management 82
Brinjal insect pests 81
  Gall midge 5, 11, 13, 14, 37, 41, 69, 70, 82, 83, 89, 262, 266
  Mealy bug 81, 89
Brinjal mealy bug 81, 82
  Biology 82
  Damage 82
  Management 82
Brontispa longissima 10, 95, 100

C

Cajanus cajan 29, 188
Capsicum annuum 82, 198, 234
Carica papaya 55, 130
Carnation 154, 245, 246, 247, 248, 259
Carnation root-knot nematode 245
  Economic importance 245
  Management 246
  Symptoms 246
Carnation root-knot and wilt complex 247
  Integrated management 247
  Interaction studies 247
Cashew 98
Cashew tea mosquito bug 98
  Damage 98
  Management 98
Cassava 159
Cassava mosaic disease 159
  Causal agent 159
  Management 159
  Spread 160
  Symptoms 159
  Transmission 160
Cauliflower 84
Cauliflower diamondback moth 85
  Management 85
Ceratocystis fimbriata 121, 196, 209, 217
Ceratovacuna lanigera 5, 39, 52

Chilli 82
Chilli gall midge 82
  Challenges in management 83
  Damage 82
  Management 83
Citrullus lanatus var. *lanatus* 149
Citrus 216
Citrus root-knot nematode 216
  Host range 216
  Management 216
  Symptoms 216
Citrus spp. 216
Climate change effect on diseases 269
  Barley yellow dwarf virus 271
  Fusarium head blight 117, 269, 271
  Indian peanut clump virus 271
  Wheat blast 108, 116, 117, 118, 119, 120, 270
  Wheat leaf rust 271
  Wheat spot blotch 270
  Wheat stem rust 271
  Wheat yellow rust 270
Climate change effect on pests 263
  Geographical distribution range shift 263
  Number of generations per year 263, 267
  Other effects 267
Climate change impact on emerging pests 262
  Increased CO_2 concentration 262
  Precipitation 263
  Risig temperature 262
Cnaphalocrosis trapezalis 47, 52
Cochliobolus sativus 270
Coconut 93
Coconut leaf beetle 10, 90, 95, 96
  Biology 96
  Damage 95
  Distribution 95
  Host range 96
  Key facts 95
  Management 96
  Spread 96
Coconut insect pests 93
  Leaf beetle 95
  Mite 93
Coconut mite 10
  Damage 94
  Key facts 93
  Management 94
Cocos nucifera 96
Coffea arabica 90, 198
Coffea canephora 90
Coffee 90
Coffee berry borer 90
  Biology 91
  Damage 91
  Ecology 91
  Economic importance 91
  Host Range 91
  Management 91
Coleus 249
Coleus forskohlii 249
Coleus root-knot nematode 249
  Management 250
  Symptoms 249
Coleus root-knot and wilt complex 251
  Interaction studies 251
  Management 251
Coleus root-knot and collar rot complex 252
  Interaction studies 252
  Management 252
  Symptoms 252
Coleus root-knot and root rot complex 254
  Integrated management 254
  Symptoms 254
Colocasia esculenta 99, 162
Conopomorpha cramerella 65, 66, 67
Coral Tree 101
Corchorus spp. 51
Cotton 25
Cotton false chinch bug 35
  Damage 35
  Management 35
Cotton flower bud maggot/gall midge 36
  Damage 37
  Key facts 36
  Management 37
Cotton mealy bug 27
  Biology 28
  Damage 28
  Economic importance 28
  Host range 29
  Management 29

Cotton mirid bug 32
- Damage 33
- Economic importance 33
- Key facts 32
- Life cycle 34
- Management 34
- Population dynamics and severity 34

Cotton insect pests 25
- False chinch bug 26, 35
- Flower bud maggot/gall midge 37
- Mealy bug 26, 26, 57, 58, 71, 82, 262
- mirid bug 23, 25, 33, 34, 99
- stink bugs 26, 34, 35
- sugarcane beetle 26, 35, 36
- tea mosquito bug 99, 101
- Three-cornered alfalfa hopper 26, 36
- Tobacco caterpillar 26, 31

Cotton stink bugs 34
- Damage 34
- Management 35

Cotton sugarcane beetle 35
- Management 36

Cotton tea mosquito bug 36
- Damage 36
- Key facts 36

Cotton three-cornered alfalfa hopper 36
- Damage 36
- Management 36

Cotton tobacco caterpillar 31
- Damage 32
- Key facts 31
- Life cycle 32
- Management 32

Cowpea hadda beetles 87
- Challenges in management 88
- Management 88

Cowpea 86

Cowpea insect pests 87
- Hadda beetles 87
- Spotted pod borer 48, 49, 50, 52, 86

Cowpea spotted pod borer 86
- Host range 86
- Life cycle 86
- Management 86

Creontiades biseratense 25, 32, 33

Cucumber 238

Cucumber root-knot nematodes 238
- Management 238
- Symptoms 238

Cucumis sativus 238

D

Dasineura gossypii 36, 37

Dendroctonus frontalis 105, 106

Dendroctonus ponderosae 16, 105, 106

Deudorix isocrates 64, 67

Dianthus caryophyllus 245

Driver of disease pathogens 269
- Agricultural changes 18
- Climate change 270
- Pathogen group 17
- Pathogen introduction 7, 16, 17, 22

E

Elaeis guineensis 97

Elephant Foot Yam 75, 152, 168, 169, 172

Elephant foot yam collar rot 169
- Management 168
- Symptoms 168

Embelica officinalis 64

Emerging pests - policy considerations 278
- Capacity development 279
- Disaster prevention 279
- Protecting livelihoods 279

Emerging pests - technical/policy responses 275
- Containment 275
- Data for projecting risk 275
- Early warning and prevention strategies 276
- Ecosystem processes 278
- Eradication 276
- Food trade industry 276
- Global frameworks 278
- Government constraints 277
- Government priorities 277
- Impact reduction 276
- Information options 276

Epilachna dodecastigma 87, 89

Erythrina spp. 101

Erythrina gall wasp 11, 101, 106
- Anticipated effect of climate change 102
- Life cycle 102

Management 102
Present distribution 103
Spread 104
Symptoms 103
Eucalyptus 103
Eucalyptus gall wasp 103
Biology 104
Damage 103
Host range 104
Management 104
Present distribution 103
Spread 104
Symptoms 103
Eucalyptus teriticornis 103
Euetheola rugiceps 35
Euschistus servus 34

F

Field bean 88
Field bean plume moth 88
Biology 89
Damage 89
Management 89
Food Security 18
Access to food 18
Availability of food 18
Stability of food 19
Utilization of food 19
Frankliniella occidentalis 79, 80, 152, 153
Fruit flies 73
Challenges in management 74
Management 74
Fusarium chlamydosporum 251, 252, 259
Fusarium culmorum 269
Fusarium graminearum 269
Fusarium oxysporum f. sp. *dianthi* 240, 247, 259
Fusarium oxysporum f. sp. lycopersici 233, 239
Fusarium oxysporum f. sp. *psidii* 125, 127, 133
Fusarium oxysporum 6, 121, 124, 125, 127, 128, 133, 205, 209, 217, 233, 239, 240, 247, 259
Fusarium solani 124, 125, 128, 129, 192, 197, 203, 204, 205, 217, 258, 259
Fusarium spp. 133, 240

G

Gauva 54
Geococcus citrinus 60, 67
Gerbera 154, 248, 249
Gerbera spp. 248
Gerbera root-knot nematode 248
Economic importance and losses 249
Management 249
Symptoms 249
Gerbera root-knot and foot rot complex 249
Integrated management 249
Interaction studies 249
Giant African snail 75
Challenges in management 76
Management 75
Ginger 157
Ginger bacterial wilt 157
Causal agent 157
Management 157
Survival and spread 157
Symptoms 157
Globodera pallida 220, 254
Globodera rostochiensis 3, 218, 220, 239
Glycine max 18, 39, 112, 198
Gossypium spp. 25
Groundnut 109
Groundnut bud necrosis virus 109
Causal virus 109
Management 110
Symptoms 109
Transmission 109
Groundnut root-knot nematodes 191
Economic importance 191
Geographical distribution in Gujarat 192
Interaction with other micro-organisms 192
Management 192
Symptoms 192
Groundnut yellow spot *Tospovirus* 112
Geographical distribution 112
Host range 112
Management 113
Particle morphology 113
Symptoms 112
Transmission 112
Guava decline disease complex 203
Interaction studies 203
Management 205
Guava Fusarium wilt 125
Causal organisms 127

Economic importance 126
Management 127
Symptoms 126
Guava root-knot nematode 196
Geographical distribution 197
Host range 198
Management 198
Spread 198
Symptoms 197
Guava spiraling whitefly 54
Biology 54
Damage 54
Host range 54
Key facts 54
Management 55

H

Helianthus annus 39
Helicoverpa armigera 5, 6, 12, 13, 48, 52, 68, 69, 70, 264, 267, 268, 272
Helopeltis antonii 12, 90, 98, 100
Helopeltis bryadi 36
Henosepilachna vigitioctopunctata 87, 88
Hereopsylla cubana 11, 51, 52
Heterodera avenae 8, 174, 186, 195
Heterodera cajani 174, 195
Hordeum vulgare 186
Hypothenemus hampei 10, 90, 91, 100, 261

I

Impatiens necrotic spot virus 152
Host range 154
Management 154
Symptoms 153
Transmission and spread 153

J

Jute 51
Jute mealy bug 51
Key facts 51

L

Lab lab purpureus var. *lignosus* 88
Leptocybe invasa 10, 101, 103, 106
Leucaena leucocephala 51
Liriomyza trifolii 10, 13, 26, 69, 70, 71, 80, 261
Litchi 65
Litchi chinensis 65
Litchi fruit borer 66
Damage 66
Management 66
Litchi insect pests 65
Fruit borer 66
Red weeveil 65
Litchi red weeveil 65
Damage 65
Management 65
Lycopersicon esculentum 76, 138, 198, 232

M

Macrophamina phaseolina 215, 217
Magnaporthe grisea 270
Magnaporthe oryzae 108, 116, 120
Management of emerging insect pests 20
Cosmopolitan pests 21
Internal feeders 21, 74
Polyphagous pests 14, 69, 262
Soil pests 21
Sporadic pests 21
Sucking pests 15, 20, 21, 40, 69
Mangifera indica 61
Mango 61
Mango leaf weevil 63
Biology 64
Damage 64
Key facts 63
Management 64
Mango fruitfly 61
Damage 61
Distribution 61
Management 61
Mango insect pests 61
Fruitfly 61
Leaf weevil 63
Red banded fruit borer 62
Mango red banded fruit borer 62
Biology 63
Damage 62
Key facts 62
Management 63
Maruca vitrata 48, 49, 52, 86
Meloidogyne arenaria 174, 191, 194, 195, 200
Meloidogyne enterolobii 8, 196, 197, 199, 200, 203, 204, 205, 217
Meloidogyne graminicola 8, 9, 174, 177, 178, 182, 183, 195, 272

Meloidogyne incognita 8, 121, 168, 196, 200, 205, 207, 209, 213, 214, 215, 217, 232, 233, 234, 236, 238, 239, 240, 243, 245, 247, 248, 249, 251, 252, 254, 255, 258, 259, 272
Meloidogyne indica 8, 196, 216, 217
Meloidogyne javanica 8, 200
Morus spp. 213
Mulberry 213
Mulberry root-knot nematodes 213
  Management 213
  Predisposing factors 213
  Symptoms 213
Mulberry root-knot & root rot complex 215
  Interaction studies 215
Musa spp 60
Myllocerus spp. 26, 46, 52
Myzus persicae 83, 131, 267

N

Nematodes on polyhouse flower crops 243
  Application of enriched organics 244
  Enrichment of organic amendments 244
  Preparation of beds 244
Nematodes on polyhouse vegetables 229
  Main causes 230
  Management 230
Nezara viridula 12, 26, 34, 266
Nilaparvata lugens 41, 52, 261, 267
Nysius raphanus 35

O

Oak 105
Oak processionary moth 105
  Anticipated effect of climate change 105
  Present distribution 105
Oil Palm 97
Oil palm rhinoceros beetle 97
  Biology 97
  Damage 97
  Management 97
Onion 145
Onion iris yellow spot virus 145
  Distribution 145
  Economic impact 145
  Host range 145
  Management 147
  Pathway causes 147
  Pathway vectors 147
  Survival and spread 147
  Symptoms 146
  Transmission 147
Oryctes rhinoceros 14, 90, 97, 100
Oryza sativa 40, 174

P

Papaya 55
Papaya mealy bug 55
  Biology 56
  Damage 56
  Host plants 57
  Key facts 55
  Management 57
Papaya ringspot virus 130
  History 130
  Management 131
  Symptoms 131
  Transmisssion 131
  Types and distribution 130
  Virus structure 131
Paracoccus marginatus 10, 13, 53, 55, 56, 67
Pearl millet 46, 47, 52, 110
Pennisetum glaucum 46
Peronospora sparsa 152, 154, 172
Phenacoccus solenopsis 5, 13, 25, 26, 27, 51, 52, 70, 81, 89
Phytophagous mites 72
  Challenges in management 73
  Management 73
Phytophthora colocasiae 152, 162, 172
Phytophthora infestans 6, 7, 134, 138, 151
Phytophthora parasitica 240, 243, 249, 259
Pigeon pea 188
Pigeon pea cyst nematode 188
  Biotypes 189
  Economic losses 188
  Geographical distribution 188
  Host range 188
  Interaction with other micro-organisms 189
  Life cycle 189
  Management 189
  Spread 189
  Symptoms 188

Pine 105
Pine insect pests 105
  Mountain pine beetle 105
  Southern pine beetle 105
Pine processionary moth 105
  Anticipated effect of climate change 105
  Present distribution 105
Pinus spp. 105
Plutella xylostella 68, 84, 85, 262, 264
Polianthes tuberosa 184, 240
Polyphagotarsonemus latus 68, 72, 73
Pomegranate 121
Pomegranate bacterial blight 121
  Causes 122
  Impact 122
  Management 123
  Survival and spread 123
  Symptoms 122
Pomegranate disease complex 124
  Causal agents 124
  Impact 124
  Management 125
  Symptoms 125
Pomegranate root-knot nematode 205
  Management 206
  Symptoms 206
Pomegranate wilt disease complex 208
  Distribution 208
  Interaction of fungi with nematodes 209
  Management 210
  Predisposing factors 209
  Spread and survival 210
  Symptoms 209
Potato 83
Potato aphid 83
  Damage 83
  Management 84
Potato cyst nematodes 218
  Challenges in management 221
  Ecology 220
  Host range 220
  Life cycle 219
  Management 222
  Physiological races 220
  Spread and survival 220
  Symptoms 219
Potato late blight 134
  Epidemiology 135
  Management 136
  Symptoms 135
Psidium guajava 54, 125, 127, 128, 129, 196, 198, 205
Puccinia graminis 271
Puccinia striiformis 270
Puccinia tritici 271
Pulse crops gram caterpillar 48
  Management 48
Pulse crop pests 48
  Gram caterpillar 48
  Pod wasp 50
  Spotted pod borer 49
Pulse crops pod wasp 50
  Damage 50
  Management 50
Pulse crops spotted pod borer 49
  Damage 49
  Economic importance 49
  Key facts 48
  Management 49
Punica granatum 121, 205
Pyrilla perpusilla 46, 47, 52, 262

Q

Quadrastichus erythrinae 11, 101, 102, 106, 266
Quercus spp. 105

R

Ralstonia solanacearum 152, 157, 172, 235, 236, 239, 243, 269
Rhizoctonia bataticola 124, 133, 252, 253, 259
Rhizoctonia solani 123, 208, 216
Rhynchaenus mangiferae 63, 64, 67
Rice 41, 174
Rice brown plant hopper 41
  Biology 41
  Damage 41
  Effect of elevated CO_2 41
  Key facts 41
  Management 42
Rice insect pests 41
  Brown plant hopper 41
  Swarming caterpillar 42
Rice root-knot nematode 174
  Distribution 175

Economic importance 176
Effect of *M. graminicola* on wheat 178
Impact on nematode in the rice-wheat 177
Management 178
Predisposing factors 178
Spread 177
Symptoms 176
Rice swarming caterpillar 43
Damage 43
Host range 44
Key facts 43
Management 44
Predisposing factors 178
Rice white tip nematode 174
Biology 184
Economic impact 184
Host range 184
Management 184
Means of dispersal 183
Symptoms 183
Rosa spp. 154
Rose 154
Rose downy mildew 154
Management 155
Predisposing factors 155
Symptoms 154

S

Saccharum officinarum 39
Sapota 58
Sapota seed borer 58
Biology 59
Damage 59
Key facts 58
Management 59
Sclerotium rolfsii 152, 168, 172, 252, 259
Serpentine leaf miner 71
Challenges in management 72
Key facts 71
Management 72
Solanum melongena 198
Solanum tuberosum 218
Solenopsis mealy bug 70
Challenges in management 71
Damage 71
Management 71
Sorghum 46
Sorghum bicolor 46
Sorghum/pearl millet grey weevil 46
Damage 46
Management 46
Sorghum/pearl millet leaf hopper 46
Damage 46
Key facts 46
Management 46
Sorghum/pearl millet leaf roller
Biology 47
Damage 47
Key facts 46
Management 47
Sorghum/pearl millet pests 45
Grey weevil 45
Leaf hopper 46
Leaf roller 46
Soybean 38
Soybean tobacco caterpillar 38
Damage 38
Management 39
Sphenarches caffer 70, 88, 89
Spissistilus festinus 35
Spodoptera litura 5, 14, 31, 38, 39, 52, 69, 267
Spodoptera mauritia 43, 52
Subabul 51, 52
Subabul *Leucaena psyllid* 51
Damage 52
Key facts 51
Management 52
Sugarcane 39
Sugarcane woolly aphid 39
Damage 40
Management 40
Sunflower 39
Sunflower tobacco caterpillar 39
Damage 39
Management 39

T

Tanaostigmodes cajaninae 50, 52
Taro 99
Taro corm borer 99
Biology 99
Damage 99

Host range 99
Management 100
Taro leaf blight 162
Management 163
Predisposing factors 163
Survival and spread 163
Symptoms 162
Tetranychus spp. 72
Thaumetopoea pityocampa 105
Thaumetopoea processionea 105
Tilletia indica 113
Tomato 232
Tomato insect pests 77
Pinworm 77
Serpentine leaf miner 80
Silver leaf whitefly 76
Western flower thrips 79
Tomato pinworm 77
Damage 77
Key facts 77
Management 78
Tomato root-knot nematodes 232
Management 232
Symptoms 232
Tomato root-knot and wilt complex 233
Interaction studies 233
Management 233
Tomato serpentine leaf miner 80
Biology 80
Damage 80
Key facts 80
Management 81
Tomato silver leaf whitefly 76
Damage 76
Key facts 76
Management 76
Tomato spotted wilt virus 142
Host range 142
Management 143
Symptoms 142
Transmission 143
Tomato western flower thrips 79
Biology 79
Damage 79
Key facts 79
Management 80
Tomato yellow leaf curl virus 137
Management 140
Symptoms 138
Transmission 139
Triticum spp. 113, 186
Trymalitis margarias 11, 58, 261
Tuberose 240
Tuberose foliar nematode 240
Host range 241
Management 241
Survival and spread 241
Symptoms 241
Tuta absoluta 10, 77, 78

V

Vigna unguiculata 86, 112, 150, 190, 199

W

Watermelon 149
Watermelon bud necrosis virus 149
Causal agent 149
Management 150
Predisposing factors 150
Symptoms 149
Transmisssion 150
Wheat 113
Wheat blast 116
Disease cycle 117
Distribution 116
Economic importance 117
Epidemiology 118
Future lines of research 120
Host range 118
Management 119
Source of inoculum 119
Survival and spread 118
Symptoms 117
Wheat/barley cereal cyst nematode 186
Biotypes 187
Economic losses 186
Host range 186
Life cycle 187
Management 187
Spread 187
Survival 187
Symptoms 186

Wheat karnal bunt 113
- Management 114
- Spread and dispersal 114
- Symptoms 113

Whiteflies 74
- Challenges in management 75
- Management 75

Withania somnifera 206, 240, 255, 256, 257, 258

X

Xanthamonas axanopodis pv. *punicae* 121
Xyleborus fornicates 121, 208, 216
Xyleborus perforans 216

Z

Zingiber officinale 157